CURRENT LABORATORY TECHNIQUES IN RABIES DIAGNOSIS, RESEARCH AND PREVENTION

VOLUME ONE

CURRENT LABORATORY TECHNIQUES IN RABIES DIAGNOSIS, RESEARCH AND PREVENTION

VOLUME ONE

Edited by

CHARLES RUPPRECHT

and

THIRUMENI NAGARAJAN

AMSTERDAM • BOSTON • HEIDELBERG • LONDON
NEW YORK • OXFORD • PARIS • SAN DIEGO
SAN FRANCISCO • SINGAPORE • SYDNEY • TOKYO
Academic Press is an imprint of Elsevier

Academic Press is an imprint of Elsevier
525 B Street, Suite 1800, San Diego, CA 92101-4495, USA
32 Jamestown Road, London NW1 7BY, UK
225 Wyman Street, Waltham, MA 02451, USA

Notices

Knowledge and best practice in this field are constantly changing. As new research and experience broaden our understanding, changes in research methods, professional practices, or medical treatment may become necessary.

Practitioners and researchers must always rely on their own experience and knowledge in evaluating and using any information, methods, compounds, or experiments described herein. In using such information or methods they should be mindful of their own safety and the safety of others, including parties for whom they have a professional responsibility.

To the fullest extent of the law, neither the Publisher nor the authors, contributors, or editors, assume any liability for any injury and/or damage to persons or property as a matter of products liability, negligence or otherwise, or from any use or operation of any methods, products, instructions, or ideas contained in the material herein.

British Library Cataloguing-in-Publication Data
A catalogue record for this book is available from the British Library

Library of Congress Cataloging-in-Publication Data
A catalog record for this book is available from the Library of Congress

ISBN: 978-0-12-800014-4

For information on all Academic Press publications
visit our website at store.elsevier.com

Typeset by MPS Limited, Chennai, India
www.adi-mps.com

14 15 16 17 18 19 10 9 8 7 6 5 4 3 2 1

CONTENTS

Foreword *xiii*
List of Contributors *xv*

PART ONE: INTRODUCTION

1. Basic Facts about Lyssaviruses **3**
Ivan V. Kuzmin
1.1 Introduction 3
1.2 Virion and Genome Organization 4
1.3 Phylogeny and Serologic Cross-Reactivity of Lyssaviruses 6
1.4 Host Range 12
1.5 Pathobiology 15
References 17

PART TWO: RABIES DIAGNOSIS

Section A: Demonstration of Viral Subunits and Antigens **25**

2. Demonstration of Lyssavirus Antigens by a Direct Rapid Immunohistochemical Test **27**
Andre Coetzer, Louis H. Nel, and Charles Rupprecht
2.1 Introduction 27
2.2 Materials 29
2.3 Methods 30
2.4 Discussion 33
References 36

3. Demonstration of Rabies Virus Antigens by a Latex Agglutination Test **37**
Songsri Kasempimolporn, Narongsak Chaiyabutr, and Visith Sitprija
3.1 Introduction 37
3.2 Materials 38
3.3 Methods 38
3.4 Discussion 40
References 42

4. Rabies Diagnosis: Demonstration of Viral Antigens by Flow Cytometry **43**

Juliano Bordignon and Carlos Roberto Zanetti

4.1 Introduction 43
4.2 Materials 44
4.3 Methods 45
4.4 Discussion 47
Acknowledgments 50
References 51

5. Demonstration of Rabies Virus Antigens by an Immunochromatographic Strip Test **53**

Songsri Kasempimolporn, Narongsak Chaiyabutr, and Visith Sitprija

5.1 Introduction 53
5.2 Materials 54
5.3 Methods 55
5.4 Discussion 57
References 59

Section B: Demonstration of Viral Nucleic Acids **61**

6. Demonstration of African Lyssavirus RNA with Real-Time Polymerase Chain Reaction **63**

Jessica Coertse, Wanda Markotter, and Louis H. Nel

6.1 Introduction 63
6.2 Materials 64
6.3 Methods 65
6.4 Discussion 68
References 72

7. Real-Time Quantitative Polymerase Chain Reaction for the Demonstration of Lyssavirus Nucleic Acid **75**

Conrad M. Freuling, Bernd Hoffmann, Melina Fischer, Lorraine M. McElhinney, Denise A. Marston, Anthony R. Fooks, and Thomas F. Müller

7.1 Real-Time Quantitative Polymerase Chain Reaction for the Demonstration of Lyssavirus Nucleic Acid 76
7.2 Methodology 76
7.3 Discussion 82
Acknowledgments 83
References 83

8. Reverse Transcription-Loop-Mediated Isothermal Amplification System for the Detection of Rabies Virus **85**

Takuya Itou, Wanda Markotter, and Louis H. Nel

8.1 Introduction 85
8.2 Materials 86
8.3 Methods 90
8.4 Discussion 93
Acknowledgments 94
References 94

9. Detection of Viral Nucleic Acids by *In Situ* Hybridization **97**

Susan A. Nadin-Davis

9.1 Introduction 97
9.2 Materials 98
9.3 Methods 100
9.4 Discussion 103
Acknowledgments 105
References 106

10. Genetic Characterization via Pyrosequencing **107**

Paola De Benedictis and Cristian De Battisti

10.1 Introduction 107
10.2 Materials 108
10.3 Methods 109
10.4 Discussion 111
Acknowledgments 115
References 115

Section C: Demonstration of Viral Antibodies and Immune Complexes **117**

11. Demonstration of Immune Complexes by Capture Enzyme-Linked Immunosorbent Assay **119**

Shampur N. Madhusudana, Kader Muhamuda, and Mani Reeta

11.1 Introduction 119
11.2 Materials 121
11.3 Methods 122
11.4 Discussion 122
Acknowledgments 123
References 124

12. **Demonstration of Viral Antibodies by an Immunochromatographic Strip Test** **127**

Akira Nishizono

12.1 Introduction 127

12.2 Materials 128

12.3 Methods 129

12.4 Discussion 130

Acknowledgments 131

References 131

13. **Rabies Diagnosis: Demonstration of Viral Antibodies by Flow Cytometry** **133**

Juliano Bordignon and Carlos Roberto Zanetti

13.1 Introduction 133

13.2 Materials 134

13.3 Methods 135

13.4 Discussion 136

Acknowledgments 139

References 140

14. **Demonstration of Rabies Antibody by a Latex Agglutination Test** **141**

Songsri Kasempimolporn, Narongsak Chaiyabutr, and Visith Sitprija

14.1 Introduction 141

14.2 Materials 142

14.3 Methods 142

14.4 Discussion 143

References 145

15. **Demonstration of Viral Antibodies by Pseudotype Virus Neutralization** **147**

Edward Wright

15.1 Introduction 147

15.2 Materials 148

15.3 Methods 149

15.4 Discussion 153

Acknowledgments 154

References 155

Section D: Typing/Differentiation of Lyssaviruses 157

16. Sanger Sequencing of Lyssaviruses 159
Lorraine M. McElhinney, Denise A. Marston, Richard J. Ellis, Conrad M. Freuling, Thomas F. Müller, and Anthony R. Fooks
16.1 Background 159
16.2 Methodology 161
16.3 Discussion 169
Acknowledgments 170
References 170

17. Next Generation Sequencing of Lyssaviruses 171
Denise A. Marston, Lorraine M. McElhinney, Emma Wise, Richard J. Ellis, Conrad M. Freuling, Thomas F. Müller, and Anthony R. Fooks
17.1 Background 171
17.2 Methodology 172
17.3 Discussion 181
Acknowledgments 183
References 183

18. Genetic Characterization of Rabies Viruses by *In Situ* Hybridization 185
Susan A. Nadin-Davis
18.1 Introduction 185
18.2 Materials 186
18.3 Methods 187
18.4 Discussion 189
Acknowledgments 191
References 191

19. Oligonucleotide Microarray: Applications for Lyssavirus Speciation 193
Akbar Dastjerdi, Anthony R. Fooks, and Nicholas Johnson
19.1 Introduction 193
19.2 Methodology 195
19.3 Discussion 201
Acknowledgments 202
References 202

20. Demonstration of Lyssavirus Nucleic Acids by Pyrosequencing **205**

Paola De Benedictis and Cristian De Battisti

20.1 Introduction 205
20.2 Materials 206
20.3 Methods 206
20.4 Discussion 210
Acknowledgments 214
References 216

Section E: Demonstration of Viral-Induced Changes in Tissues/Organs **219**

21. Rabies Diagnosis: MR Imaging **221**

Jiraporn Laothamatas, Witaya Sungkarat, and Thiravat Hemachudha

21.1 Introduction 221
21.2 Materials 222
21.3 Methods 223
21.4 Discussion 230
Acknowledgments 230
References 231

PART THREE: RABIES BIOLOGICS

Section F: Rabies Vaccines for Humans or Other Animals **235**

22. Attenuated Vaccines for Veterinary Use **237**

Ad Vos, Grit Neumann, Boris Hundt, and Andreas Neubert

22.1 Introduction 237
22.2 Materials 238
22.3 Methods 240
22.4 Discussion 243
References 244

23. Pox Viral Vectored Vaccines for Rabies **245**

Jacqueline Weyer and Louis H. Nel

23.1 Introduction 245
23.2 Materials 246
23.3 Methods 248
23.4 Discussion 251
References 252

24. Recombinant Rabies Virus Vaccines **255**
Milosz Faber
24.1 Introduction 255
24.2 Materials, Methods, and Results 256
24.3 Discussion 260
References 261

25. Adenoviral Vectors as Vaccine Carriers for Prevention of Rabies **265**
Hildegund C.J. Ertl
25.1 Introduction 265
25.2 Basic Characteristics of Ad Viruses 266
25.3 Types of Ad Vectors 266
25.4 Transgene Product for Rabies Vaccines 267
25.5 Construction of Ad Vectors 267
25.6 Ad Virus Rescue and Expansion 268
25.7 Quality Control of Ad Vectors 270
25.8 Discussion 272
References 273

Section G: Rabies Antibodies/Fragments **275**

26. Production of Human Monoclonal Antibodies using Mammalian Expression Systems **277**
Wilfred E. Marissen
26.1 Introduction 278
26.2 Materials 278
26.3 Methods 279
Acknowledgments 288
References 289

27. Generation of Immune Globulin Single Variable Domains by Display Technologies **291**
Hans de Haard, Anna Hultberg, Theo Verrips, Peter Vanlandschoot, Michael Saunders, Catelijne Stortelers, and Steven Van Gucht
27.1 Introduction 292
27.2 Materials 293
27.3 Methods 294
27.4 Discussion 298
27.5 Experimental Tips 298

Acknowledgments 300
References 301

PART FOUR: APPENDIX

Appendix A 305
Thirumeni Nagarajan and Charles Rupprecht

Appendix B 307
Thirumeni Nagarajan and Charles Rupprecht

Appendix C 309
Evelyne Picard-Meyer, Emmanuelle Robardet, and Florence Cliquet

Index *321*

FOREWORD

This first edition of *Current Laboratory Techniques in Rabies Diagnosis, Research, and Prevention* is the endproduct of the cumulative efforts, joint expertise, and persistent determination of dedicated professionals in the field of rabies prevention and control. Their aim in writing and publishing this book is to provide the most up-to-date information on techniques to detect lyssaviruses, such as rabies virus, and also to provide further resources for those working in the field of rabies management and research throughout the world.

Writing an updated book was deemed necessary by global rabies experts due to the numerous technical advances in rabies methodology that have occurred during the past 20 years. Additionally, over the past decade, there has been an increasing focus on supporting large-scale integrated rabies prevention and control programs that necessitate improved capacity for laboratory-based surveillance. By integrating newer diagnostic tools described in this book into routine rabies diagnostic procedures, public health professionals will have access to a more comprehensive picture of the epidemiology of lyssaviruses in specific regions as well as globally. Thus, this new edition will provide public health professionals involved in rabies programs with new tools to not only monitor existing rabies endemic situations, but also to predict regions where future outbreaks may occur, thereby limiting or preventing rabies virus from expanding into these high risk areas.

With a few notable exceptions, there is generally a wide gap between the laboratory capacity in most regions of the world where canine-mediated human rabies still causes tens of thousands of deaths every year, and in countries where dog-to-dog transmission of rabies has been eliminated or is on the verge of elimination. Unfortunately, the ability to diagnose rabies in many canine rabies-endemic countries remains limited or nonexistent. Without the capability to conduct rabies diagnoses on a routine basis in endemic countries, it is impossible to determine the risk of infection after an animal bite, and therefore all human exposures to suspect rabid animals must be considered high risk and should be considered for post-exposure prophylaxis (PEP). By incorporating the ability to diagnose rabies in animals that expose humans, governments begin to construct the first building blocks in the establishment of reliable risk-assessment systems that ultimately save human lives, while decreasing the overall cost of rabies biologics. Without the establishment of standardized risk-assessment systems that include reliable diagnostic laboratories, the cost of providing PEP to all bite victims has the potential to escalate exponentially and eventually overwhelm public health budgets already facing less than adequate resources. Therefore, establishing the capacity to conduct rabies diagnoses should be considered an investment in basic public health, not only from the standpoint of saving human lives, but also as an investment in eventually reducing the financial burden of providing vaccines to those at risk, and as an investment in building sustainable rabies prevention and control programs.

The wealth of comprehensive information contained in *Current Laboratory Techniques in Rabies Diagnosis, Research, and Prevention* may not be immediately applicable to newly established diagnostic laboratories with limited budgets for purchasing expensive equipment, but the meticulous information describing routine procedures for detecting the presence of lyssaviruses will provide essential information for laboratories that are aiming to build effective surveillance systems for monitoring the epidemiology of rabies and the success of the control programs in their own country.

On behalf of all of us working in the field of public health, our special thanks go out to all of the contributors of *Current Laboratory Techniques in Rabies Diagnosis, Research, and Prevention.* It is hoped that the information contained herein will encourage governments and public health experts to establish reliable diagnostic facilities in regions where animal exposure continues to pose a threat to human and animal health, and to inspire researchers to design even more effective systems in the near future.

Deborah J. Briggs
Executive Director, Global Alliance for Rabies Control,
Manhattan, KS, USA

LIST OF CONTRIBUTORS

Juliano Bordignon
Department of Virology, Instituto Carlos Chagas, ICC/Fiocruz, Curitiba, Paraná, Brazil; Virology, Instituto Carlos Chagas, ICC/Fiocruz, Curitiba, Paraná, Brazil

Narongsak Chaiyabutr
Queen Saovabha Memorial Institute, Thai Red Cross Society, Bangkok, Thailand

Florence Cliquet
French Agency for Food, Environmental and Occupational Health and Safety (Anses), Nancy Laboratory for Rabies and Wildlife, Community Reference Laboratory for Rabies, Community Reference Laboratory for Rabies Serology, OIE Reference Laboratory for Rabies, Technopôle Agricole et Vétérinaire, WHO Collaborating Centre for Research and Management in Zoonoses Control, Malzéville, France

Jessica Coertse
Department of Microbiology and Plant Pathology, University of Pretoria, Pretoria, South Africa

Andre Coetzer
Department of Microbiology and Plant Pathology, University of Pretoria, Pretoria, Gauteng, South Africa

Akbar Dastjerdi
Animal Health and Veterinary Laboratories Agency, Addlestone, Surrey, United Kingdom

Cristian De Battisti
FAO Reference Centre for Rabies, Istituto Zooprofilattico Sperimentale delle Venezie, Legnaro (Padova), Italy

Paola De Benedictis
FAO Reference Centre for Rabies, Istituto Zooprofilattico Sperimentale delle Venezie, Legnaro (Padova), Italy

Hans de Haard
arGEN-X, Zwijnaarde, Belgium

Richard J. Ellis
Central Sequencing Unit, Animal Health and Veterinary Laboratories Agency, Addlestone, Surrey, United Kingdom

Hildegund C.J. Ertl
The Wistar Institute, Philadelphia, PA, USA

Milosz Faber
Department of Microbiology and Immunology, Thomas Jefferson University, Philadelphia, Pennsylvania, USA

Melina Fischer
FLI, Institute of Diagnostic Virology, Greifswald-Insel Riems, Germany

Anthony R. Fooks
University of Liverpool, Department of Clinical Infection, Microbiology and Immunology, Liverpool, United Kingdom; Wildlife Zoonosis and Vector-Borne Disease Research Group, Animal Health and Veterinary Laboratories Agency, Addlestone, Surrey, United Kingdom; Institute of Infection and Global Health, University of Liverpool, Liverpool, United Kingdom

Conrad M. Freuling
Friedrich-Loeffler-Institut (FLI), Federal Research Institute for Animal Health, Institute of Molecular Virology and Cell Biology, Greifswald-Insel Riems, Germany

Thiravat Hemachudha
WHO Collaborating Center for Research and Training on Viral Zoonoses, and Department of Medicine, Faculty of Medicine, Chulalongkorn University, Bangkok, Thailand

Bernd Hoffmann
FLI, Institute of Diagnostic Virology, Greifswald-Insel Riems, Germany

Anna Hultberg
arGEN-X, Zwijnaarde, Belgium

Boris Hundt
IDT Biologika GmbH, Dessau-Rosslau, Germany

Takuya Itou
Nihon University Veterinary Research Center, Fujisawa, Kanagawa, Japan

Nicholas Johnson
Animal Health and Veterinary Laboratories Agency, Addlestone, Surrey, United Kingdom

Songsri Kasempimolporn
Queen Saovabha Memorial Institute, Thai Red Cross Society, Bangkok, Thailand

Ivan V. Kuzmin
Global Alliance for Rabies Control, Manhattan, KS, USA

Jiraporn Laothamatas
Advanced Diagnostic Imaging Center, and Department of Radiology, Faculty of Medicine Ramathibodi Hospital, Mahidol University, Bangkok, Thailand

Shampur N. Madhusudana
Department of Neurovirology, WHO Collaborating Centre for Reference and Research on Rabies, National Institute of Mental Health and Neurosciences, Bangalore, India

Wilfred E. Marissen
Crucell Holland B.V., Leiden, The Netherlands

Wanda Markotter
Department of Microbiology and Plant Pathology, University of Pretoria, Pretoria, South Africa

Denise A. Marston
Wildlife Zoonosis and Vector-Borne Disease Research Group, Animal Health and Veterinary Laboratories Agency, Addlestone, Surrey, United Kingdom

Lorraine M. McElhinney
Wildlife Zoonosis and Vector-Borne Disease Research Group, Animal Health and Veterinary Laboratories Agency, Addlestone, Surrey, United Kingdom

Kader Muhamuda
Research Associate, Department of Microbiology and Molecular Genetics, University of Pittsburgh School of Medicine, Pittsburgh, PA, USA

Thomas F. Müller
Friedrich-Loeffler-Institut, Federal Research Institute for Animal Health, Institute of Molecular Virology and Cell Biology, Greifswald-Insel Riems, Germany

Susan A. Nadin-Davis
Centre of Expertise for Rabies, Canadian Food Inspection Agency, Ottawa Laboratory Fallowfield, Ottawa, Canada

Thirumeni Nagarajan
Research and Development, Biological E. Limited, Shameerpet, Hyderabad, Andhra Pradesh, India

Louis H. Nel
Centre for Emerging and Zoonotic Diseases, National Institute for Communicable Diseases, Sandringham, Gauteng, South Africa; Department of Microbiology and Plant Pathology, University of Pretoria, Pretoria, Gauteng, South Africa

Andreas Neubert
IDT Biologika GmbH, Dessau-Rosslau, Germany

Grit Neumann
IDT Biologika GmbH, Dessau-Rosslau, Germany

Akira Nishizono
Department of Microbiology, Faculty of Medicine, Oita University, Japan

Evelyne Picard-Meyer
French Agency for Food, Environmental and Occupational Health and Safety (Anses), Nancy Laboratory for Rabies and Wildlife, Community Reference Laboratory for Rabies, Community Reference Laboratory for Rabies Serology, OIE Reference Laboratory for Rabies, Technopôle Agricole et Vétérinaire, WHO Collaborating Centre for Research and Management in Zoonoses Control, Malzéville, France

Mani Reeta
Department of Neurovirology, WHO Collaborating Centre for Reference and Research on Rabies, National Institute of Mental Health and Neurosciences, Bangalore, India

Emmanuelle Robardet
French Agency for Food, Environmental and Occupational Health and Safety (Anses), Nancy Laboratory for Rabies and Wildlife, Community Reference Laboratory for Rabies,

Community Reference Laboratory for Rabies Serology, OIE Reference Laboratory for Rabies, Technopôle Agricole et Vétérinaire, WHO Collaborating Centre for Research and Management in Zoonoses Control, Malzéville, France

Charles Rupprecht
Ross University School of Veterinary Medicine, Basseterre, St. Kitts, West Indies

Michael Saunders
arGEN-X, Zwijnaarde, Belgium

Visith Sitprija
Queen Saovabha Memorial Institute, Thai Red Cross Society, Bangkok, Thailand

Catelijne Stortelers
Ablynx, Zwijnaarde, Belgium National Reference Centre of Rabies, Scientific Institute of Public Helath, Brussels, Belgium

Witaya Sungkarat
Advanced Diagnostic Imaging Center, and Department of Radiology, Faculty of Medicine Ramathibodi Hospital, Mahidol University, Bangkok, Thailand

Steven Van Gucht
arGEN-X, Zwijnaarde, Belgium

Peter Vanlandschoot
Ablynx, Zwijnaarde, Belgium National Reference Centre of Rabies, Scientific Institute of Public Helath, Brussels, Belgium

Theo Verrips
Utrecht University, Utrecht, The Netherlands

Ad Vos
IDT Biologika GmbH, Dessau-Rosslau, Germany

Jacqueline Weyer
Centre for Emerging and Zoonotic Diseases, National Institute for Communicable Diseases, Sandringham, Gauteng, South Africa

Emma Wise
Wildlife Zoonosis and Vector-Borne Disease Research Group, Animal Health and Veterinary Laboratories Agency, Addlestone, Surrey, United Kingdom

Edward Wright
Viral Pseudotype Unit, Faculty of Science and Technology, University of Westminster, London, United Kingdom

Carlos Roberto Zanetti
Department of Microbiology, Immunology and Parasitology, Federal University of Santa Catarina, Florianópolis, Santa Catarina, Brazil

PART ONE

Introduction

CHAPTER ONE

Basic Facts about Lyssaviruses

Ivan V. Kuzmin
Global Alliance for Rabies Control, Manhattan, KS, USA

Chapter Contents

1.1 Introduction 3
1.2 Virion and Genome Organization 4
1.3 Phylogeny and Serologic Cross-Reactivity of Lyssaviruses 6
1.4 Host Range 12
1.5 Pathobiology 15
References 17

1.1 INTRODUCTION

Rabies is one of the most feared infectious diseases known to humankind since ancient times.[1] The name of the disease may originate from the Sanskrit word "rabhas" (violence). Rabies (an acute progressive encephalomyelitis) is caused by negative-sense single-stranded RNA (ssRNA) viruses from the genus *Lyssavirus*, family *Rhabdoviridae*.[2] It is neither scientifically nor clinically correct to say that lyssaviruses other than the rabies virus (RABV) cause a "rabies-like" disease, as the disease is the same. Despite the significant progress in rabies prevention and control achieved during recent decades, the disease still causes more than 55,000 human deaths every year.[3] Nearly 99% of these cases are associated with dog-transmitted rabies in the developing countries of Asia and Africa. Nevertheless, even in many developed countries where canine rabies has been eliminated, the disease is still present in wildlife, which poses public health concerns and a possibility of re-introduction of dog rabies.[4] As with other diseases, efficient strategies for rabies control, prevention, prophylaxis, and treatment depend on the availability of effective diagnostic tests, laboratory-based surveillance, and safe and potent biologics. This introductory chapter is dedicated to a basic description of causative agents of rabies, including their taxonomy, morphology, serologic cross-reactivity, and generic pathobiological and ecological patterns that are significant for the development of reliable diagnostic tests, and for the creation of efficient protective biologics.

C. Rupprecht & T. Nagarajan (Eds): Current Laboratory Techniques in Rabies Diagnosis, Research, and Prevention, Volume 1.
Doi: http://dx.doi.org/10.1016/B978-0-12-800014-4.00001-9

1.2 VIRION AND GENOME ORGANIZATION

Lyssaviruses have relatively large bullet-shaped virions, 130–250 nm in length and 60–100 nm in diameter (Figure 1.1). The lipid bilayer envelope is acquired from host cell membranes during budding. The envelope is studded with glycoprotein (G) spikes organized in trimers. The internal nucleocapsid core consists of a ribonucleoprotein (RNP) complex comprising the genomic RNA tightly bound to the nucleoprotein (N), together with an RNA-dependent RNA polymerase (L) and phosphoprotein (P). The nucleocapsid is active for transcription and replication: the N-RNA template is processed by the L protein, which contains most enzymatic activities, and its cofactor the P protein. The nucleocapsid has a helical symmetry, is uncoiled and filamentous, about 700×20 nm in size. In the virion, the matrix protein (M) condenses the nucleocapsid, interacts with the N-RNA complex, and associates with the host-derived lipid bilayer containing the transmembrane G protein.[5]

The lyssavirus genome is represented by a non-segmented, linear, negative-sense ssRNA about 12 kb in length. It includes five major genes that are arranged in the conserved linear order 3′-N-P-M-G-L-5′. Each of the individual genes is flanked by transcription initiation and termination/polyadenylation signals that are largely conserved among members of

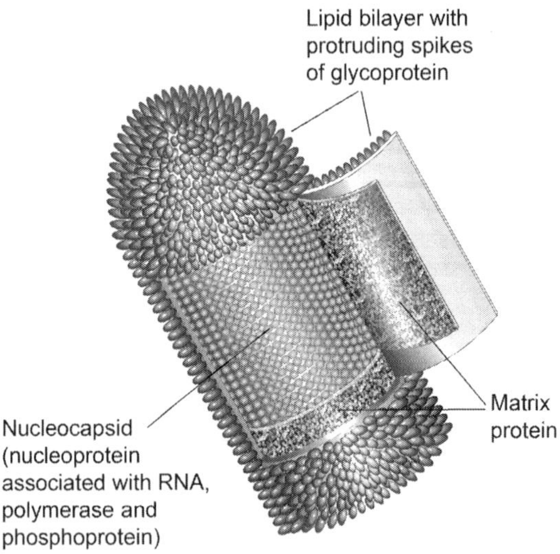

Figure 1.1 Schematic illustration of a lyssavirus virion.

the same species. Transcription units are separated by short untranscribed intergenic regions. The ends of the genomic RNA, termed 3′ leader and 5′ trailer sequences, exhibit terminal complementarity and contain promoter sequences that initiate replication of the genome and antigenome, respectively.[5]

Lyssaviruses have five structural proteins designated L, G, N, P, and M.[5] The L (220–240 kDa) has multiple domains and performs the functions required for genome transcription and replication, including RNA-dependent RNA polymerase, mRNA 5′ capping enzyme, cap methyltransferase, 3′ poly (A) polymerase, and protein kinase activities. The G (65–90 kDa) is assembled into trimers to form the virion surface spikes. The G protein binds to host cell receptors, induces virus endocytosis, and mediates fusion of viral and endosomal membranes. The G protein induces production of virus-neutralizing antibodies and elicits cell-mediated immune responses. The N (47–62 kDa) is the major component of the nucleocapsid, and actively interacts with RNA, and L and P proteins. The P (20–30 kDa) plays multiple roles during transcription and replication as a non-catalytic cofactor of the viral polymerase. It mediates the physical link and proper positioning of the L protein on the N-RNA template, and acts as a chaperone during synthesis of N by forming N-P complexes that prevent N from self-aggregating and binding to cellular RNA.[5] In addition, P was shown to have multiple binding sites for different viral or host proteins to allow their assembly into multi-molecular complexes. The P protein forms elongated and flexible dimers, and has a modular organization composed of three structured domains separated by intrinsically disordered regions. The highly acidic N-terminal domain has binding sites for the viral L and N proteins, host cell kinases, and importins. The central domain has an L protein binding site and dimerization domain, and the C-terminal basic domain is involved in N-RNA binding.[6] The P may also interact with the host cellular transport systems such as the dynein motor complex, nucleo-cytoplasmic transporters, and microtubules[7–9] to facilitate intracellular movement of viral components. Furthermore, P interferes with the innate immune response by inhibiting different steps of the host cell interferon response. The M is an inner component of the virion (20–30 kDa). It appears to be involved in regulation of the genome RNA transcription. The M binds to nucleocapsids and the cytoplasmic domain of the G, thereby facilitating the budding process, and also mediates such pathobiological effects as cell rounding, intracellular membrane redistribution, and apoptosis.[5]

1.3 PHYLOGENY AND SEROLOGIC CROSS-REACTIVITY OF LYSSAVIRUSES

Until the 1950s, it was believed that RABV was unique. The discoveries of serologically related viruses in Nigeria, Lagos bat virus (LBV) from a Pteropodid bat,[10] and Mokola virus from a shrew[11] demonstrated that the structure of this virus group was more complex, and the terms "rabies-related viruses" (RRVs) and "rabies serogroup" were introduced.[11] Another serologically related virus, Duvenhage virus, was isolated from a man who died of rabies after a bite from an insectivorous bat in 1970 in South Africa;[12] it was recognized as a fourth viral serotype.[13] As viruses regularly isolated from bats in Europe since the 1950s were related serologically to Duvenhage virus, they were initially included in the Duvenhage serotype.[14,15]

Later, application of monoclonal antibodies (MAbs) refined the classification of the "rabies serogroup":[16] European bat lyssaviruses were not only distinguished from the African Duvenhage virus,[17] but separated into two distinct serotypes, temporally termed "biotypes".[18] This differentiation was later supported by gene sequencing and phylogenetic analysis.[19] Extensive phylogenetic studies on the diversity of RRVs led to the creation of the novel operational term "genotype", which has been broadly used in the scientific literature for over a decade. New genotypes were discovered, and quantitative criteria for their demarcation were proposed.[19–21]

To accommodate the growing variety of RRVs, the genus *Lyssavirus* was established under the auspices of the International Committee on the Taxonomy of Viruses (ICTV). The name of the genus originated from Greek mythology: Lyssa (Λυσσα) was a goddess, or spirit of rage, fury, madness, or frenzy.[22] The existing "genotypes" served as a basis for lyssavirus taxonomy, but were refined to satisfy the official ICTV taxonomy rules, which operate with more complex entities such as viral species. The present taxonomy of lyssaviruses includes 14 species. One additional representative, Lleida bat virus, is known by a fragment of genome sequence only[23] and therefore does not have taxonomic status (Table 1.1, Figure 1.2).

Demarcation criteria for lyssavirus species include: (1) Genetic distances, with the threshold of 80–82% nucleotide identity for the complete N gene, or 80–81% nucleotide identity for the concatenated coding regions of N+P+M+G+L genes. Globally, all isolates belonging to the same species have higher identity values than the threshold, except the

Table 1.1 Viruses Currently Included in the Genus *Lyssavirus*

Recognized and Proposed Species	Phylogroup	Natural Host	Geographical Range	Comments
Rabies virus; RABV (type species)	I	Bats (*Chiroptera*) of multiple species, terrestrial mammals (predominantly *Carnivora*)	Terrestrial mammals – worldwide (except Australia, Antarctica and several insular territories); bats – New World only	Responsible for the vast majority of human rabies cases in the world. All currently available human and veterinary vaccine strains originate from this species.
Australian bat lyssavirus; ABLV	I	Pteropodid bats (at least 4 species of *Pteropus* genus) and insectivorous bats (*Saccolaimus albiventris*)	Australia (perhaps, with several surrounding islands)	Given limited surveillance, host range among insectivorous bats may be greater. Three human cases documented.
European bat lyssavirus, type 1; EBLV-1	I	Insectivorous bats (predominantly *Eptesicus serotinus*)	The major part of Europe, from Spain to the Ukraine.	Given the limited surveillance in eastern Europe and Asia, may be distributed more broadly, along the reservoir species range. Spill-over infections in wild and companion animals, as well as a human case have been documented.
European bat lyssavirus, type 2; EBLV-2	I	Insectivorous bats (predominantly *Myotis daubentonii* and *M. dasycneme*)	North-western Europe	Two human cases have been documented.
Khujand virus; KHUV	I	Insectivorous bat *Myotis mystacinus*	Central Asia	Known by a single isolate. Given the limited surveillance in eastern Europe and Asia, may be distributed more broadly. No human cases have been documented.

(*Continued*)

Table 1.1 (Continued)

Recognized and Proposed Species	Phylogroup	Natural Host	Geographical Range	Comments
Aravan virus; ARAV	I	Insectivorous bat *Myotis blythi*	Central Asia	Known by a single isolate. Given the limited surveillance in eastern Europe and Asia, may be distributed more broadly. No human cases have been documented.
Bokeloh bat lyssavirus; BBLV	I	Insectivorous bat *Myotis nattereri*	Western Europe	Known by three isolates from Germany and France. No human cases have been documented.
Irkut virus; IRKV	I	Insectivorous bat *Murina leucogaster*	Eastern Asia	Known by three isolates, from bats and from a human who developed rabies after bat bite.
Duvenhage virus; DUVV	I	Insectivorous bats	Sub-Saharan Africa	Known by 4 isolates, 3 of which came from humans, bitten by bats, and one from a bat, of the presumably Miniopterus species.
Lagos bat virus; LBV	II	Pteropodid bats of several genera (*Eidolon helvum, Rousettus aegyptiacus, Epomophorus spp.*, etc.)	Sub-Saharan Africa	Constitutes several lineages with long genetic distances. Potentially, in the future may be subdivided into 2-3 separate species. Spill-over infections reported in wild and companion animals. No human cases documented to date.

Mokola virus; MOKV	II	Unknown	Sub-Saharan Africa	Twice isolated from shrews, once from a rodent. The majority of other isolates were obtained from companion animals, such as cats and dogs, as the result of spill-over infections. Two human cases have been reported.
Shimoni bat virus; SHIBV	II	Insectivorous bat *Hipposideros commersoni* (*H. vittatus*)	Kenya	Known by a single isolate. Serologic surveys suggest that *H. commersoni (H. vittatus)* is the likely reservoir. No human cases have been documented.
West Caucasian bat virus; WCBV	NA*	Insectivorous bats from genus *Miniopterus*	South-eastern Europe	Known by a single isolate. However, serologic surveys suggest that WCBV (or other serologically-related virus) may be present in Miniopterus bats in Africa (Kenya). No human cases have been documented.
Ikoma lyssavirus; IKOV	NA	African civet (*Civettictis civetta*)	Tanzania	Known by a single isolate. The natural host is questionable. Given phylogenetic relatedness to the West Caucasian bat virus, the index case in African civet may result from a spill-over infection of bat origin. No human cases have been documented.
Lleida bat lyssavirus; LLEBV	NA	Insectivorous bat *Miniopterus schreibersii*	Spain	Known by a single partial genome sequence with unsuccessful virus isolation efforts. No human cases have been documented.

*NA: Neither a member of Phylogroup I nor II. Other Phylogroup not assigned.

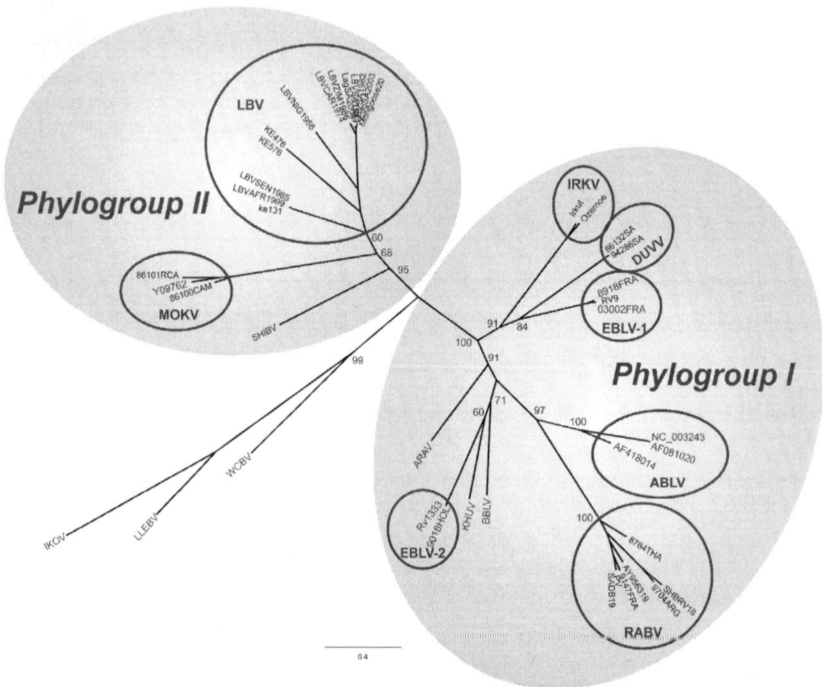

Figure 1.2 Maximum likelihood phylogenetic tree of lyssaviruses based on partial nucleoprotein gene sequences (850 nucleotides). Branch lengths are drawn to scale, and bootstrap values (1,000 replicates) are shown for key nodes.

viruses currently included in the *Lagos bat virus* (LBV), species. However, as these viruses are segregated into a monophyletic cluster, in the absence of other demarcation characters, there is currently no objective plan to subdivide LBV into several species, as was suggested by several authors.[24,25] (2) Topology and consistency of phylogenetic trees, obtained with various evolutionary models. (3) Antigenic patterns in reactions with anti-nucleocapsid MAbs and serologic cross-reactivity using polyclonal antisera. (4) Whenever available, additional characteristics, such as ecological properties, host and geographic range, and pathological features are considered.[5] In the future, such essential information may help to elucidate the diversity of viruses currently included in the LBV species.

Based on genetic distances and serologic cross-reactivity, the *Lyssavirus* genus has been subdivided into two phylogroups.[26] The demarcation was based on phylogenetic relationships: 74% or more amino acid sequence identity within the G ectodomain between viruses of one phylogroup, and less than 64% sequence identity between viruses from different phylogroups; cross-neutralization within phylogroups, but absence of cross-neutralization between members of different phylogroups; the presence of K/R333 in the G of Phylogroup I viruses and D333 in the G of Phylogroup II viruses (with following implication for a potential of reduced pathogenicity for the latter). In general, serologic cross-reactivity somewhat correlated with genetic distances, and sequences that shared more than 72% amino acid sequence identity within their G ectodomains cross-neutralized each other.[26]

Phylogroup I includes the species *Rabies virus* (RABV), *European bat lyssaviruses, type 1* (EBLV-1) and *type 2* (EBLV-2), *Duvenhage virus* (DUVV), *Australian bat lyssavirus* (ABLV), *Aravan virus* (ARAV), *Khujand virus* (KHUV), *Bokeloh bat lyssavirus* (BBLV) and *Irkut virus* (IRKV). Phylogroup II includes *Lagos bat virus* (LBV), *Mokola virus* (MOKV), and *Shimoni bat virus* (SHIBV). The remaining viruses, *West Caucasian bat virus* (WCBV), *Ikoma lyssavirus* (IKOV), and Lleida bat virus (LLEBV), cannot be included in either of these phylogroups. Although the latter three viruses are related phylogenetically to each other (Figure 1.2), genetic distances between them are greater than the distances between Phylogroups I and II, and it is unclear whether WCBV, IKOV, and LLEBV would cross-neutralize each other. Therefore, it is not possible to assign these three viruses to a new Phylogroup III based on the existing demarcation criteria.

The limited serologic cross-reactivity between lyssaviruses from different phylogroups is important from a public health and veterinary standpoint. All commercially available rabies biologics are based on RABV. They elicit efficient protection against Phylogroup I lyssaviruses, but not against other members of the genus.[26–29] Moreover, antigenic and genetic distinctions should be considered in applications of diagnostic tests to ensure their ability to detect all varieties of lyssaviruses, as described in the relevant chapters of this book.

Several approaches were explored in the development of biologics capable of eliciting protection against non-Phylogroup I lyssaviruses. The first potential MOKV vaccine, a recombinant baculovirus expressing MOKV G, elicited protection against the homologous virus, but not against a heterologous challenge with RABV.[30] Several vectors and promoters were implemented in other DNA vaccines against MOKV, which expressed either G gene, N gene, or both. Neither vaccine was fully protective in a single immunization, although booster doses increased their efficacy significantly, except the vaccine based on N gene.[31]

Other studies explored DNA vaccines that expressed MOKV and RABV chimeric G genes.[27,28] The success of this strategy lies in the observation that lyssavirus G can be divided into two parts, separated by a flexible bridge. Each of these parts contains one of the two important antigenic domains in the elicitation of a protective immune response, the antigenic sites II (NH_2 part) and III (COOH part). Such chimeric vaccines induced VNAs against RABV, EBLV-1, EBLV-2, MOKV and LBV, e.g. against a variety of Phylogroup I and II viruses available at that time. The only virus that for some reason was neutralized weakly was DUVV.[27,28]

Experimental vaccinia viruses encoding G genes of RABV, MOKV and WCBV, either singly or in dual combinations, were constructed in another study.[32] Constructs expressing a single G gene protected animals against challenge with homologous virus. Similarly, recombinants expressing G genes from two lyssaviruses induced protection against both homologous viruses, but no significant cross-reactivity was observed (e.g. the RABV+MOKV vaccine elicited protection against RABV, MOKV and LBV but did not protect from WCBV, whereas the RABV+WCBV vaccine protected mice from RABV and WCBV only).

With recent progress in reverse genetics and recombinant vaccine construction[33] an opportunity exists to develop modern next generation biologics, which would incorporate G genes from several different viruses in a selected backbone. Obviously, there are limitations for the number of G genes inserted into a viral genome. However, insertion of at least two additional G genes in the RABV genome was attempted successfully.[33] Construction of three chimeric G genes, and a replacement of the original G gene with these three constructs in a lyssavirus backbone, may cover the antigenic spectrum of lyssaviruses detected to date. However, it is unclear whether all the inserted G genes will be expressed efficiently and represented on the virion surface. Likely, larger DNA-viruses should be considered for the creation of recombinant "pan-lyssavirus" vaccines. Issues relating to the construction of such biologics are covered in other relevant chapters of this book.

1.4 HOST RANGE

Although all mammals are susceptible to lyssaviruses, rabies in nature is maintained primarily by carnivores and bats. Given the diversity of lyssaviruses found in bats compared to those in carnivores (Table 1.1), it is commonly believed that bats are the primary evolutionary hosts of lyssaviruses.[34–35] In addition, a diversity of lyssaviruses has been documented in the Old World, whereas only RABV was found in the New World to date. This suggested that lyssaviruses evolved in the Old World.[2] Yet, the existing knowledge does not explain why RABV is present in American bats in the absence of this virus in Old World bats. By inference from phylogeny, RABV should be present in Old World bats. The absence of such records may be explained in part by poor surveillance. For example, although bat lyssaviruses have been documented over the globe, no such records exist to date for unique species in south-eastern Asia, the Middle East, and Northern Africa, although isolation of "RABVs" from bats was reported historically from Thailand[36] and India[37] and more recently was supported by detection of anti-lyssavirus antibodies in bat sera throughout Asia.[38–41]

Typically, each lyssavirus species and each phylogenetic lineage within a species is associated with a specific host. These relationships are best studied for RABV which is distributed globally, with the exception of several insular territories, Australia, and Antarctica.[3] The number of phylogenetic lineages of RABV documented in New World bats corresponds roughly to the number of bat species tested so far, which suggests a long history of co-adaptation of each viral variant and its reservoir host. Moreover, some broadly distributed bat species, such as *Eptesicus fuscus* and *Tadarida brasiliensis*, harbour several lineages of RABV.[42]

At least several well-established lineages of RABV, circulating in North American terrestrial mammals (i.e. the raccoon lineage, south-central skunk lineage, and other skunk lineages from Mexico) are linked phylogenetically to bat RABVs.[34,43] Among wild carnivores, skunks of different species appear very permissive to various RABV variants. In addition to the south-central and Mexican skunk RABVs which are associated phylogenetically with bat viruses, there are several other lineages (i.e. north-central skunk, California skunk, and dog-related skunk RABV lineages from Mexico) which belong to a "cosmopolitan" RABV lineage. Skunks in Eastern Canada have acquired the Arctic RABV variant,[43,44] and skunks along the east coast of North America are frequently infected by the raccoon RABV.[45–46] Moreover, several outbreaks caused by bat RABV were documented in skunks in Arizona.[47]

In contrast, the North American raccoon (*Procyon lotor*) seems well adapted and constrained only to one specific RABV variant, which is believed to have emerged during the 1940s in Florida,[34,43] and since then has rapidly spread along the east coast of North America.[48] Intensive oral rabies vaccination along the border of the enzootic area currently prevents the spread of raccoon rabies to the western parts of the continent.[4]

Other American lineages of RABV circulating in dogs, coyotes (*Canis latrans*) and several species of foxes, belong to the "cosmopolitan" group. The "cosmopolitan" RABV has spread with infected dogs to many parts of the world during the period of European colonization.[34,43,44] Rabies control in dogs resulted in the elimination of dog-mediated rabies in the USA and Canada, although it still occurs in Latin America. Mongooses in several Caribbean islands acquired the "cosmopolitan" RABV from dogs, and maintain its circulation.[49–50] In addition to the representatives of "cosmopolitan" RABV lineage, Arctic RABV circulates among Arctic foxes (*Vulpes lagopus*), red foxes (*V. vulpes*) and striped skunks (*Mephitis mephitis*) in northern territories of the continent.[51,52] Oral rabies vaccination of wild canids significantly reduced RABV circulation in their populations in North America.

In the Old World, RABV has been documented in carnivores only as concerns reservoirs, with the exception of epizootics in kudu antelopes (*Tragelaphus strepsiceros*) in southern Africa,[53] predominantly in canids and herpestids.[4] The "cosmopolitan" lineage of RABV circulates in dogs and wild canids in vast territories of Eurasia and Africa.[43,44,54] In addition, distinct "autochthonous" lineages of RABV circulate in dogs in several areas of Africa and south-east Asia.[43] Arctic RABV circulates circumpolarly among Arctic and red foxes. Phylogenetic inferences suggest that the Arctic RABV lineage originated from so-called Arctic-like lineages, currently present in the Indian subcontinent, central and eastern Asia.[51,52] Raccoon dogs (*Nyctereutes procyonoides*), besides maintaining the circulation of Arctic-like RABV in their native habitats in eastern Asia, readily participate in the circulation of "cosmopolitan" RABV in the areas enzootic for fox rabies where this species was successfully introduced. The wolf (*Canis lupus*) was historically a major wildlife rabies reservoir in Eurasia. However, with population reductions, the significance of this species in circulation of RABV has diminished.[55,56]

Following the rational use of oral vaccination programs, rabies had been eliminated from the red fox populations in Western Europe by the end of the 20th century.[57] Nevertheless, rabies still remains present in red foxes, corsac foxes, jackals and raccoon dogs in Eastern Europe and Asia, which carries a risk of re-introduction to the countries that are currently considered "rabies free".[55,56,58]

Mustelids were almost never reported as important rabies reservoirs, although the European badger (*Meles meles*) might be relatively frequently infected in the areas affected by fox rabies, but these represent spill-over infections rather than sustained circulation.[56] The only documented exception is the Chinese ferret-badger (*Melogale moschata*), which was suggested as a reservoir of a specific RABV lineage of dog origin.[59] Recently, ferret-badger rabies was documented in Taiwan after several decades of absence of terrestrial rabies.[60] Mongoose (e.g. *Herpestidae*) maintain circulation of dog-related RABV in southern Asia, but in Africa they harbour a specific RABV variant.[61,62] Domestic and wild species of cats serve as vectors, and a source of exposure to humans, but have never been documented as capable reservoirs that can maintain sustained circulation of RABV in the absence of another major host. Although rodents may be infected like other mammals, reservoir species have not been documented.

Geographic and host ranges of other lyssaviruses are more restricted compared to RABV. A brief description is provided in Table 1.1.

1.5 PATHOBIOLOGY

Lyssaviruses are quintessential neurotropic pathogens. The virus is generally delivered into a wound via a bite. Very rarely an exposure may occur via contamination of mucous surfaces, or via inhalation of highly-concentrated aerosols generated in laboratory settings or in the unique environment of caves inhabited by multi-million-member colonies of *Tadarida brasiliensis* bats.[2] The virus can replicate at the inoculation site, as was shown for skeletal muscle fibres.[63] The virus typically penetrates neurons through the neuromuscular junction.[64] The attachment to the host cell receptors is mediated by G spikes protruding from the virion envelope. Once bound, the virus enters the neuron by endocytosis, and the endosomal vesicle is transported by retrograde axonal flow dependent on the microtubule network, using dynein motors. The pH decrease within the endosome provokes conformational changes of the G spikes, which mediates fusion between the endosomal and viral membranes, and the (-) strand RNA genome encapsidated by the N protein is liberated into the cytoplasm.[64,65] Transcription and replication occur at the neuron perikaryon.

The N-RNA template is transcribed by the viral polymerase (RdRp) and its cofactor, the P protein. This generates in cascades the monocistronic (+) strand RNAs, corresponding to the leader RNA (Le^+), and the five mRNAs.[66] All but one of the monocistronic mRNAs produce a single protein from a single ORF initiated at the first AUG codon. The P mRNA produces 3 or 4 proteins, initiated from downstream in-frame AUG codons.[67] Because the RdRp is pausing at each termination signal to polyadenylate the proximal mRNA, it eludes the intergenic region and only reinitiates at the distal initiation/capping signal, and the size of the untranscribed intergene regions influences the rate of re-initiation. Consequently, there is a progressive decrease in mRNA amounts from the 3′ encoded to the 5′ encoded cistrons. The N and P proteins, encoded by the two 3′ proximal mRNAs, are produced in greater quantities. The N gene is one of the most conserved lyssavirus genes whereas the P gene is variable. For these reasons, the N gene is most commonly targeted by molecular diagnostic tests to ensure their sensitivity and reliability.

The viral progeny is released at the synaptic junctions and propagates up to the central nervous system (CNS) by using motor or sensory neurons and following neuronal connections. This avoids or limits internal surveillance by the host immune system, resulting in the absence of an early antibody response. The duration of an asymptomatic incubation period is 2 months on average,[64] but may be very variable, from a week to several months and even years.[68] During the incubation period no rabies diagnosis can be performed on the animal or human tissues due to very limited viral load, uncertain location of the virus in peripheral nerves, and absence of immune response.

Having been delivered to the CNS, the virus disseminates rapidly. Nearly all regions of the CNS may be affected, but the *medulla oblongata* is a principal area from which the virus is transported to the salivary glands along the innervation pathways which allows the transmission to other susceptible hosts via a bite.[69] Therefore, the most reliable diagnosis of rabies can be done during clinical period of the disease (or post-mortem) on CNS tissues, and the brainstem is the most important region for diagnostic tests.

The clinical period is rapid and severe (usually one to two weeks). The prodromal period is lacking specific clinical signs and is limited to fever, malaise, sometimes flu-like signs. Patients may feel paresthesia and tingling in the inoculation site. With the development of encephalitis, cerebral symptoms manifest. Humans often develop hallucinations and delirium. Some symptoms, which have been considered as 'classic' but are only observed in 50% or less of patients, are hydrophobia, aerophobia, photophobia, and phonophobia. Immune response can usually be detected after several days of disease onset. As the N is present in infected cells in the largest amounts, anti-N antibodies (non-neutralizing) can be detected first by such methods as indirect fluorescent antibody (IFA) test or enzyme-linked immunosorbent assay (ELISA). The virus-neutralizing anti-G antibodies can usually be detected a few days later by any of the dedicated tests, if at all. The antibodies can be detected in both serum and cerebrospinal fluid (CSF). The presence of anti-lyssavirus antibodies in serum may occur in vaccinated patients whereas the presence of such antibodies in CSF is indicative for CNS infection (rabies). In a majority of cases the immune response cannot clear lyssavirus infection. Once symptoms appear, death occurs usually within 1–10 days due to cardiac or respiratory arrest. Intensive care may prolong the disease for 1–1.5 months.[70,71] A few cases of recovery from clinical rabies have been documented in humans, and other animals.[72–76]

Before the fatal outcome, the infected animal transmits the infection to a number of susceptible individuals (usually via bite) to ensure sustained viral circulation. In addition to salivary glands, during the late stage of the disease, the virus spreads from the CNS to the periphery via neuronal pathways, and can be detected in various organs and tissues (which is critical for organ transplantation issues). Detection of RABV in saliva and in nerves, and surrounding hair follicles of skin, are the essential tests for ante-mortem diagnosis in humans.[77–79]

REFERENCES

1. Baer GM, Neville J, Turner GS. *Rabbis and rabies: A pictorial history through the ages.* Mexico City, Mexico: Laboratorios Baer; 1996.
2. Kuzmin IV, Tordo N. Genus Lyssavirus. In: Dietzgen RD, Kuzmin IV, editors. *Rhabdoviruses: Molecular taxonomy, evolution, genomics, ecology, host–vector interactions, cytopathology and control.* London, UK: Caister Academic Press; 2012. p. 37–57.
3. WHO expert consultation on rabies *Second report. WHO Technical Report Series; no. 982.* Geneva, Switzerland: World Health Organization; 2013.
4. Rupprecht CE, Slate D. Rabies prevention and control: Advances and challenges. In: Dietzgen RD, Kuzmin IV, editors. *Rhabdoviruses: Molecular taxonomy, evolution, genomics, ecology, host–vector interactions, cytopathology and control.* London, UK: Caister Academic Press; 2012. p. 215–52.
5. Dietzgen RG, Calisher CH, Kurath G, Kuzmin IV, Rodriguez LL, Stone DM, et al. Family Rhabdoviridae. In: King AMQ, Adams MJ, Carstens EB, Lefkowitz EL, editors. *Virus Taxonomy: Ninth report of the International Committee on Taxonomy of Viruses.* Oxford, UK: Elsevier; 2011. p. 686–714.
6. Leyrat C, Ribeiro Jr. EA, Gérard FCA, Ivanbov I, Ruigrok RWH, Jamin M. Structure, interactions with host cell and functions of rhabdovirus phosphoprotein. *Future Virol* 2011;**6**:465–81.

7. Das SC, Nayak D, Zhou Y, Pattnaik AK. Visualization of intracellular transport of vesicular stomatitis virus nucleocapsids in living cells. *J Virol* 2006;**80**:6368–77.
8. Albertini AAV, Schoehn G, Weissenhorn W. Ruigrok RWH. Structural aspects of rabies virus replication. *Cell Mol Life Sci* 2008;**65**:282–94.
9. Min BE, Martin K, Wang R, Tafelmeyer P, Bridges M, Goodin M. A host–factor interaction and localization map for a plant-adapted rhabdovirus implicates cytoplasm-tethered transcription activators in cell-to-cell movement. *Mol Plant Microbe Interact* 2010;**23**:1420–32.
10. Boulger LR, Porterfield JS. Isolation of a virus from Nigerian fruit bats. *Trans Roy Soc Trop Med Hyg* 1958;**52**:421–4.
11. Shope RE, Murphy FA, Harrison AK, Causey OR, Kemp GE, Simpson DI, et al. Two African viruses serologically and morphologically related to rabies virus. *J Virol* 1970;**6**:690–2.
12. Meredith CD, Rossouw AP, van Praag Koch H. An unusual case of human rabies thought to be of chiropteran origin. *S Afr Med J* 1971;**45**:767–9.
13. Schneider LG, Dietzschold B, Dierks RE, Matthaeus W, Enzmann PJ, Strohmaier K. Rabies group-specific ribonucleoprotein antigen and a test system for grouping and typing of rhabdoviruses. *J Virol* 1973;**11**:748–55.
14. Schneider LG. Antigenic variants of rabies virus. *Comp Immunol Microbiol Infect Dis* 1982;**5**:101–7.
15. Schneider LG, Barnard BJH, Schneider HP. Application of monoclonal antibodies for epidemiological investigations and oral vaccination studies: I. African viruses. In: Kuwert E, Merieux C, Koprowski H, Bogel K, editors. *Rabies in the tropics*. Berlin, Germany: Springer Verlag; 1985. p. 49–53.
16. Wiktor TJ, Koprowski H. Monoclonal antibodies against rabies virus produced by somatic cell hybridization: detection of antigenic variants. *Proc Natl Acad Sci USA* 1978;**75**:3938–42.
17. Dietzschold B, Rupprecht CE, Tollis M, Lafon M, Mattei J, Wiktor TJ, et al. Antigenic diversity of the glycoprotein and nucleocapsid proteins of rabies and rabies-related viruses: implications for epidemiology and control of rabies. *Rev Infect Dis* 1988;**10**(S4):785–98.
18. King A, Davis P, Lawrie A. The rabies viruses of bats. *Vet Microbiol* 1990;**23**:165–74.
19. Bourhy H, Kissi B, Tordo N. Molecular diversity of the Lyssavirus genus. *Virology* 1993;**194**:70–81.
20. Fraser GC, Hooper PT, Lunt RA, Gould AR, Gleeson LJ, Hyatt AD, et al. Encephalitis caused by a lyssavirus in fruit bats in Australia. *Emerg Infect Dis* 1996;**2**:327–31.
21. Kuzmin IV Hughes GJ, Botvinkin AD, Orciari LA, Rupprecht CE. Phylogenetic relationships of Irkut and West Caucasian bat viruses within the Lyssavirus genus and suggested quantitative criteria based on the N gene sequence for lyssavirus genotype definition. *Virus Res* 2005;**111**:28–43.
22. Theodorides J. *Histoire de la Rage "Cave Canem."*. Paris, France: Masson; 1986.
23. Aréchiga Ceballos N, Vázquez Morón S, Berciano JM, Nicolás O, Aznar López C, Juste J, et al. Novel lyssavirus in bat, Spain. *Emerg Infect Dis* 2013;**19**:793–5.
24. Markotter W, Kuzmin I, Rupprecht CE, Nel LH. Phylogeny of Lagos bat virus: challenges for lyssavirus taxonomy. *Virus Res* 2008;**135**:10–21.
25. Delmas O, Holmes EC, Talbi C, Larrous F, Dacheux L, Bouchier C, et al. Genomic diversity and evolution of the lyssaviruses. *PLoS One* 2008;**3**:e2057.
26. Badrane H, Bahloul C, Perrin P, Tordo N. Evidence of two lyssavirus phylogroups with distinct pathogenicity and immunogenicity. *J Virol* 2001;**75**:3268–76.
27. Bahloul C, Jacob Y, Tordo N, Perrin P. DNA-based immunisation for exploring the enlargement of immunological cross-reactivity against the lyssaviruses. *Vaccine* 1998;**16**:417–25.

28. Jallet C, Jacob Y, Bahloul C, Drings A, Desmezieres E, Tordo N, et al. Chimeric lyssavirus glycoproteins with increased immunological potential. *J Virol* 1999;**73**:225–33.
29. Hanlon CA, Kuzmin IV, Blanton JD, Weldon WC, Manangan JS, Rupprecht CE. Efficacy of rabies biologics against new lyssaviruses from Eurasia. *Virus Res* 2005;**111**:44–54.
30. Tordo N, Bourhy H, Sather S, Ollo R. Structure and expression in baculovirus of the Mokola virus glycoprotein: an efficient recombinant vaccine. *Virology* 1993;**94**:59–69.
31. Nel LH, Niezgoda M, Hanlon CA, Morril PA, Yager PA, Rupprecht CE. A comparison of DNA vaccines for the rabies-related virus, Mokola. *Vaccine* 2003;**21**:2598–606.
32. Weyer J, Kuzmin IV, Rupprecht CE, Nel LH. Cross-protective and cross-reactive immune responses to recombinant vaccinia viruses expressing full-length lyssavirus glycoprotein genes. *Epidemiol Infect* 2008;**136**:670–8.
33. Faber M, Li J, Kean RB, Hooper DC, Alugupalli KR, Dietzschold B. Effective preexposure and postexposure prophylaxis of rabies with a highly attenuated recombinant rabies virus. *Proc Natl Acad Sci USA* 2009;**106**:11300–5.
34. Badrane H, Tordo N. Host switching in lyssavirus history from the Chiroptera to the Carnivora orders. *J Virol* 2001;**75**:8096–104.
35. Rupprecht CE, Turmelle A, Kuzmin IV. A perspective on lyssavirus emergence and perpetuation. *Curr Opin Virol* 2011;**1**:662–70.
36. Smith PC, Lawhaswasd K, Vick WE, Stanton JS. Isolation of rabies virus from fruit bats in Thailand. *Nature* 1968;**216**:384.
37. Pal SR, Arora BM, Chuttani PN, Broer S, Choudhury S, Joshi RM. Rabies virus infection of a flying fox bat, Pteropus poliocephalus, in Chandigarh, Northern India. *Trop Geogr Med* 1980;**32**:265–7.
38. Arguin PM, Murray-Lillibridge K, Miranda ME, Smith JS, Calaor AB, Rupprecht CE. Serologic evidence of lyssavirus infections among bats, the Philippines. *Emerg Infect Dis* 2002;**8**:258–62.
39. Reynes JM, Molia S, Audry L, Hout S, Ngin S, Walston J, et al. Serologic evidence of lyssavirus infection in bats, Cambodia. *Emerg Infect Dis* 2004;**10**:2231–4.
40. Kuzmin IV, Niezgoda M, Carroll DS, Keeler N, Hossain MJ, Breiman RF, et al. Lyssavirus surveillance in bats, Bangladesh. *Emerg Infect Dis* 2006;**12**:486–8.
41. Lumlertdacha B, Boongird K, Wanghongsa S, Wacharapluesadee S, Chanhome L, Khawplod P, et al. Survey for bat lyssaviruses, Thailand. *Emerg Infect Dis* 2005;**11**:232–6.
42. Streicker DG, Lemey P, Velasco-Villa A, Rupprecht CE. Rates of viral evolution are linked to host geography in bat rabies. *PLoS Pathog* 2012;**8**:e1002720.
43. Nadin-Davis SA, Real LA. Molecular phylogenetics of the lyssaviruses – insights from a coalescent approach. *Adv Virus Res* 2011;**79**:203–38.
44. Smith JS, Orciari LA, Yager PA, Seidel HD, Warner CK. Epidemiologic and historical relationships among 87 rabies virus isolates as determined by limited sequence analysis. *J Infect Dis* 1992;**166**:296–307.
45. Blanton JD, Palmer D, Rupprecht CE. Rabies surveillance in the United States during 2009. *J Am Vet Med Assoc* 2010;**237**:646–57.
46. Blanton JD, Dyer J, McBrayer J, Rupprecht CE. Rabies surveillance in the United States during 2011. *J Am Vet Med Assoc* 2012;**241**:712–22.
47. Kuzmin IV, Shi M, Orciari LA, Yager PA, Velasco-Villa A, Kuzmina NA, et al. Molecular inferences suggest multiple hosts shifts of rabies viruses from bats to mesocarnivores in Arizona during 2001–2009. *PLoS Pathog* 2012;**8**:e1002786.
48. Biek R, Henderson JC, Waller LA, Rupprecht CE, Real LA. A high-resolution genetic signature of demographic and spatial expansion in epizootic rabies virus. *Proc Natl Acad Sci U S A* 2007;**104**:7993–8.

49. Nadin-Davis SA, Torres G, Ribas Mde L, Guzman M, De La Paz RC, et al. A molecular epidemiological study of rabies in Cuba. *Epidemiol Infect* 2006;**134**:1313–24.
50. Nadin-Davis SA, Velez J, Malaga C, Wandeler AI. A molecular epidemiological study of rabies in Puerto Rico. *Virus Res* 2008;**131**:8–15.
51. Nadin-Davis SA, Muldoon F, Wandeler AI. Persistence of genetic variants of the arctic fox strain of Rabies virus in southern Ontario. *Can J Vet Res* 2006;**70**:11–19.
52. Kuzmin IV, Hughes GJ, Botvinkin AD, Gribencha SG, Rupprecht CE. Arctic and Arctic-like rabies viruses: distribution, phylogeny and evolutionary history. *Epidemiol Infect* 2008;**136**:509–19.
53. Scott TP, Fischer M, Khaiseb S, Freuling C, Höper D, Hoffmann B, et al. Complete genome and molecular epidemiological data infer the maintenance of rabies among kudu (Tragelaphus strepsiceros) in Namibia. *PLoS One* 2013;**8**:e58739.
54. Bourhy H, Reynes JM, Dunham EJ, Dacheux L, Larrous F, Huong VT, et al. The origin and phylogeography of dog rabies virus. *J Gen Virol* 2008;**89**:2673–81.
55. Bourhy H, Kissi B, Audry L, Smreczak M, Sadkowska-Todys M, Kulonen K, et al. Ecology and evolution of rabies virus in Europe. *J Gen Virol* 1999;**80**:2545–57.
56. Kuzmin IV, Botvinkin AD, McElhinney LM, Smith JS, Orciari LA, Hughes GJ, et al. Molecular epidemiology of terrestrial rabies in the former Soviet Union. *J Wildl Dis* 2004;**40**:617–31.
57. Pastoret PP, Brochier B. Epidemiology and control of fox rabies in Europe. *Vaccine* 1999;**17**:1750–4.
58. Singer A, Kauhala K, Holmala K, Smith GC. Rabies in northeastern Europe – the threat from invasive raccoon dogs. *J Wildl Dis* 2009;**45**:1121–37.
59. Liu Y, Zhang S, Wu X, Zhao J, Hou Y, Zhang F, et al. Ferret badger rabies origin and its revisited importance as potential source of rabies transmission in Southeast China. *BMC Infect Dis* 2010;**10**:234.
60. Rabies – Taiwan (04): ferret-badger, more cases. ProMed Mail. Archive Number: 20130729.1851999.
61. Nel LH, Sabeta CT, von Teichman B, Jaftha JB, Rupprecht CE, Bingham J. Mongoose rabies in southern Africa: a re-evaluation based on molecular epidemiology. *Virus Res* 2005;**109**:165–73.
62. Van Zyl N, Markotter W, Nel LH. Evolutionary history of African mongoose rabies. *Virus Res* 2010;**150**:93–102.
63. Charlton KM, Nadin-Davis S, Casey GA, Wandeler AI. The long incubation period in rabies: delayed progression of infection in muscle at the site of exposure. *Acta Neuropathol* 1997;**94**:73–7.
64. Tordo N, Ceccaldi PE, Gaudin Y, Wandeler A. Rhabdovirus: Rabies. In: Collier L, Mahy B, editors. *Topley & Wilson's microbiology and microbial infections* 10th ed. London, UK: Hodder Arnold; 2005. p. 1104–36.
65. Albertini A, Bressanelli S, Lepault J, Gaudin Y. Structure and working of viral fusion machinery. *Curr Top Membr* 2011;**68**:49–80.
66. Albertini AA, Ruigrok RW, Blondel D. Rabies virus transcription and replication. *Adv Virus Res* 2011;**79**:1–22.
67. Chenik M, Chebli K, Blondel D. Translation initiation at alternate in-frame AUG codons in the rabies virus phosphoprotein mRNA is mediated by a ribosomal leaky scanning mechanism. *J Virol* 1995;**69**:707–12.
68. Boland TA, McGuone D, Jindal J, Rocha M, Cumming M, Rupprecht CE, et al. Phylogenetic and epidemiologic evidence of multi-year incubation in human rabies. *Ann Neurol* 2013 in press.
69. Hanlon CA, Smith JS, Anderson GR. Recommendations of a national working group on prevention and control of rabies in the United States. Article II: Laboratory diagnosis

of rabies. The National Working Group on Rabies Prevention and Control. *J Am Vet Med Assoc* 1999;**215**:1444–6.
70. Hemachudha T, Laothamatas J, Rupprecht CE. Human rabies: a disease of complex neuropathogenetic mechanisms and diagnostic challenges. *Lancet Neurol* 2002;**1**:101–9.
71. Hemachudha T, Ugolini G, Wacharapluesadee S, Sungkarat W, Shuangshoti S, Laothamatas J. Human rabies: neuropathogenesis, diagnosis, and management. *Lancet Neurol* 2013;**12**:498–513.
72. Willoughby Jr. RE, Tieves KS, Hoffman GM, Ghanayem NS, Amlie-Lefond CM, Schwabe MJ, et al. Survival after treatment of rabies with induction of coma. *N Engl J Med* 2005;**352**:2508–14.
73. Jackson AC. Therapy of human rabies. *Adv Virus Res* 2011;**79**:365–75.
74. Hamir AN, Niezgoda M, Rupprecht CE. Recovery from and clearance of rabies virus in a domestic ferret. *J Am Assoc Lab Anim Sci* 2011;**50**:248–51.
75. Davis AD, Gordy PA, Bowen RA. Unique characteristics of bat rabies viruses in big brown bats (Eptesicus fuscus). *Arch Virol* 2013;**158**:809–20.
76. Centers for Disease Control and Prevention (CDC) Recovery of a patient from clinical rabies--California, 2011. *MMWR Morb Mortal Wkly Rep* 2012;**61**:61–5.
77. Nagaraj T, Vasanth JP, Desai A, Kamat A, Madhusudana SN, Ravi V. Ante mortem diagnosis of human rabies using saliva samples: comparison of real time and conventional RT-PCR techniques. *J Clin Virol* 2006;**36**:17–23.
78. Dacheux L, Reynes JM, Buchy P, Sivuth O, Diop BM, Rousset D, et al. A reliable diagnosis of human rabies based on analysis of skin biopsy specimens. *Clin Infect Dis* 2009;**47**:1410–7.
79. Nadin-Davis SA, Sheen M, Wandeler AI. Development of real-time reverse transcriptase polymerase chain reaction methods for human rabies diagnosis. *J Med Virol* 2009;**81**:1484–97.

Rabies Diagnosis

Demonstration of Viral Subunits and Antigens

CHAPTER TWO

Demonstration of Lyssavirus Antigens by a Direct Rapid Immunohistochemical Test

Andre Coetzer[1], Louis H. Nel[1], and Charles Rupprecht[2]
[1]Department of Microbiology and Plant Pathology, University of Pretoria, Pretoria, Gauteng, South Africa
[2]Ross University School of Veterinary Medicine, Basseterre, St. Kitts, West Indies

Chapter Contents

2.1 Introduction 27
2.2 Materials 29
  2.2.1 Reagents 29
  2.2.2 Equipment 29
  2.2.3 Biological Material: Deterioration or Decomposition 29
2.3 Methods 30
  2.3.1 Performing Diagnostic Assay 30
  2.3.2 Reading and Recording of Results 32
    2.3.2.1 Staining Intensity 32
    2.3.2.2 Antigen Distribution 33
  2.3.3 Test Results are Reported if the Following Observations are Made 33
2.4 Discussion 33
References 36

2.1 INTRODUCTION

During the early part of the 21st century, while at the U.S. Centers for Disease Control and Prevention (CDC), Rupprecht and Niezgoda developed a novel diagnostic assay for lyssavirus diagnosis known as the Direct Rapid Immunohistochemical Test (DRIT). The development of the DRIT assay (as it is currently known) is based on a simplified version of the standard avidin–biotin complex immunohistochemical (IHC) diagnostic assay.[1,2] The main differences between DRIT and standard IHC tests is that the original IHC test takes many hours to conduct, requires

C. Rupprecht & T. Nagarajan (Eds): Current Laboratory Techniques in Rabies Diagnosis, Research, and Prevention, Volume 1.
Doi: http://dx.doi.org/10.1016/B978-0-12-800014-4.00002-0

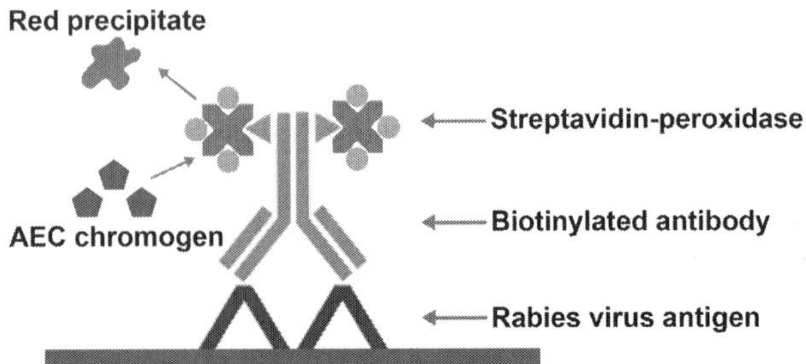

Figure 2.1 A DRIT positive diagnosis: A red precipitate is formed, due to conversion of the AEC chromogen in the presence of the streptavidin–anti-rabies antibody–biotin complex bound to the viral antigen present in the tissue impression.

a linker antibody conjugated to a biotin moiety, and is performed upon cut sections of fixed tissue, while the DRIT is accomplished in less than 1 hour, utilizes either a polyclonal (PAb) or monoclonal antibody (MAb) that is directly labeled with a biotin moiety, and is performed upon a simple touch impression of brain tissue, allowing an anatomic-pathologic appreciation of viral inclusions in neurons. Additionally, the need for a linker antibody is negated by the subsequent addition of a streptavidin that has a high affinity for the biotin moiety. The streptavidin is tagged to a reporter enzyme that is used to catalyze the formation of a colored precipitate once the antibodies are bound to the rabies virus (RABV) antigens in the presence of an appropriate substrate (hydrogen peroxide, H_2O_2) and chromogen (amino-ethylcarbazole, AEC) (Figure 2.1).

To date, a number of studies using the DRIT assay have been published, and these were unanimous in showing that the diagnostic utility of the DRIT assay appeared equal to that of the gold standard direct fluorescent antibody test (DFA).[3–8] By comparison, in preliminary studies, the DRIT matches the diagnostic sensitivity and specificity of the DFA and offers certain key advantages. Significantly, the DRIT diagnostic process relies on a compound light microscope rather than a fluorescent microscope, which is more expensive to procure and maintain.[4] Furthermore, the DRIT results are easy to interpret, which could potentially reduce the number of diagnostic errors.

The objective of this chapter is to provide an in-depth description of the standard operating procedure (SOP) of the DRIT assay, and to also provide a comprehensive list of all the reagents and equipment required to perform this assay in an appropriate diagnostic facility.

2.2 MATERIALS

2.2.1 Reagents

We have provided the names of the supplier(s) whose products we have tested and are familiar with in practice, but comparable products from other suppliers may also be available.

- Formalin, 10% neutral buffered (*Sigma-Aldrich, Cat # HT501128*)
- 1× Phosphate buffered saline (PBS) (Fisher Scientific, Cat # SH30256,02)
- Tween 80 (Sigma-Aldrich, Cat # P1754)
- H_2O_2 (3%) (from any commercial supplier)
- Biotinylated antibody preparation (possible suppliers will be discussed)
- Streptavidin-peroxidase (Kirkegaard & Perry Laboratories Inc, Cat # 71-00-38)
- AEC staining kit (Sigma-Aldrich, Cat # AEC101)
- Gills formulation #2 (Sigma-Aldrich, Cat # GHS232)
- Gel/mount (Fisher Scientific, Cat # BM-M01)
- Distilled water (Fisher Scientific, Cat # S75232)

2.2.2 Equipment

- Single-edge, frosted microscope slides
- Tissue-Tek slide staining kit or multiple individual glass staining dishes
- Slide rack
- Tissue culture plate lids or commercially supplied humidity chamber
- Pipettor, 10–100 µL
- Pipettor, 100–1,000 µL
- Universal pipettor tips, 5–100 µL
- Universal pipettor tips, 100–1,000 µL
- Plastic dropper bottle
- Corning cover glass
- Light microscope with 20× and 40× objectives

2.2.3 Biological Material: Deterioration or Decomposition

The DRIT diagnostic assay relies on the detection of viral antigens that are present in the central nervous system (CNS) tissue. As such, the physical condition of the tissue is very important, and fresh tissue showing no signs of decomposition should ideally be used. The criteria upon which tissue is graded is as follows:[2]

- Good: Tissue is fresh and shows no signs of decomposition.

- Fair: Slight tissue decomposition can be observed, but the anatomic regions can still be clearly identified. The tissue might show signs of discoloration around the periphery, but the tissue is still stable and fresh around the core of the sample.
- Poor: The tissue shows signs of discoloration (substantial green color), liquefaction, desiccation, or an unrecognized gross anatomy.

If negative results are obtained from a tissue sample in "poor" condition, the test report should state only that the tissue condition of the sample is such that the test cannot rule out the presence of RABV in the specimen because the negative predictive value is unknown when the specimen conditions do not meet the minimum test criteria. As such, a negative finding should not be mentioned because it is often misinterpreted as a negative diagnosis (which could be a "false negative" regardless of test). Positive results may be reported as such.[2]

2.3 METHODS

2.3.1 Performing Diagnostic Assay

1. Create a tissue impression of a fresh section of cerebellum or medulla oblongata (or any other available brain tissue) by lightly pressing a degreased microscope slide on the brain pieces. The impressions are subsequently dried on a piece of absorbent tissue paper and left to dry at room temperature (RT) for 5 minutes. Positive and negative controls are included in every staining session, and are fixed and stained according to the same SOP.
2. Immerse the tissue impressions in 10% neutral buffered formalin for 10 minutes (Dish I).
3. Remove tissue impressions from the 10% neutral buffered formalin and dip-rinse slides several times in fresh tris phosphate buffered solution (TPBS) buffer (Dish II).
4. Immerse the tissue impressions in 3% H_2O_2 for 10 minutes (Dish III).
5. Remove excess H_2O_2 by dip-rinsing the slides in fresh TPBS buffer (Dish IV).
6. Immerse tissue impressions in fresh TPBS buffer (Dish V) and work with one slide at a time while leaving the remaining tissue impressions submerged in the TPBS buffer.
7. Remove the slide from the immersed TPBS buffer and shake off the buffer. After shaking off the buffer, the excess buffer can be blotted from the sides surrounding the tissue impression.

8. Place tissue impression in a humidity chamber (the tissue culture plate lid can be lined with a moistened paper towel and used to replace commercially available humidity chambers). Add the biotinylated antibody preparation drop-by-drop until the entire tissue impression is covered.
9. Incubate all the tissue impressions in the humidity chamber for 10 minutes at RT (the incubation can be done on a bench top in the laboratory).
10. Shake off the biotinylated antibody after incubation and immerse the tissue impressions in TPBS buffer (Dish V; this TPBS buffer can be used until Step 16 in the diagnostic process). Work with one slide at a time while leaving the remaining tissue impressions submerged in the TPBS buffer.
11. Repeat Step 7.
12. Place the tissue impression in a humidity chamber and add the ready-to-use streptavidin-peroxidase to each tissue impression until it is completely covered.
13. Incubate all the tissue impressions in the humidity chamber for 10 minutes at RT.
14. Shake off the streptavidin-peroxidase after incubation and immerse the tissue impressions in TPBS buffer (Dish V). Work with one slide at a time while leaving the remaining tissue impressions submerged in the TPBS buffer.
15. Repeat Step 7.
16. Place the tissue impression in a humidity chamber and add the working solution of AEC chromogen to each tissue impression until it is completely covered.
17. Incubate all the tissue impressions in the humidity chamber for 10 minutes at RT.
18. Shake off the excess AEC chromogen after incubation and immerse the tissue impressions in distilled water (Dish VI) for 5 minutes.
19. Immerse the tissue impression in a counterstain of Gills #2 Hematoxylin (diluted 1:2 with distilled water) for 2 minutes (Dish VII).
20. Immediately remove the excess counterstain from the tissue impressions by dip-rinsing the slides in distilled water (Dish VIII). Perform a second dip-rinse in fresh distilled water (Dish IX) to ensure that all the excess counterstain is removed from the tissue impression.
21. Immerse slides in fresh distilled water (Dish X) and work with one slide at a time.

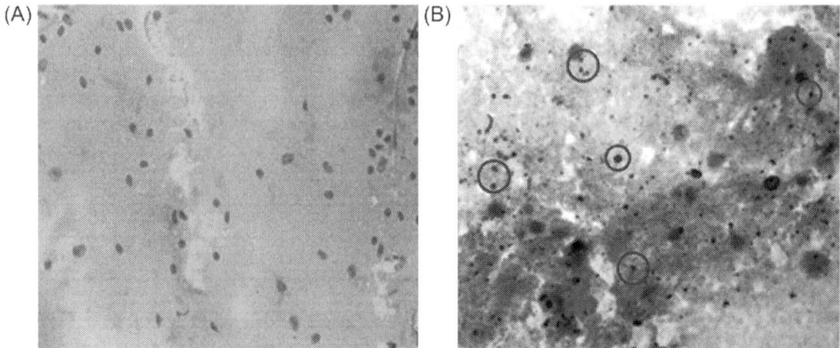

Figure 2.2 (A) DRIT negative result: There are no rose-red inclusions on the blue neuronal background indicating that the biotinylated antibody did not react with any viral antigen. (B) DRIT positive result: The red inclusions (encircled) are visible on the blue neuronal background. The highlighted inclusions are visible in various shapes and sizes, and are almost too numerous to count properly.

22. Shake off distilled water and blot excess distilled water from the sides surrounding the tissue impression.
23. Mount the slides with a water-soluble mounting medium and cover glass, taking care not to allow the tissue impressions to dry out.
24. View the slides by light microscopy. Use a 20× objective to scan the field, and a 40× objective for a closer inspection of rose-red inclusion bodies present on the blue neuronal background.
25. Record results.

2.3.2 Reading and Recording of Results

Negative results: A sample is considered negative if more than 40 fields are viewed at a magnification of at least 200× and no rose-red inclusion bodies are visible on the neuronal background (Figure 2.2A).

Positive results: Both the staining intensity and antigen distribution is based on the presence of the rose-red inclusion bodies than can manifest in the form of round or oval masses of varying sizes (Figure 2.2B).

2.3.2.1 Staining Intensity

The staining intensity is graded from +1 to +4, and the positive control slides should always contain a tissue impression with a staining intensity of +4. If a slight loss of color is observed, the staining intensity is graded as +3, while noticeably dull stains are graded as +2 or +1.

2.3.2.2 *Antigen Distribution*

The antigen distribution is graded by the amount of observed antigen and is scored as follows:

- +4, large numbers of both small and large inclusions of varying shapes are visible in almost every field of the impression.
- +3, numerous inclusions of varying sizes and shapes are found in almost every microscopic field.
- +2, 10–50% of the observed fields contain a few inclusions of varying sizes and shapes.
- +1, very few (usually one or two) inclusions of varying sizes and shapes are visible in less than 10% of the observed fields.

2.3.3 Test Results are Reported if the Following Observations are Made

Control slides: The positive control slide should contain an accumulation of both large and small rose-red inclusions with a staining intensity of +4 and an antigen distribution of either +3 or +4. No staining should be present on the negative control slide.

Test slides: Samples that are positive should have a staining intensity of +3 or +4 and an antigen distribution of +2 to +4 in slides made from both the brain stem and cerebellum. Samples that are negative should have no specific staining whatsoever.

2.4 DISCUSSION

Although the DRIT diagnostic assay is a relatively new development, it has nevertheless found application in a number of studies in different geographical locations and settings. This provides a good measure of confidence in the technique, and from our experience, we would not hesitate to recommend that rabies diagnostic laboratories consider this method as a primary method for the postmortem diagnosis of rabies in animals. Although most studies have focused upon RABV, all other lyssaviruses have also been recognized by use of this technique.[8]

To date, at least two biotinylated antibody preparations have been evaluated in comparative studies between the DRIT and DFA. These two antibody preparations will be discussed further in this chapter. Until recently, the CDC was the only supplier of a biotinylated antibody preparation for the purpose of performing the DRIT. The biotinylated antibody preparation, produced by the CDC, is supplied as a ready-to-use cocktail consisting of two biotinylated MAbs. The ready-to-use MAb cocktail is readily provided by the CDC once the two MAbs have been biotinylated and mixed in the correct predetermined ratios. A recent study showed that alternative antibody preparations could be biotinylated and used in the SOP of the diagnostic assay.[8] In the aforementioned study, a PAb preparation was biotinylated using a commercial biotinylation kit and applied to the pre-existing SOP for the DRIT. During the course of the study, the DRIT, using the biotinylated PAb preparation, was applied to 250 CNS tissue samples originating from southern Africa, and it was found that the DRIT had a diagnostic efficacy that was marginally higher than that of the DFA test.[8]

Although the SOP as described in this chapter is based on a pre-existing SOP published elsewhere, it has been simplified from the original protocol by using a commercially supplied AEC staining kit.[2,8] Not only does the use of the ready-to-use commercial staining kit minimize the handling of possible carcinogenic chemicals, but the kit also allows the working solution to be easily prepared by simply mixing three prepared vials immediately before use.

Apart from the advantages that the DRIT holds over the DFA test in terms of cost and ease of interpreting the results, the DRIT is also suited for application in developing countries where chemical preservatives are used to maintain the tissue integrity during transport and shipment to diagnostic facilities. The transport of samples in buffered formalin, although still performed in some developing countries, is not recommended for samples that are destined to be diagnosed with the DFA due to the stringent enzymatic treatment of the sample that is needed before diagnosis can be performed.[9] However, in contrast to the DFA, where acetone is the primary fixative, buffered formalin is used only briefly in the DRIT, as an excellent preserver of tissue impressions, and as a superior chemical for viral inactivation. Moreover, the DRIT can be performed easily on 50% glycerol/PBS stored tissue samples after washing. The use of glycerol saline as a chemical preservative for tissue transport when cold chains are not available is advocated as the samples can be subjected to diagnosis without any enzymatic treatment. A recent study has shown that although the storage of samples in glycerol saline is slightly detrimental to the observed immunogenic patterns upon applying either the DFA or DRIT, the DRIT is hindered to a lesser extent by the specific chemical preservation of the tissue.[4]

To achieve accurate and consistent results when performing either the DFA test or DRIT, certain factors need to be controlled effectively. False results may occur due to the use of incorrect antibody concentrations, inadequately adjusted equipment, non-specific background staining, failure to use effective control slides, or color blindness on the part of the operator.[10] Care should thus be taken to ensure that all the equipment is serviced regularly and that diagnostic technicians undergo routine proficiency testing. When performing the DRIT, the working concentration of the biotinylated antibody preparation should be determined whenever a new batch is used or when different equipment will be used to view antigenic distribution or staining intensity. These steps will in turn limit the occurrence of incorrect diagnosis. Despite the fact that numerous factors have to be controlled to achieve effective diagnosis with DRIT, the test still maintains a diagnostic sensitivity and specificity of nearly 100% while requiring only a simple compound light microscope to perform the test. The DRIT is thus an ideal option currently available to diagnostic facilities that cannot perform the DFA test easily, but still wish to rely on the direct detection of viral antigens in neuronal tissue samples, without the need for tissue homogenization. Such applications should lead to improvements of decentralized, laboratory-based surveillance in developing countries where the burden of rabies is the greatest.

Additional Information

Reagent Preparation and Usage

1. Formalin, 10%: ready-to-use. [*Change out after 2 runs or once a week.*]
2. TPBS buffer: Add 990 mL 1 × PBS and 10 mL Tween 80. Shake well until the Tween 80 is completely dissolved. [*Change out after each run.*]
3. 3% H_2O_2: Ready-to-use. [*Change out after each run.*]
4. Biotinylated polyclonal antibody: Apply working dilution as determined by means of a titration series of the stock solution of biotinylated antibody.
5. Streptavidin-peroxidase: Ready-to-use.
6. AEC staining kit: The staining kit contains three dropper bottles (vials) filled with the required reagents. Create the working solution immediately before use as follows: *Mix 4 mL distilled water, 2 drops of vial 1 (Acetate buffer, 2.5 M, pH 5.0), 1 drop of vial 2 (3-amino-9-ethylcarbazole in N,N-dimethylformamide), and 1 drop of vial 3 (3% H_2O_2 in distilled water).*
7. Gills formulation #2: Diluted 1:2 in distilled water. [*Change out every week.*]
8. Gel/mount: Ready-to-use.
9. Distilled water: Ready-to-use. [*Change out after each run.*]

REFERENCES

1. Fekadu M, Greer PW, Chandler FW, Sanderlin DW. Use of the avidin-biotin peroxidase system to detect rabies antigen in formalin-fixed paraffin-embedded tissues. *J Virol Methods* 1988;**19**(2):91–6.
2. Niezgoda M, Rupprecht CE. Standard operating procedures for the direct, Rapid Immunohistochemistry test for the detection of rabies virus antigen. *Natl Lab Train Netw Course* 2006:1–16.
3. Lembo T, Niezgoda M, Velasco-Villa A, Cleaveland S, Ernest E, Rupprecht CE. Evaluation of a direct, Rapid Immunohistochemical Test for rabies diagnosis. *Emerg Infect Dis* 2006;**12**(2):310–3.
4. Durr S, Naissengar S, Mindekem R, Diguimbye C, Niezgoda M, Kuzmin I, et al. Rabies diagnosis for developing countries. *PLoS Negl Trop Dis* 2008;**2**(3):e206.
5. Tao XY, Niezgoda M, Du JL, Li H, Wang XG, Huang Y, et al. The primary application of direct rapid immunohistochemical test to rabies diagnosis in China. *Chin J Exp Clin Virol* 2008;**22**(3):168–70.
6. Madhusudana SN, Subha S, Thankappan U, Ashwin YB. Evaluation of a direct rapid immunohistochemical test (dRIT) for rapid diagnosis of rabies in animals and humans. *Virol Sin* 2012;**27**(5):299–302.
7. Saturday GA, King R, Fuhrmann L. Validation and operational application of a rapid method for rabies antigen detection. *US Army Med Dep J* 2009:42–5.
8. Coetzer A, Markotter W, Sabeta CT, Nel LH. Comparison of biotinylated monoclonal and polyclonal antibodies in an evaluation of a direct rapid immunohistochemical test for the routine diagnosis of rabies in southern Africa. *University of Pretoria* 2013:1–155.
9. Reid FL, Hall NH, Smith JS, Baer GM. Increased Immunofluorescent Staining of Rabies-Infected, formalin-fixed brain tissue after pepsin and trypsin digestion. *J Clin Microbiol* 1983;**18**(4):968–71.
10. Dean DJ, Abelseth MK, Atanasiu P. The fluorescent antibody test. In: Meslin F-X, Kaplan MM, Koprowski H, editors. *Laboratory techniques in rabies* 4th ed.. Geneva: World Health Organization; 1996. p. 88–9.

CHAPTER THREE

Demonstration of Rabies Virus Antigens by a Latex Agglutination Test

Songsri Kasempimolporn, Narongsak Chaiyabutr, and Visith Sitprija
Queen Saovabha Memorial Institute, Thai Red Cross Society, Bangkok, Thailand

Chapter Contents

3.1 Introduction 37
3.2 Materials 38
3.2.1 Reagents 38
3.2.2 Biological Materials 38
3.2.3 Equipment 38
3.3 Methods 38
3.3.1 Production of Anti-Rabies Serum 38
3.3.2 Preparation of Anti-Rabies IgG 39
3.3.3 Latex Coating 39
3.3.4 Test Saliva 39
3.3.5 Slide Agglutination 39
3.4 Discussion 40
3.4.1 Experimental Tips 40
3.4.2 Technical Notes 40
3.4.3 Critical Parameters 41
References 42

3.1 INTRODUCTION

Attempts to diagnose rabies have been made by examining the saliva, if no other tissue is available for testing. Diagnosis, looking for evidence of the virus in saliva, can be carried out using several methods. In dogs, rabies virus (RABV) appears in the saliva before and during the appearance of clinical signs, but the virus may be shed intermittently.[1–4] Sensitive but sophisticated and expensive techniques such as polymerase chain reaction (PCR) can detect viral RNA in cerebrospinal fluid (CSF)

C. Rupprecht & T. Nagarajan (Eds): Current Laboratory Techniques in Rabies Diagnosis, Research, and Prevention, Volume 1.
Doi: http://dx.doi.org/10.1016/B978-0-12-800014-4.00003-2

and saliva samples, which are unsuitable for fluorescent antibody examination. RABV can be detected in a much higher frequency in saliva than in CSF.[5] A more convenient and affordable diagnostic test that can be completed within minutes is desirable. The development of a simple and inexpensive test for the diagnosis of rabies using saliva based on latex agglutination has been described.[6] RABV antigen in saliva could be detected by agglutination on a glass slide using latex particles coated with anti-rabies antibody. Specific agglutination is due to bridge formation of the viruses on and between the latex particles, which could be judged macroscopically. Saliva samples could be used as an alternative to brain specimens for RABV antigen testing, which, amongst other advantages, offers fewer occupational and disposal risks. The latex agglutination test (LAT), because of its simplicity and lack of requirement for instrumentation, might be used to rapidly screen animals suspected of having rabies. Because the virus may be absent or present at very low levels in saliva, samples taken for diagnosis cannot definitively rule out rabies.

3.2 MATERIALS

3.2.1 Reagents

- Polystyrene latex beads (diameter, 0.3 µm; 10% suspension (w/v))
- 0.01 M Phosphate buffered saline (PBS), pH 7.2
- 1% bovine serum albumin (BSA) in PBS
- Latex diluent (PBS with 0.1% BSA and 0.005% polyvinylpyrrolidone)

3.2.2 Biological Materials

- Purified anti-rabies immune globulin (IgG)
- Saliva samples

3.2.3 Equipment

- Double-concave slides or clear pressed glass with 10 depressed cavities

3.3 METHODS

3.3.1 Production of Anti-Rabies Serum

1. Immunize horses (or rabbits) with a series of purified cell culture rabies vaccine.[7]

2. Test bleed the horses on days 14, 16, and 18 after the last injection, separate the serum, and store at −20°C.
3. Titrate pooled aliquots from all the immune serum in the rapid fluorescent focus inhibition test (RFFIT).[8]

3.3.2 Preparation of Anti-Rabies IgG

1. Fractionate IgG from serum by DEAE-cellulose column chromatography.[9]
2. Pool the eluted IgG and concentrate by placing in a dialysis bag (3.5 kDa MW cut-off) covered with polyethylene glycol 6000.
3. Dialyze against PBS and determine the protein content by Bradford's method.[10]

3.3.3 Latex Coating

1. Wash a 10% suspension of latex beads with PBS by centrifugation at 12,500 × g for 15 min at 4°C.
2. Re-suspend the latex to 1% suspension (w/v) in PBS.
3. Mix a 1% suspension of latex beads with an equal volume of anti-rabies IgG at the optimum concentration.
4. Gently shake at room temperature for 2 h.
5. Wash the latex twice with PBS by centrifugation at 12,500 × g for 15 min at 4°C.
6. Block any remaining latex protein-binding sites with 1% BSA in PBS at room temperature for 30 min.
7. Repeat the washing step.
8. Re-suspend the sensitized latex beads in latex diluent at a concentration of 0.4% and store at 4°C.

3.3.4 Test Saliva

1. Collect saliva from carcasses by immersing a sponge swab into 2 mL of PBS and then swabbing the anterior surface of the tongue and the cheek mucosa for 15 to 20 s.
2. Put the swab into a test tube. Squeeze the swab firmly against the wall of the tube to release the contents. Discard the swab.
3. Centrifuge at 12,500 × g for 20 min and keep the supernatant at 4°C.

3.3.5 Slide Agglutination

1. Add two 25-μL aliquots of coated latex beads on a double-concave slide, one with 25 μL of test saliva, and one with 25 μL of PBS as a negative control.

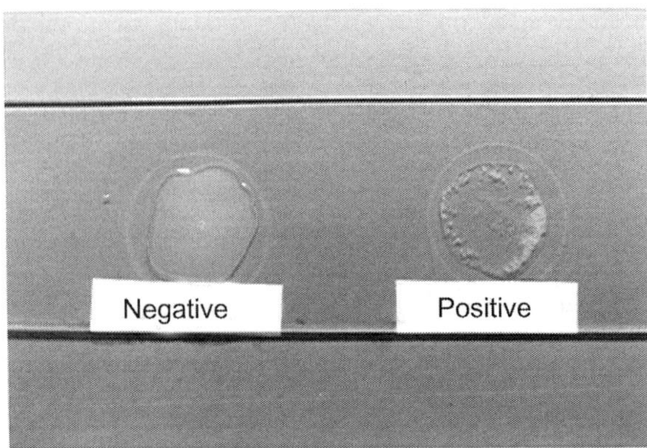

Figure 3.1 Latex agglutination test for detection of rabies virus in saliva.

2. Rock gently for 5–15 min and read agglutination visually against a dark background. A typical pattern of agglutination is shown in Figure 3.1.

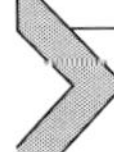

3.4 DISCUSSION

3.4.1 Experimental Tips

1. Coating the beads with IgG is not difficult as it binds readily to polystyrene latex beads. The optimum concentration of IgG for coating must be determined; a concentration of 1–3 mg/mL is usually suitable. Preliminary experiments should be performed with some saliva samples from animals that are previously known to be positive or negative for RABV infection. IgG concentration affects the sensitivity. When the higher IgG concentration is used, a higher sensitivity is obtained. However, the excessive amount of IgG on the surface of latex beads can lead to a prozone effect that causes false negative results.
2. The most desirable latex particle size is 0.3 μm because particles of this size agglutinate the virus specifically as compared with a larger size (0.8 μm). Latex coated with IgG could be kept at 4°C for up to 3 months.
3. BSA is used to block any remaining latex protein-binding sites. Because BSA competes for binding with IgG, excessive blocking reaction time can force IgG on the latex particle surface to be replaced by BSA.

3.4.2 Technical Notes

1. RABV is well distributed in many parts of the body (including salivary glands) during the final stage of infection. Therefore, the sensitivity of

LAT varies depending on the time of saliva collection relative to the onset of clinical signs. Saliva collected during the late stage of disease yields more positive results.

2. The LAT can be performed on a live rabies-suspected animal.[11] However, the testing of serial saliva samples taken at different time intervals is necessary. Furthermore, samples taken for ante-mortem diagnosis cannot definitively rule out rabies.
3. Some components in saliva affect the specific agglutination of sensitized latex. However, the components associated with false-positive reactions are not known. Samples that are classified as positive can also be forwarded to a well-equipped rabies laboratory for confirmatory diagnosis.

3.4.3 Critical Parameters

1. There are different parameters that can be adjusted to optimize an immunoreaction and to obtain the best response. The success of a LAT largely depends on the nature of the adsorbed antibodies. Immunoreactivity differences can be found if IgG are obtained from different animal species or purchased from different commercially available antibodies. Therefore, one needs to evaluate and optimize the antibody before choosing it to develop a LAT.
2. Non-occupied parts of the antibody-latex particles are usually covered with a second macromolecule (i.e. BSA) to suppress non-specific interactions of the complementary of the antigen. In addition, BSA is able to increase the colloidal stability of IgG-coated latex particles. The higher the amount of adsorbed BSA, the higher the stability. However, with very high values, desorption of antibody may occur, and subsequently loss of the immunoreactivity. Therefore, a balance of adsorbed antibody/BSA ratio must be achieved.
3. Some of the parameters or factors that must be taken into account are the following: latex particle size, adsorption conditions which determine the amount and mode of attachment of the antibody, reaction time, agglutination conditions (pH, ionic strength, temperature), etc.[12]

Additional Information

Sources of Supplies

- Polystyrene latex beads (Sigma, St Louis, MO).
- Double-concave glass slides (Various sources).
- Clear pressed glass with 10 depressed cavities (Marienfeld, Germany).

REFERENCES

1. Vaughn Jr JB, Gerhardt P, Newell KW. Excretion of street rabies virus in the saliva of dogs. *JAMA* 1965;**193**:363–8.
2. Schneider LG. Spread of virus from the central nervous system. In: Baer GM, editor. *The natural history of rabies*, Vol. 1. New York: Academic Press; 1975. p. 273–301.
3. Fekadu M, Shaddock JH, Bear GM. Excretion of rabies virus in the saliva of dogs. *J Infect Dis* 1982;**145**:715–9.
4. Rupprecht CE. Rhabdoviruses: rabies virus [Chapter 61]. In: Baron S, editor. *Medical microbiology* 4th ed.. Texas: University of Texas Medical Branch at Galveston; 1996.
5. Saengseesom W, Mitmoonpitak C, Kasempomolporn S, Sitprija V. Real-time PCR analysis of dog cerebrospinal fluid and saliva samples for ante-mortem diagnosis of rabies. *Southeast Asian J Trop Med Public Health* 2007;**38**:53–7.
6. Kasempimolporn S, Saengseesom W, Lumlertdacha B, Sitprija V. Detection of rabies virus in dog saliva using a latex agglutination test. *J Clin Microbiol* 2000;**38**:3098–9.
7. Luekrajang T, Wangsai J, Phanuphak P. Production of antirabies serum of equine origin. In: Meslin FX, Kaplan MM, Koprowski H, editors. *Laboratory techniques in rabies* 4th ed.. Geneva: World Health Organization; 1996. p. 401–4.
8. Smith JS, Yager PA, Bear GM. A rapid reproducible test for determining rabies neutralizing antibody. *Bull WHO* 1973;**48**:535–41.
9. Levy HB, Sober HA. A simple chromatographic method for preparation of gammaglobulin. *Proc Soc Exp Biol Med* 1990;**103**:250–2.
10. Bradford MM. A rapid and sensitive method for the quantitation of microgram quantities of protein utilizing the principle of protein-dye binding. *Anal Biochem* 1976;**72**:248–54.
11. Kasempimolporn S, Sichanasai B, Saengseesom W, Puempumpanich S, Chatraporn S, Sitprija V. Prevalence of rabies virus infection and rabies antibody in stray dogs: a survey in Bangkok, Thailand. *Prev Vet Med* 2007;**78**:325–32.
12. Ortega-Vinuesa JL, Bastos-Gonzalez D. A review of factors affecting the performances of latex agglutination tests. *J Biomater Sci Polymer Edn* 2001;**12**:379–408.

CHAPTER FOUR

Rabies Diagnosis: Demonstration of Viral Antigens by Flow Cytometry

Juliano Bordignon[1] and Carlos Roberto Zanetti[2]
[1]Virology, Instituto Carlos Chagas, ICC/Fiocruz, Curitiba, Paraná, Brazil
[2]Microbiology, Immunology and Parasitology, Federal University of Santa Catarina, Florianópolis, Santa Catarina, Brazil

Chapter Contents

4.1 Introduction 43
4.2 Materials 44
4.2.1 Reagents 44
4.2.2 Equipment 45
4.2.3 Biological Materials 45
4.3 Methods 45
4.3.1 Infection and Incubation 45
4.3.2 Cell Staining 45
4.3.3 Flow Cytometric Analysis of Infection 46
4.3.4 Interpretation of Results 46
4.4 Discussion 47
4.4.1 Experimental Tips 48
4.4.2 Critical Parameters and Troubleshooting 48
4.4.3 Precautions 49
4.4.4 Alternative Materials and Methods 49
4.4.5 Time Considerations 49
4.4.6 Limitations 50
4.4.7 Future Considerations 50
Acknowledgments 50
References 51

4.1 INTRODUCTION

Given the severity of rabies, the diagnosis of viral infection must be both rapid and conclusive, with laboratory results guiding patient treatment and epidemiological assessments. The technique of choice for rabies diagnosis is immunofluorescence assays, or immunohistochemistry (identification

C. Rupprecht & T. Nagarajan (Eds): Current Laboratory Techniques in Rabies Diagnosis, Research, and Prevention, Volume 1.
Doi: http://dx.doi.org/10.1016/B978-0-12-800014-4.00004-4

of Negri bodies) on suspect tissues.[1] These techniques are both rapid and inexpensive, but like any test, under some circumstances both may yield false-negative results, making confirmation by another technique inevitable. Rabies virus (RABV) isolation in cell culture is one pertinent confirmatory technique for rabies diagnosis, and is at least as sensitive as the mouse inoculation test (MIT).[2] In addition, cell culture techniques make it possible to isolate the strain of the virus actually causing the disease, and this is potentially important for epidemiological and research purposes. Virus isolation in cells is faster than MIT and does not entail the ethical issues associated with *in vivo* isolation in mice. Nevertheless, the isolation of RABV in cell cultures requires several days (3 to 5) to allow the production of sufficient RABV for detection by immunofluorescence or another technique.[2] Flow cytometry is an automatic, fast, easy, and reproducible technique for the detection and quantification of RABV antigens in cell culture. Several authors have demonstrated the detection and quantification of cells infected with other viruses, such as rotavirus, HIV-1, and herpes simplex virus, by flow cytometry.[3–5]

Flow cytometry has emerged as a technique of choice for automated analyses of cells in suspension.[6] Through optical means, based on the laser-mediated activation of fluorescent dyes, this technique makes it possible to analyze single particles or cells. Flow cytometry identifies differences in cell size (based on forward scatter, FSC) and complexity (based on side scatter, SSC). Furthermore, if the cell is first stained with conjugated antibodies, this technique can be used to detect antigens in cell culture. The laser excites the fluorophore conjugated to the antibody, making it possible to detect the corresponding antigen in a cell. Even intracellular and nuclear antigens can be detected, provided that the cells are correctly fixed and permeabilized. We describe here a protocol for monitoring infection with fixed (Pasteur virus, PV) and street RABV strains in the BHK-21 and C6 cells lines by flow cytometry.

4.2 MATERIALS

4.2.1 Reagents

- 4% *p*-formaldehyde.
- Fluorescent anti-rabies conjugate.
- 24-well cell culture dishes.
- DMEM-F12 and MEM cell culture media.

- Fetal calf serum (FCS).
- 1× phosphate buffer (PBS) pH 7.4.
- Coulter IsotonII® diluter.
- FACS lysis solution (Becton, Dickinson & Company).
- 100 mM glutamine solution.
- Penicillin/streptomycin (10 IU/μg/mL) and amphotericin (25 μg/mL).
- Trypsin–EDTA solution (Sigma Cat #4049).

4.2.2 Equipment

- Flow cytometer.
- Cell culture facility with a CO_2 incubator, freezers and refrigerators, a centrifuge, laminar flow cabinets, and other devices commonly used in cell culture laboratories.

4.2.3 Biological Materials

- BHK-21 (C-13; ATCC CCL10), C6 (ATCC CCL107),[7,8] and mouse neuroblastoma cells (MNA)[9] have been used for the detection of RABV infection in cell culture. The use of other cell lines might require slight modifications to the protocol.
- Standard and titrated RABV strains, such as PV or challenge standard virus (CVS), are to be used as a positive control for RABV infection.
- Clinical or field samples.

4.3 METHODS

4.3.1 Infection and Incubation

BHK-21 (C-13; ATCC CCL10) and C6 (ATCC CCL107) cells in supplemented DMEM-F12 medium (10% FCS, 2 mM glutamine, 100 IU/mL penicillin, 100 μg/mL streptomycin and 250 ng/mL amphotericin B/mL) are used to seed 24-well plates (3.0×10^5 cells/well). The plates are incubated overnight at 37°C and 5% CO_2, and the cells are then infected with 1 $CCID_{50}$ of PV or CVS as positive control, for 1 h, at 37°C, under an atmosphere containing 5% CO_2. After infection, the cells are washed once (1× PBS) and incubated for 72 h in supplemented DMEM-F12 medium.

4.3.2 Cell Staining

After incubation, the cells are washed with 1× PBS and disaggregated by incubation in trypsin-EDTA solution for 5 minutes. The detached cells are recovered and suspended in 5.0 mL of MEM supplemented with 2% FCS. The cells are then washed twice with 1× PBS plus 0.05% Tween-20 (wash buffer), fixed, and permeabilized by incubation for 10 minutes in a 1:1 solution (v/v) of 4% *p*-formaldehyde and FACS lysissolution (Becton, Dickinson and company, Cat #349202, 1:10 in distilled water). The cells are washed twice in wash buffer and centrifuged for 5 minutes at ~1,900× *g* at room temperature. The cell pellet is re-suspended in 100 μL solution of a 1:20 dilution of RABV nucleocapsid-specific polyclonal antibody conjugated with FITC (Sanofi, France) in 1× PBS and incubated for 45 minutes in the dark at 37°C. Finally, the cells are washed twice in wash buffer, centrifuged, and the final pellet is recovered in Coulter IsotonII diluter and stored in the dark at 4°C until flow cytometric analysis.

4.3.3 Flow Cytometric Analysis of Infection

For the analysis of infected cells, the flow cytometer must be set up for the detection of cells on the basis of size (FSC) and complexity (SSC). Gating on fixed cells, with the exclusion of cell debris and doublets, must be set up (Figure 4.1). The equipment settings for the detection of cells and fluorescence may differ between different types of equipment (analog vs. digital, for example) and between the fluorophores conjugated to the antibody or the cell line used. A trained technician should be responsible for defining the best settings and the gates for the analysis of infection. The next step involves the use of unstained cells to set thresholds for cell autofluorescence. An isotype control (an unrelated antibody conjugated to the same fluorophore as the specific antibody) should also be used to control for nonspecific binding. Once the flow cytometer has been set up, uninfected (and stained) and infected cells (samples and positive controls) can be analyzed. At least 10,000 events should be counted during the flow cytometric analysis for each sample. The number of positive cells in controls or samples is determined on the basis of a gate set up with uninfected controls. Two important pieces of information can be obtained from this analysis: the percentage of cells positive for RABV antigens, and the intensity of staining. The percentage of cells positive for RABV antigens provides information about the presence of the virus in the cells, whereas staining intensity provides an indication of the level of viral replication.

4.3.4 Interpretation of Results

With flow cytometry, we have been able to distinguish between uninfected and infected cells (for BHK-21 and C6 cells) as quickly as 12 hours

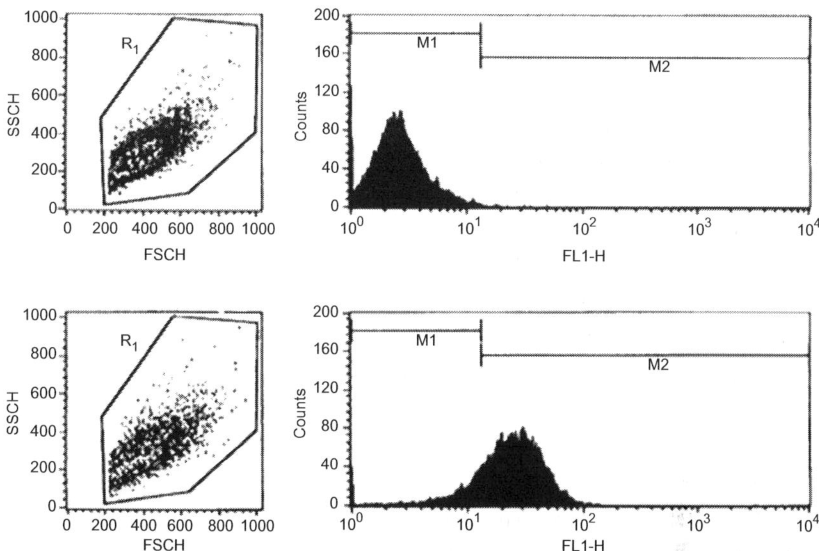

Figure 4.1 Detection of intracytoplasmic RABV antigens by flow cytometry. Upper panel: Uninfected control cells. Lower panel: 5×10^5 BHK-21 cells were infected with virus strain PV and infection was monitored at 72 hpi. The cells were disaggregated with trypsin, washed, fixed and permeabilized, incubated with FITC-conjugated antibodies against the rabies ribonucleoprotein (RNP) for 30 minutes, and analyzed for fluorescence by flow cytometry. Each histogram represents 10,000 events. For details on the gate strategy, see Bordignon et al. (2002).[8]

post-infection (hpi), for cells infected with fixed or wild RABV strains. At this time point, 4.7 to 7.1% of the cells were positive for RABV antigens. After 24 hpi, infection rates exceeded 50% of the cells, peaking at 48–72 hpi, with approximately 80% of the cells positive for RABV antigens. Vengatesan et al. (2006) also reported high rates of infection for MNA cells at 72 hpi.[9]

4.4 DISCUSSION

In the last few years, flow cytometry has emerged as a method of choice for the analysis of cells in suspension.[6] Flow cytometry can be used for the rapid and reproducible analysis of single particles by optical methods based on the laser-mediated activation of fluorescent dyes. The possibility of carrying out multi-parameter examinations of individual cells has made flow cytometry an invaluable tool for both qualitative and quantitative analyses. Flow cytometry has been used for the indirect detection of antigens from a number of viruses, including HIV-1,[4] herpes simplex virus,[5] and rotavirus.[3] Using the protocol described above, we have been able to demonstrate the infection of C6 and BHK-21 cells with RABV after 12 hpi (4.7–7.1%). However, the number of RABV-positive cells detected peaked at 48 to 72 hpi for both cell types (88% for C6/PV and 71% for BHK-21/PV cells). Vengatesan et al. (2006) also demonstrated that the level of infection of MNA cells peaked at 48–72 hpi following inoculation with 50 FFD_{50} of the PV strain (75.28±4.67 % and 84.43±0.93 %, respectively).[9]

Vengatesan et al. (2006) also reported a change in the complexity of infected cells, increasing SSC.[9] They suggested that this change might be induced by the formation of inclusion bodies within cells. This observation has not yet been confirmed, but might facilitate the identification of RABV-infected cells without the need for fluorescence staining.

Although flow cytometry successfully detected RABV antigens, it is important to bear in mind that the field strains used by Bordignon et al. (2002) were adapted to cell line conditions before use.[8] It is also of the utmost importance to test the technique on samples taken directly from the nervous tissues of animals suspected for rabies. Finally, the detection of RABV antigens within cells by flow cytometry is a rapid, reliable, and quantitative technique that could potentially shorten the time required for the confirmation of viral infection in cell cultures. Flow cytometry is also potentially useful for quantitative applications, such as the testing of antiviral drugs and vaccine challenges. The detection of RABV antigens by flow cytometry is also less subjective than the manual observations involved in the fluorescence antibody test (FAT).[9]

4.4.1 Experimental Tips

The use of a freshly prepared 4% *p*-formaldehyde solution for the fixing/permeabilization of cells is recommended to prevent cell damage/lysis. The repeated freezing and thawing of this solution should be avoided.

4.4.2 Critical Parameters and Troubleshooting

The main problems in the flow cytometry analysis of infected cells are the definition of the cell population analyzed (FSC×SSC), and determining the correct settings for the equipment. Two key elements are relevant here: the availability of a trained technician to ensure that the equipment is set up correctly, and the use of appropriate controls, such as unstained cells and isotype controls.

The cells must also be permeabilized for the detection of intracellular antigens (such as those of RABV). A 4% solution of *p*-formaldehyde is very effective for the fixing and permeabilization of cells (together with FACS lysis solution, diluted 1:10 in water).

4.4.3 Precautions

All the usual safety considerations must be taken into account when using flow cytometry for RABV detection. However, once the samples have been fixed with *p*-formaldehyde, no additional special precautions are required for the analysis of samples in a flow cytometer with multiple users.

4.4.4 Alternative Materials and Methods

Recently described methods and reagents for cell fixation/permeabilization, such as Cytofix/Cytoperm kits from Becton, Dickinson & Company (see Additional Information), could be used as an alternative to the protocol described above. Vengatesan et al. (2006) obtained better results with Cytofix/Cytoperm than with the fixation buffer described in the protocol above.[9] An unconjugated primary antibody could be used for cell staining, but this would necessitate incubation with a secondary antibody conjugated to a fluorescent dye (e.g. FITC-conjugated anti-mouse antibody if the primary antibody was produced in mouse). This would increase the time required for cell staining, but would have no effect on the efficiency of antigen detection. Finally, at the end of the protocol, the cells could also be suspended in sterile 1× PBS before flow cytometric analysis, rather than Isoton II (Beckman Coulter). If the cells are not analyzed immediately, they should be re-suspended in sterile 1× PBS mixed in a 1:1 ratio with 3% *p*-formaldehyde and stored in the dark at 4°C. Cells treated in this way can be analyzed after three to four days, with no loss of fluorescence.

4.4.5 Time Considerations

The principal advantage of flow cytometry for RABV detection is that this technique yields results more rapidly than traditional immunofluorescence assays on rabies-infected cells. PV-infected BHK-21 and C6 cells present RABV antigens detectable by flow cytometry after as little as 12h of infection. Vengatesan et al. (2006) were able to demonstrate the presence of RABV antigens in N2A cells after 24hpi.[9] However, the definition of this time point as the standard for RABV detection would require a more complete study, with smaller amounts of virus used for challenge, and the aim of defining the sensitivity of this technique with respect to the traditional direct fluorescent antibody (DFA) method.

Sample preparation for flow cytometry takes no longer than that for traditional immunofluorescence, as the steps involved are very similar. However, the reading and analysis of samples are faster for flow cytometry than for traditional immunofluorescence methods involving microscopy. Flow cytometry also provides a faster, more reproducible quantification of infection than microscopy.

4.4.6 Limitations

Despite these major advantages of flow cytometry over other techniques for RABV antigen detection, this method is subject to certain limitations. Flow cytometry requires expensive equipment, which may hinder its use for routine rabies diagnosis. Furthermore, a highly trained technician is required to ensure that reliable results are obtained for flow cytometry. In recent years, new equipment that is less expensive and easier to operate and maintain has become available from some companies (Accuri™ C6 from BD and Guava easyCyte™ 8HT from Merck Millipore), potentially favoring the more widespread use of flow cytometry for RABV detection.

4.4.7 Future Considerations

It has not been possible to analyze RABV particles directly by flow cytometry due to limitations concerning the detection of small particles. However, new techniques in which the virus is conjugated to magnetic nanoparticles are making it possible to identify the different antigens in the viral structures more accurately.[10] Improvements in traditional flow cytometry methods for cells should also increase our understanding of the variation of antigens presented by virus particles, making it possible to improve antigen detection and vaccine development.

ACKNOWLEDGMENTS

We thank Prof. André Báfica from the Laboratory of Immunobiology of the Federal University of Santa Catarina for critical reading and helpful discussions concerning this chapter.

Additional Information

Examples of major sources of supplies/equipment:

- Becton, Dickinson & Company (www.bdbiosciences.com).
- Millipore (www.millipore.com).
- MiltenyiBiotec (www.miltenyibiotec.com).
- Beckman Coulter (www.beckmancoulter.com).
- Bio-Rad (www.bio-rad.com).

REFERENCES

1. Trimarchi CV, Debbie JG. Standardization and quantitation of immunofluorescence in the rabies fluorescent-antibody test. *Appl Microbiol* 1972;**24**(4):609–12.
2. Woldehiwet Z. Clinical laboratory advances in the detection of rabies virus. *Clinica Chimica Acta* 2005;**351**:49–63.
3. Barardi CRM, Emsile KR, Vesey G, Williams KL. Development of a rapid and sensitive quantitative assay for rotavirus based on flow cytometry. *J Virol Methods* 1998;**74**:31–8.
4. Yang G, Olson JC, Pu R, Vyas GN. Flow cytometric detection of human immunodeficiency virus type 1 proviral DNA by polymerase chain reaction incorporating digoxigenin- or fluorescein-labeled dUTP. *Cytometry* 1995;**21**:197–202.
5. McSharry JJ, Constantino R, McSharry MB, Venezia RA, Lehman JM. Rapid detection of herpes simplex virus in clinical samples by flow cytometry after amplification in tissue culture. *J Clin Microbiol* 1990;**28**:1864–6.
6. Jensen BD, Horan PK. Flow cytometry: rapid isolation and analysis of single cells. *Methods Enzymol* 1989;**171**:549–81.
7. Bordignon J, Piza AT, Alvarez-Silva M, Caporale GMM, Carrieri ML, Kotait I, et al. Isolation and replication of rabies virus in C6 rat glioma cells (Clone CCL107). *Biologicals* 2001;**29**:67–73.
8. Bordignon J, Pires SC, Caporale GMM, Carrieri ML, Kotait I, Lima HC, et al. Flow cytometry assay for intracellular rabies virus detection. *J Virol Meth* 2002;**105**:181–6.
9. Vengatesan D, Raj GD, Raja A, Ramadass P, Gunaseelan L. Detection of rabies virus antigen or antibody using flow cytometry. *Cytometry* 2006;**70B**:335–43.
10. Arakelyan A, Fitzgerald W, Margolis L, Grivel JC. Nanoparticle-based flow virometry for the analysis of individual virions. *J Clin Investig* 2013;**123**(9):3716–27.

CHAPTER FIVE

Demonstration of Rabies Virus Antigens by an Immunochromatographic Strip Test

Songsri Kasempimolporn, Narongsak Chaiyabutr, and Visith Sitprija
Queen Saovabha Memorial Institute, Thai Red Cross Society, Bangkok, Thailand

Chapter Contents

5.1 Introduction 53
5.2 Materials 54
  5.2.1 Reagents 54
  5.2.2 Biological Materials 54
  5.2.3 Materials and Equipment 55
5.3 Methods 55
  5.3.1 Colloidal Gold Conjugation 55
  5.3.2 Test Strip Preparation 55
  5.3.3 Application of the Test Strip 56
5.4 Discussion 57
  5.4.1 Experimental Tips 57
  5.4.2 Technical Notes 57
  5.4.3 Future Considerations 58
References 59

5.1 INTRODUCTION

A number of immunochromatographic assays in a lateral flow format have been described for rabies virus (RABV) antigen detection in clinical samples, particularly brain tissue.[1–4] These tests employ two monoclonal antibodies (MAbs) which recognize the nucleoprotein (N) of RABV. RABV can be detected in a high frequency of samples collected antemortem.[5,6] Field observations suggest that on very rare occasions, a dog may shed RABV for slightly longer than the 10-day observation period prior to clinical development of rabies.[7–9] A convenient and affordable diagnostic

C. Rupprecht & T. Nagarajan (Eds): Current Laboratory Techniques in Rabies Diagnosis, Research, and Prevention, Volume 1.
Doi: http://dx.doi.org/10.1016/B978-0-12-800014-4.00005-6

test is desirable to provide prompt results for the purpose of the implementation of rabies prevention and control measures.

This chapter describes an immunochromatographic strip test for the rapid detection of RABV antigens using a combination of polyclonal anti-rabies antibody (PAb) and MAb specific to RABV glycoprotein (G).[10] PAb is labeled with the colloidal gold as the detector reagent, and MAb is immobilized on the nitrocellulose membrane as the capture reagent. The antibody–colloidal gold conjugate participates in the binding reaction to form a complex with the RABV if it is present in the sample. Because this complex binds to the capture antibody, the presence of a clear red-purple colored line on the strip identifies RABV infection. The interpretation of the result is visual within minutes. Saliva could be used as an alternative to brain specimens for RABV testing that can be readily used as the sample for the strip test, which makes it an ideal field diagnostic procedure. However, the strip test can be limited in terms of the amount of antigen it can detect. Clearly, a negative result should not be interpreted as indicating a rabies-negative animal. As the strip test can be used outside the laboratory and be applicable as an on-site testing assay, it represents a powerful screening tool for epidemiological surveys and disease control.

5.2 MATERIALS

5.2.1 Reagents

- Colloidal gold particles (20 nm in diameter).
- 0.2 M sodium carbonate.
- 0.01 M Phosphate buffer, pH 7.4 (see Additional Information).
- 10% bovine serum albumin (BSA) in 2 mM sodium borate buffer (pH 9.0).
- 2 mM sodium borate (pH 9.0) containing 1% BSA and 0.1% sodium azide.
- Storage buffer (20 mM sodium borate buffer (pH 8.0) containing 2% (w/v) sucrose, 5% (w/v) trehalose, and 0.1%sodium azide).

5.2.2 Biological Materials

- Purified rabbit anti-rabies immune globulin (IgG) (see Chapter 3).
- Purified mouse MAb specific to RABV G (anti-G MAb).[11]
- Goat anti-rabbit IgG
- Saliva samples (see Chapter 3)

5.2.3 Materials and Equipment

- Glass fiber membrane.
- Nitrocellulose membrane.
- Cellulosic paper.
- Plastic cassette.
- Automated dispensers.

5.3 METHODS

5.3.1 Colloidal Gold Conjugation

1. Adjust the pH of colloidal gold solution to 8.0–9.0 with 0.2 M sodium carbonate.
2. Add 2 mL of polyclonal anti-rabies IgG solution at the optimum concentration in 0.01 M phosphate buffer drop-by-drop to 10 mL of the pH-adjusted gold solution, stirring for 30 min at room temperature.
3. Add 10% BSA solution to the mixture to yield a final 1% (v/v) BSA concentration, gently rocking for over 10 min.
4. Wash by repeated centrifugation (14,000 × *g*) at 10°C for 30 min with 2 mM sodium borate containing 1% BSA.
5. Re-suspend the precipitate in 0.5 mL of the storage buffer and store at 4°C.

5.3.2 Test Strip Preparation

1. Dispense anti-G MAb at the concentration of 0.75 μg/strip, and anti-rabbit IgG at the concentration of 0.70 μg/strip, respectively, at the test and control lines on a nitrocellulose membrane (25 mm wide and 300 mm long) using an automated micro-spray dispenser. The test and control lines are situated 0.5 cm apart at the center of the membrane. After drying for 1 h at 40°C, the membrane is stored in a desiccator at room temperature.
2. Dispense the gold-conjugated antibody solution onto a glass fiber membrane (5 mm × 300 mm) at 10 μL/cm using an automated spray dispenser to produce the conjugate pad. After drying for 1 h at 56°C, the pad is stored in a desiccator at room temperature.
3. Cut a glass fiber sample pad in a size of 15 mm × 300 mm, and likewise a cellulose absorbent pad.
4. Assemble the sample pad, conjugate pad, blotted membrane, and absorbent pad on the backing card (60 mm × 300 mm) in the order shown

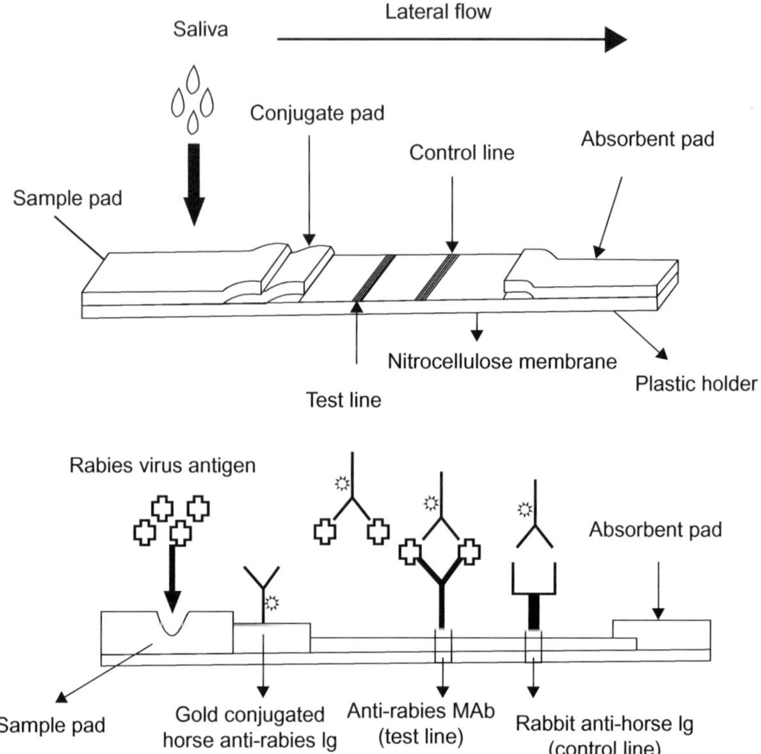

Figure 5.1 Schematic diagram of lateral flow immunochromatographic strip test.

in Figure 5.1. All pads overlap to enable migration of samples along the membrane phase. Cut the laminated card to 3.5 mm width strips with a cutter.

5. Place the strip in a plastic housing and store in a desiccated foil pouch until ready to test.

5.3.3 Application of the Test Strip

1. Apply a 100-μL portion of a saliva sample to a sample application window of the test device.
2. Interpret the test result within 10 min. A positive result is the appearance of red-purple colored lines at both the test zone and the control zone. The sample is considered negative if only one line appears in the control zone of the membrane (Figure 5.2). If there is no color present in the control line, the test should be considered as invalid.

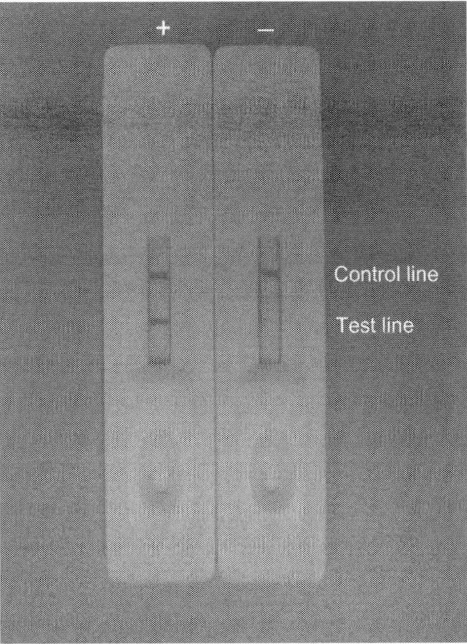

Figure 5.2 Test strip to detect rabies virus antigens showing typical positive and negative results.

5.4 DISCUSSION

5.4.1 Experimental Tips

1. The optimal size of the gold particles is somewhat dependent on the proteins and matrices to be used in the assay. Commonly, particles used in lateral flow assays are in the range of 20–40 nm.
2. It is necessary to determine the optimal concentration of antibody for stabilizing the gold. Unstable gold will precipitate and/or turn black in color. Stable gold will retain its ruby color. In addition, the optimal pH for conjugation can also be determined. Most antibodies coat well at pH 7.8 at a protein concentration of 10 μg/mL of colloidal gold.
3. For applications using brain extracts, sample cleanup may be necessary to discard cells and/or debris from obstructing the immunostrip. The sample has to be in the liquid state.

5.4.2 Technical Notes

1. The lateral flow system consists of overlapping components (conjugate pad, membrane, and wick), assembled such that the test fluid is

transported from the conjugate pad to the membrane by capillary action, and is finally retained on the wick. Two antibodies binding distinct epitopes present on the test antigen are used: one (detection antibody) conjugated with a signal generator (e.g. colloidal gold) and the other (capture antibody) immobilized onto solid phase. An additional antibody specific to the detection antibody can be used to produce a control signal.

2. Salivary secretion of the virus is intermittent, and the testing of saliva samples taken at different time intervals is critical. The amount of virus and the duration of virus shedding in the saliva are of paramount importance for rabies diagnosis. Any doubtful or negative results need to be confirmed by animal history, clinical signs, and established tests.
3. Saliva samples contain various components, such as a small number of immune active and epithelial cells, digestive enzymes, and bacteria. These components may interfere with the antigen-antibody reaction. Suspect samples can be forwarded to a well-equipped rabies laboratory for confirmatory diagnosis.
4. Because the inactivation of IgG is caused mainly by a change in the structural conformation during dehydration and storage, its prevention can be accomplished by handling the molecules in the presence of a stabilizer such as BSA, casein, sugar, or trehalose. Shelf life in the presence of a stabilizer is approximately 2 years at room temperature.

5.4.3 Future Considerations

The strip test may be a practical method in the field. If used appropriately, the strip test can be a valuable adjunct to traditional methods for antemortem rabies diagnosis. To use this test, adequate validation according to international standards is a priority. The procedures still require standardization and stringent quality control.

Additional Information

Sources of Supplies/Equipment

- Colloidal gold (Heron Diagnostics, Thailand)
- Glass fiber (Rapid 24®, Whatman, GE Healthcare, Uppsala, Sweden)
- Nitrocellulose membrane (FF80HP®, Whatman, GE Healthcare, Uppsala, Sweden)
- Cellulosic paper (CF4®, Whatman, GE Healthcare, Uppsala, Sweden)
- Backing card (G&L, San Jose, CA)

- Goat anti-rabbit immunoglobulin (Sigma, Louis, MO)
- Automated spray dispenser (Imagene, Hanover, NH)
- Automated micro-spray dispenser (Biodot, Irvine, CA)
- Plastic cassette (Mcmould Entico, Thailand)

Preparation of Buffer

- Phosphate buffer, pH 7.4, 0.01 M.
- Dissolve 1.38 g of sodium dihydrogen phosphate ($NaH_2PO_4 \cdot H_2O$) to a final volume of 1,000 mL.
- Dissolve 1.42 g of disodium hydrogen phosphate (Na_2HPO_4) to a final volume of 1,000 mL.
- Start with 500 mL of Na_2HPO_4 and add $NaH_2PO_4 \cdot H_2O$ while stirring until the desired pH is reached.

REFERENCES

1. Kang BK, Oh JS, Lee CS, Park BK, Park YN, Hong KS, et al. Evaluation of a rapid immunodiagnostic test kit for rabies virus. *J Virol Methods* 2007;**145**:30–6.
2. Nishizono A, Khawplod P, Ahmed K, Goto K, Shiota S, Mifune K, et al. A simple and rapid immunochromatographic test kit for rabies diagnosis. *Micribiol Immunol* 2008;**52**:243–9.
3. Markotter W, York D, Sabeta CT, Shumba W, Zula G, Le Roux K, et al. Evaluation of a rapid immunodiagnostic test kit for detection of African lyssaviruses from brain material. *Onderstepoort J Vet Res* 2009;**76**:257–62.
4. Servat A, Picard-Meyer E, Robardet E, Muzniece Z, Must K, Cliquet F. Evaluation of a rapid immunochromatographic diagnostic test for the detection of rabies from brain material of European mammals. *Biologicals* 2012;**40**:61–6.
5. Crepin P, Audry L, Rotivel Y, Gacoin A, Caroff C, Bourhy H. Intravitam diagnosis of human rabies by PCR using saliva and cerebrospinal fluid. *J Clin Microbiol* 1998;**36**:1117–21.
6. Saengseesom W, Mitmoonpitak C, Kasempimolporn S, Sitprija V, Real-time PCR. analysis of dog cerebrospinal fluid and saliva samples for ante-mortem diagnosis of rabies. *Southeast Asian J Trop Med Public Health* 2007;**38**:53–7.
7. Vaughn Jr JB, Gerhardt P, Newell KW. Excretion of street rabies virus in the saliva of dogs. *JAMA* 1965;**193**:363–8.
8. Schneider LG. Spread of virus from the central nervous system. In: Baer GM, editor. *The natural history of rabies*. New York: Academic Press; 1975. p. 273–301.
9. Somayajulu MV, Reddy GV. Live dogs and dead man. *J Assoc Phys India* 1989;**37**:6–17.
10. Kasempimolporn S, Saengseesom W, Huadsakul S, Boonchang S, Sitprija V. Evaluation of a rapid immunochromatographic test strip for detection of rabies virus in dog saliva samples. *J Vet Diag Invest* 2011;**23**:1197–201.
11. Lafon M. Techniques for the production, screening and characterization of monoclonal antibodies. In: Meslin FX, Kaplan MM, Koprowski H, editors. *Laboratory techniques in rabies* 4th ed.. Geneva: World Health Organization; 1996. p. 133–44.

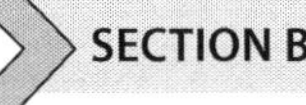

SECTION B

Demonstration of Viral Nucleic Acids

CHAPTER SIX

Demonstration of African Lyssavirus RNA with Real-Time Polymerase Chain Reaction

Jessica Coertse, Wanda Markotter, and Louis H. Nel
Department of Microbiology and Plant Pathology, University of Pretoria, Pretoria, South Africa

Chapter Contents

6.1 Introduction 63
6.2 Materials 64
  6.2.1 Reagents 64
  6.2.2 Equipment 64
  6.2.3 Biological Material 65
6.3 Methods 65
  6.3.1 Preparation of Reagent Master Mix 65
  6.3.2 Cycling Protocol 66
    6.3.2.1 Analysis of Results 67
6.4 Discussion 68
References 72

6.1 INTRODUCTION

Molecular methods were first proposed as an alternative method for diagnosis and epidemiological studies of lyssaviruses in the 1990s.[1] Shortly thereafter, a heminested reverse-transcriptase polymerase chain reaction (hnRT-PCR) assay was described utilizing a cocktail of seven primers for amplification of rabies virus (RABV), Lagos bat virus (LBV), Mokola virus (MOKV), Duvenhage virus (DUVV), and European bat Lyssavirus types 1 and 2 (EBLV-1 and EBLV-2).[2] During the last decade, the molecular method of choice for lyssavirus detection has shifted from RT-PCR to real-time PCR methods as they are less time-consuming, more sensitive, and less prone to contamination. Real-time PCR differs from conventional PCR with regards to the manner in which the amplicon is detected – oligoprobes with fluorescent potential are monitored during

C. Rupprecht & T. Nagarajan (Eds): Current Laboratory Techniques in Rabies Diagnosis, Research, and Prevention, Volume 1.
Doi: http://dx.doi.org/10.1016/B978-0-12-800014-4.00006-8

amplification versus post-amplification detection using UV-irradiated agarose gel electrophoresis with a DNA intercalating dye.[3]

Various systems are available for real-time PCR technology; however, only two chemistries are used widely for rabies. The sequence-independent chemistry require a minimum of two primers and a non-specific fluorescent dye such as SYBR Green I, while sequence-dependent methods usually use a 5′ nuclease hydrolysis probe (TaqMan™ probe) labeled with a fluorescent reporter and quencher in combination with two primers.[4]

Since the description of the first real-time RT-PCR assay utilizing a cocktail containing several primers and probes in a two-step protocol,[5] numerous real-time PCR assays have been described[6,7] that simplify the reaction set-up and subsequent analysis of results. The objective of this chapter is to provide an in-depth description of a real-time RT-PCR assay. This assay has a lower limit of detection of 10 RNA copies and is specific to African variants of RABV, LBV, DUVV, and MOKV. This one-step protocol allows for amplification and detection of viral RNA within 2 hours in a single capillary using a single primer-probe set.

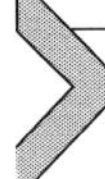

6.2 MATERIALS

6.2.1 Reagents

- One Step PrimeScript™ RT-PCR Kit (Takara Bio Inc, Cat. # RR064A)
- 541lys primer (10 pmol): 5′ – CACMGSNAAYTAYAARACNAA – 3′
- 550B primer (10 pmol): 5′ – GTRCTCCARTTAGCRCACAT – 3′
- 620lyssa probe (10 pmol): 5′ – (FAM)CATCACACCTTGATGACAA CTCACAA(BHQ-1) – 3′
- [FAM: 6-carboxyfluorescein, BHQ-1: Black Hole Quencher 1]

6.2.2 Equipment

- Pre-cooled (4°C) LightCycler® centrifuge adapter block
- LightCycler capillaries (20 μL) (Roche, Cat. # 49292920001)
- LightCycler capping tool
- Micropipettes and corresponding aerosol barrier tips
- Microcentrifuge tube
- LightCycler 1.5 thermal cycler system including software
- Centrifuge

6.2.3 Biological Material

- *Sample RNA*: Total RNA should be extracted from tissues or biological fluids using TRIzol reagent (Invitrogen, USA) according to the manufacturer's instructions.
- *Standard RNA*: *In vitro* transcribed RNA (CVS)[8] at a concentration of 10^5 copies/μL.

6.3 METHODS

6.3.1 Preparation of Reagent Master Mix

1. Place all reagents on ice until thawed.
2. Centrifuge (approximately 15 seconds) all reagents prior to use to ensure that the entire content of the tube is at the bottom – this reduces the risk of contamination or loss of reagents.
3. The master mix should be prepared in a dedicated bio-flow cabinet or PCR "clean room" to avoid contamination.
4. In a microcentrifuge tube, prepare the reagent master mix by adding the following components in the order mentioned (Table 6.1).
5. Always add an additional reaction to minimize any possible variation of the reactions due to pipetting errors or inaccurate dispensing of reagents [Total number of reactions = number of samples + no template control (NTC) + standard (STD) + 1].
6. Gently mix the reagent master mix and centrifuge the tube for 5–15 seconds at 600 *g*.
7. Place glass capillaries into the pre-cooled centrifuge adapters without touching the sides of the capillary.
8. Pipette 19 μL of the reagent master mix into each capillary.

Table 6.1 List of Reagents and Volumes to be Added to the Master Mix

Reagent	**Amount Per Reaction (μL)**
2× One Step RT-PCR Buffer III	10
Nuclease free water	5.85
TaKaRa Ex Taq™ HS (5 U/μL)	0.4
PrimeScript™ RT enzyme Mix II	0.4
541lys primer (10 pmol)	1
550B primer (10 pmol)	1
620lyssa probe (10 pmol)	0.35
Total	19

9. Add 1 μL of nuclease free water into a single capillary (NTC) and cap immediately with the capping tool.
10. Close the lid of the LightCycler centrifuge adapter block containing the capillaries with the reagent master mix and move the block to a dedicated area for the addition of sample RNA.
11. Add 1 μL of RNA template to a capillary and cap immediately with the capping tool.
12. Once all the sample RNA has been added to capillaries and capped, 1 μL of STD RNA is added to a capillary and capped immediately with the capping tool.
13. Place the adapters containing the capillaries into a centrifuge.
14. Centrifuge for 5–15 seconds at 600 *g*.
15. Make sure that the reaction is distributed throughout the capillary and that the volumes of all the capillaries are equal.
16. Transfer the capillaries to the sample carousel and place in the LightCycler instrument.

6.3.2 Cycling Protocol

1. Open the LightCycler software and enter the cycling protocol from Table 6.2.
2. Start run, and wait for the instrument to reach temperature and to locate the samples (if a sample was not placed in the carousel or a capillary does not contain any fluorescent probe, the software will indicate

Table 6.2 Cycling Parameters for use on the LightCycler® 1.5 Thermal Cycler System

Target °C	Hold (mm:ss)	Slope (°C/s)	Acquisition Mode
Program name: Reverse transcription (Analysis mode: None)			
42°C	30:00	20	None
95°C	00:10	20	None
Program name: PCR amplification (Analysis mode: Quantification)			
40 Cycles			
95°C	00:05	20	None
42°C	00:15	20	Single
72°C	00:06	2	None
Program name: Cooling (Analysis mode: None)			
40°C	00:30	20	None

this with an error message that has to be acknowledged before the run will start).

6.3.2.1 Analysis of Results

1. Once the run is completed, results may be analyzed by selecting the "Analysis" tab.
2. Select the "Absolute quantification" option.
3. The software will prompt you to enter sample settings:
   a. Select enable controls
      i. For the sample containing no RNA (NTC), select "Negative control" under sample type.
      ii. For the sample containing the *in vitro* transcribed RNA (STD), select "Standard" under the sample type and enter "1E5" under the concentration column.
      iii. Sample names can be entered if preferred, and "Unknown" should be selected under sample type.
   b. After sample settings have been entered, select the "Analysis" tab again and select the "Absolute quantification" option again.
   c. On the analysis screen select the "Standard curve (In run)" tab and then the "Use external" option – the external standard curve is selected.
4. The results will be displayed in table format together with the amplification curves (Table 6.3 and Figure 6.1).
   a. In the results table, the Cp value and concentration/RNA copy number of the samples will be displayed.
   b. In the results table there will also be a "Calls" column that will indicate if the controls performed as expected (i.e. will show "success") or if the controls failed (i.e. "failure").
5. For results to be valid, the negative control should be negative and the standard should have a Cp value of 22–23.
   a. If the negative control should fail or be called uncertain: results should be discarded as contamination occurred.
   b. If the standard does not perform as expected: If a slightly higher or lower Cp value is observed, it is likely due to a pipetting error or inaccurate dispensing of reagents; if the Cp value is significantly lower (more than 3 cycles), it indicates degradation of the RNA, and this standard should be discarded. The quantification of the reactions is therefore inaccurate and the run should be repeated.

Table 6.3 Results Obtained After the Completion of the RT-PCR and Subsequent Absolute Quantification Analysis

						Calls		
Pos	**Name**	**Type**	**Cp**	**Conc**	**Std**	**Comb**	**Target**	**Control**
1	S1	Unknown	31.4	3E2		Positive	Positive	Success
2	S2	Unknown	25.26	1.27E4		Positive	Positive	Success
3	S3	Unknown				Negative	Negative	Success
4	S4	Unknown	>35	>4.69E1		Positive	Positive	Success
5	S5	Unknown	28.34	1.77E3		Positive	Positive	Success
6	S6	Unknown	28.77	1.37E3		Positive	Positive	Success
7	CVS STD	Standard	22.29	1E5	1E5	Success	Positive	Success
8	NTC	Negative control				Success	Negative	Success

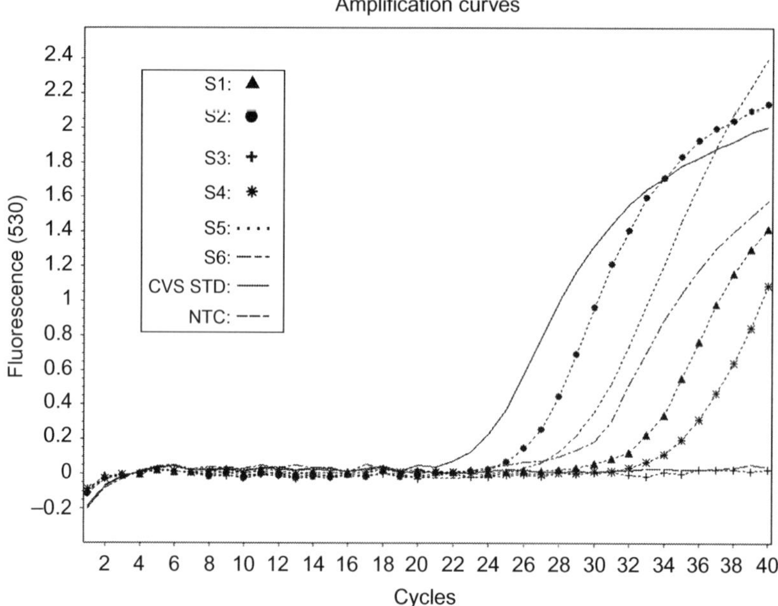

Figure 6.1 Amplification curves of samples (indicated in Table 6.3) after completion of the real-time RT-PCR protocol.

6.4 DISCUSSION

The one-step real-time RT-PCR assay described in this chapter allows for the detection and quantification of viral RNA within 2 hours with a lower limit of detection of 10 RNA copies. This assay may also be relevant in cases where samples would be problematic for conventional molecular methods such as tissue/RNA degradation[9] or from tissues fixed in formalin. Although the main purpose of this assay is for the detection of viral nucleic acids, the amplified products may also be directly sequenced for identification purposes.[10] However, because a very short region of the nucleoprotein gene is targeted, epidemiological information from such sequences is limited and should be interpreted with caution. This assay is very sensitive and therefore care should be taken when preparing the reagent master mix as contamination may occur easily. The use of challenge virus standard (CVS) strain of RABV as the standard in this assay will also allow for detection of contamination with this control via subsequent sequencing as it displays significant divergence from other lyssaviruses.

The use of *in vitro* transcribed RNA is the most common material used for absolute quantification using external standards.[11] However, RNA is prone to degradation, and therefore it is recommended that RNA standards be stored at −80°C in nuclease-free water containing carrier nucleic acid (such as tRNA). Working stock of standard RNA (containing carrier nucleic acid) may be stored at −20°C, but should be replaced every 3 months or as soon as variation in the Cp values are observed.

Alternative real-time RT-PCR reagents may also be used[10] but would need to be optimized carefully, and several assay characteristics determined before results are reported/published.[12] Furthermore, this assay was specifically designed to detect lyssavirus species circulating in Africa and has not been tested for the successful detection of other lyssaviruses elsewhere in the world, and also not optimized for newly discovered lyssaviruses (i.e. Bokelohbat lyssavirus, Ikomalyssavirus, etc.). However, all lyssaviruses described before 2009 were considered in the overall design, and detection should be successful, but would need to be evaluated. Several other real-time RT-PCR assays have been developed[13–17] and these may differ in the precise method, reagents used, and the lyssavirus species targeted (Table 6.4). Nevertheless, the principle of real-time detection of specific nucleic acid amplification from two site-specific primers applies to all these methods.

The assay described in this chapter overcomes several inherent disadvantages of conventional methods such as low dynamic range, low sensitivity, risk of contamination, and time needed to obtain a result. However, this assay can only be performed in laboratories with the necessary equipment, strict quality control procedures, and personnel with sufficient experience in molecular methods.

Table 6.4 Details of Real-time RT-PCR Assays used for Lyssavirus Detection

Reference	Real-time RT-PCR Method	RNA Extraction Method	Target Region	Detected Species	Geographical Region[a]	Reagents	LOD[b]
6	Two-step, sequence independent	TRIzol (Invitrogen)	N gene	RABV, LBV, MOKV, DUVV, EBLV 1 & 2, ABLV, KHUV, IRKV, ARAV, WCBV, SHIBV	Worldwide	SYBR® Green Hot Start (Applied Biosystems)	5–50 DNA copies and 190–1430 RNA copies
13	One-step, TaqMan	QIAmp Viral RNA Kit (Qiagen) or RNeasy kit (Qiagen)	N gene	RABV	Worldwide	QuantiTect™ Probe RT-PCR Kit (Qiagen) or RNA Ultrasense™ One-Step Quantitative RT-PCR System (Invitrogen)	10 RNA copies

8	One-step, TaqMan (quant.)	TRIzol (Invitrogen) or QIAmp Viral RNA Kit (Qiagen)	N gene	RABV, LBV, MOKV, DUVV	Africa	LightCycler® RNA Amplification Kit Hybprobe (Roche)	10 RNA copies
14	One-step, TaqMan (quant.)	TRIzol (Invitrogen)	N gene	RABV	Worldwide	SuperScript III Platinum One-Step qRT-PCR kit (Invitrogen)	0.05 $TCID_{50}$ EU
15	One-step, TaqMan	Not specified	N gene	RABV	India, Thailand, Indonesia	QuantiTect™ Probe RT-PCR Kit (Qiagen)	4 RNA copies
16	Two-step, sequence independent	TRIzol LS (Invitrogen) with glycogen	N gene	RABV	India	SYBR Green I (ABI Biosystems)	1043 DNA copies
7	One-step, TaqMan	TRIzol (Invitrogen)	N gene	RABV, EBLV-1, EBLV-2	Worldwide	Separate reagents	1 pg RNA
17	Two-step, TaqMan	TRIzol (Invitrogen)	N gene	RABV	USA	TaqMan master mix (Applied Biosystems)	Not specified

[a]Refers to the geographical location of isolates/samples successfully detected with this assay. (Most assays were designed based on all available sequence information to represent the global diversity of a specific Lyssavirus species.)
[b]Limit of detection.

Additional Information

Construction of a Standard Curve

- *In vitro* transcribed CVS RNA should be diluted to represent 10^1–10^{10} RNA copies/μL.
- Every dilution should be tested in triplicate in the same run using the conditions as described in this chapter.
- The above-mentioned run should be repeated.

Statistical Analysis

The results obtained from the two standard curve runs should be evaluated statistically using the LightCycler® software and any statistical program, such as Microsoft Excel.

- PCR efficiency (E): Determined from the slope of the log linear phase of the amplification curves. This can be done by using the LightCycler software or using the following formula: $E=10^{-1/\text{slope}}$.
- Standard deviation (SD): This is determined automatically by the LightCycler software after sample settings are entered. This is the most widely used measure of dispersion and gives an indication of the difference between a sample and the mean.
- Coefficient of variation (CV): The CV is the standard deviation expressed as a percentage of the mean and should be determined for every dilution (tested in triplicate) by dividing the SD by the mean and then converting the value to percentage.
- Reproducibility: Refers to the inter-assay variability, which is determined by the CV value observed for the same dilution between different runs.
- Precision: Refers to the intra-assay variability, which is determined by the CV value for the same dilution within the same run.
- Linear dynamic range: Indicates the lowest and highest quantifiable copy number over which the reaction is linear. This can be determined by using the "Pearson correlation" function of Microsoft Excel.

REFERENCES

1. Sacramento D, Bourhy H, Tordo N. PCR technique as an alternative method for diagnosis and molecular epidemiology of rabies virus. *Mol Cell Probes* 1991;**5**:229–40.
2. Heaton PR, Johnstone P, McElhinney LM, Cowley R, O'Sullivan E, Whitby JE. Hemi-nested PCR assay for detection of six genotypes of rabies and rabies-related viruses. *J Clin Microbiol* 1997;**35**(11):2762–6.
3. MacKay IM, Arden KE, Nitsche A, Real-time PCR. in virology. *Nucleic Acids Res* 2002;**30**(6):1292–305.
4. Fischer M, Hoffman B, Freuling CM, Müller T, Beer M. Perspectives on molecular detection methods of lyssaviruses. *Berl Münch Tierärztl Wochenschr* 2012;**125**:264–71.

5. Black EM, Lowings JP, Smith J, Heaton PR, McElhinney LM. A rapid RT-PCR method to differentiate six established genotypes of rabies and rabies-related viruses using TaqMan™ technology. *J Virol Methods* 2002;**105**:25–35.
6. Hayman DTS, Banyard AC, Wakeley PR, Harkess G, Marston D, Wood JLN, et al. A universal real-time assay for the detection of lyssaviruses. *J Virol Methods* 2011;**177**:87–93.
7. Wakeley PR, Johnson N, McElhinney LM, Marston D, Sawyer J, Fooks AR. Development of a real-time, TaqMan reverse transcription-PCR assay for detection and differentiation of Lyssavirus genotypes 1, 5 and 6. *J Clin Microbiol* 2005;**43**(6):2786–92.
8. Coertse J, Weyer J, Nel LH, Markotter W. Improved PCR methods for detection of African rabies and rabies-related lyssaviruses. *J Clin Microbiol* 2010;**48**(11):3949–55.
9. Mollentze N, Weyer J, Markotter W, le Roux K, Nel LH. Dog rabies in southern Africa: regional surveillance and phylogeographical analyses are an important component of control and elimination strategies. *Virus Genes* 2013;**27**(3):569–73.
10. Coertse J, Nel LH, Sabeta CT, Weyer J, Grobler A, Walters J, et al. A case study of rabies diagnosis from formalin-fixed brain material. *J S Afr vet Ass* 2011;**82**(4):250–3.
11. Pfaffl MW. Quantification strategies in real-time PCR. In: Bustin SA, editor. *A-Z of quantitative PCR.*. : La Jolla International University Line; 2004. p. 87–120.
12. Bustin SA, Benes V, Garson JA, Hellemans J, Huggett J, Kubista M, et al. The MIQE Guidelines: Minimum Information for publication of Quantitative real-time PCR Experiments. *Clin Chem* 2009;**55**(4):611–22.
13. Hoffman B, Freuling CM, Wakeley PR, Rasmussen TB, Leech S, Fooks AR, et al. Improved safety for molecular diagnosis of classical rabies viruses by use of a TaqMan real-time reverse transcription-PCR "double check" strategy. *J Clin Microbiol* 2010;**48**(11):3970–8.
14. Nadin-Davis SA, Sheen M, Wandeler AI. Development of a real-time reverse transcriptase polymerase chain reaction methods for human rabies diagnosis. *J Med Virol* 2009;**81**:1484–97.
15. Wacharapluesadee S, Sutipanya J, Damrongwatanapokin S, Phumesin P, Chamnapood P, Leowijuk C, et al. Development of a TaqMan real-time RT-PCR assay for the detection of rabies virus. *J Virol Methods* 2008;**151**:317–20.
16. Nagaraj T, Vasanth JP, Desai A, Kamat A, Madhusudana SN, Ravi V. Ante mortem diagnosis of human rabies using saliva samples: Comparison of real-time and conventional RT-PCR techniques. *J Clin Virol* 2006;**36**:17–23.
17. Hughes GJ, Smith JS, Hanlon CA, Rupprecht CE. Evaluation of a TaqMan PCR assay to detect rabies virus RNA: Influence of sequence variation and application to quantification of viral loads. *J Clin Microbiol* 2004;**42**(1):299–306.

CHAPTER SEVEN

Real-Time Quantitative Polymerase Chain Reaction for the Demonstration of Lyssavirus Nucleic Acid

Conrad M. Freuling[1], Bernd Hoffmann[2], Melina Fischer[2], Lorraine M. McElhinney[3,4], Denise A. Marston[3], Anthony R. Fooks[3], and Thomas F. Müller[1]

[1]Friedrich-Loeffler-Institut (FLI), Federal Research Institute for Animal Health, Institute of Molecular Virology and Cell Biology, Greifswald-Insel Riems, Germany
[2]FLI, Institute of Diagnostic Virology, Greifswald-Insel Riems, Germany
[3]Wildlife Zoonosis and Vector-Borne Disease Research Group, Animal Health and Veterinary Laboratories Agency, Addlestone, Surrey, United Kingdom
[4]Institute of Infection and Global Health, University of Liverpool, Liverpool, United Kingdom

Chapter Contents

7.1 Real-Time Quantitative Polymerase Chain Reaction for the Demonstration of Lyssavirus Nucleic Acid 76
7.2 Methodology 76
 7.2.1 Viral RNA Extraction 76
 7.2.2 Generic Pan Lyssavirus Real-Time Polymerase Chain Reactions 76
  7.2.2.1 Materials/Equipment 76
  7.2.2.2 Preparation of Mastermix 77
  7.2.2.3 Internal Controls 77
  7.2.2.4 Thermal Program 80
  7.2.2.5 Read Out 80
 7.2.3 Real-Time Polymerase Chain Reaction for Lyssavirus Species using Specific Taqman Probes 80
  7.2.3.1 Materials/Equipment 80
  7.2.3.2 Preparation of Mastermix 81
  7.2.3.3 Thermal Program 81
  7.2.3.4 Read Out 81
7.3 Discussion 82
Acknowledgments 83
References 83

C. Rupprecht & T. Nagarajan (Eds): Current Laboratory Techniques in Rabies Diagnosis, Research, and Prevention, Volume 1.
Doi: http://dx.doi.org/10.1016/B978-0-12-800014-4.00007-X

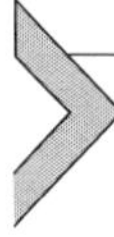

7.1 REAL-TIME QUANTITATIVE POLYMERASE CHAIN REACTION FOR THE DEMONSTRATION OF LYSSAVIRUS NUCLEIC ACID

In comparison with conventional polymerase chain reaction (PCR) assays, real-time PCRs are closed-tube systems, thus reducing the possibility of amplicon or cross-contamination that may lead to false positive results. A generally smaller fragment size causes shorter time for the PCR, and without the need for a confirmation as in gel-based assays, the real-time PCR (RT-PCR) is faster in the turnaround time and allows for a higher throughput. Also, a combination of various assays (e.g. internal control, multiplex, etc.) is possible, and some assays have shown a relatively higher sensitivity in comparison to conventional RT-PCR protocols.[1–3] Also, RT-PCR allows for a quantitation of the template RNA. Numerous real-time RT-PCR assays have been developed and were published, some of which were selectively developed for the individual virus strain,[4] while others were developed for a wider taxonomic range.[1,5]

In this chapter, we describe a diagnostic approach that uses two independent pan-lyssavirus real-time RT-PCR assays,[6,7] targeting the nucleoprotein gene (N-gene) and polymerase gene (L-gene), respectively. In parallel, or as a second step, lyssavirus species-specific real-time RT-PCR assays using probes can be applied to determine the virus species and independently confirm the results obtained by the generic pan-lyssa PCR. Whilst the real-time RT-PCR using primers JW12 and N165-146 has been shown to detect lyssaviruses including the more divergent Ikoma lyssavirus;[5,6,8] the use of an additional conserved target region offers extra diagnostic safety.[7]

7.2 METHODOLOGY

7.2.1 Viral RNA Extraction

Extract RNA following the instructions given in the chapter on conventional gel-based PCRs (see Volume 2) and store RNA samples at −80°C until use. Due to the high sensitivity of the real-time PCR, great care must be taken to ensure that any contamination is excluded.

7.2.2 Generic Pan Lyssavirus Real-Time Polymerase Chain Reactions

7.2.2.1 Materials/Equipment

Several thermal cyclers are available that allow real-time detection of fluorescence. The difference in performance when using established manufacturers appears negligible. However, consumables may be specific for the individual thermal cyclers of different companies. For example, the CFX96 quantitative PCR system (Bio-Rad Laboratories Inc., Hercules, USA) should be used with the respective Bio-Rad 96-well PCR plates. For real-time RT-PCR, numerous test kits are available commercially, including the OneStep RT-PCR Kit (Qiagen, Hilden, Germany). When establishing real-time RT-PCR in the laboratory, quality assurance should involve testing a 10-fold serial dilution series of stock RNA (1 μg/μL) over a range of 1 μg–1 pg for viral RNA stocks isolated from positive brain samples. The assay should detect RNA in the 10 pg dilution with high efficiency (>90%).

7.2.2.2 Preparation of Mastermix

The mastermix contains all the components for the reverse transcription and amplification. Sufficient mastermix should be prepared to be able to analyze all the samples and no template control(s) plus positive control(s). Samples should be run at least in duplicate, and all controls run alongside to ensure the test is running within the required parameters.

Mastermixes can be made up fresh for the test to be run (Table 7.1) or made up in batches and frozen for up to one year in suitably sized aliquots. The assays described here were optimized for limited sample material using a total reaction volume of 12.5 μL.

7.2.2.3 Internal Controls

The functionality of amplification can be tested in an extra tube using heterologous internal control (IC) systems[9] in combination with endogenous housekeeping genes[10] (Table 7.2).

Table 7.1 Lyssavirus Mastermix per Sample and Per Assay Added in this Order

Item	Components	Volume
1.	RNase-free water	5.50 μL
2.	5× RT-PCR Buffer (OneStep, Qiagen, Hilden, Germany)	2.50 μL
3.	dNTPs 10 mM	0.50 μL
4.	0.5 μL ResoLightDye (Roche, Mannheim, Germany)	0.50 μL
5.	Forward primer (see Table 7.2 for details)	0.25 μL
6.	Reverse primer (see Table 7.2 for details)	0.25 μL
7.	RT-PCR Enzyme Mix (OneStep, Qiagen, Hilden, Germany)	0.50 μL
8.	RNA template or RNase free water for no control (NTC)	2.50 μL
	Total	12.50 μL

Table 7.2 Real Time Polymerase Chain Reaction Primer and Probe Specifications

	PCR assay	Primer/Probe	Sequence (5′–3′)	Position	Concentration [μM]	Reference
Pan-lyssavirus	N-gene PCR	JW12	ATG TAA CAC CYC TAC AAT G	55–73	20	6
		N165-146	GCA GGG TAY TTR TAC TCA TA	165–146	20	
	L-gene PCR	Pan-Lyssa-7531F	TTC TTC GCT YTR ATG TCW TGG AA	7074–7096	20	7
		Pan-Lyssa-7749R	ATG RTT GTT CCA CTT YTC ATA RTC	7292–7269	20	
Lyssavirus-specific	R13 MP	JW12	ATG TAA CAC CYC TAC AAT G	55–73	10	
		N165-146	GCA GGG TAY TTR TAC TCA TA	165–146	10	
		LysGT1-FAM	FAM -ACA AGA TTG TAT TCA AAG TCA ATA ATC AG- BHQ1	81–109	2.5	
		LysGT5-HEX	HEX -AA CAR GGT TGT TTT YAA GGT CCA TAA- BHQ1	80–105	2.5	5
		LysGT6-Cy5	Cy5 -ACA RAA TTG TCT TCA ARG TCC ATA ATC AG- BHQ3	81–109	2.5	
		BBLV-1TEX	TEX -CTC TGA CAA GAT TGT CTT CAA AGT C- BHQ2	76–101	2.5	
	R14 MP	RV-N-196-F	GAT CCT GAT GAY GTA TGT TCC TA	266–288	10	
		RV-N-283-R	RGA TTC CGT AGC TRG TCC A	353–335	10	11
		RabGT1-B-FAM	FAM -CAG CAA TGC AGT TYT TTG AGG GGA C- BHQ1	297–321	2.5	

	R14 MP	EBLV1-353F	GCT CAA ACR GGA GGT CAA GA	431–450	10	
		EBLV1-440R	AGA CAR AGA AGA AGT CCW ACC A	510–489	10	7
		EBLV1-392HEX	HEX -ACC CTA CRA CAC CTG AAC ATG CAT CT- BHQ1	462–487	2.5	
	R14 MP	EBLV2-42F	RGT GTC TGT AAA RCC AGA AG	112–131	15	
		EBLV2-173R	GAC AGA ATR GAC TTA TAA GCT CT	243–221	15	13
		EBLV2 N Probe	Cy5 -TCG GAA AAA ACC CAG CAT AAC CCT- BHQ2	175–198	2.5	
	R14 MP	BBLV-2F	CCT TGG TRA ACA TTC AGA GAA CG	390–412	10	
		BBLV-2R	GGC CAC AGT TGG ATC CCT TG	475–456	10	14
		BBLV-2TEX_as	TEX -TCC TCC GGT CAA GGC CCA RTT GCC- BHQ2	422–445	2.5	
	IC_h	EGFP1-F	GAC CAC TAC CAG CAG AAC AC	637–656	2.5	
		EGFP2-R	GAA CTC CAG CAG GAC CAT G	768–750	2.5	9
		EGFP-Probe 1	FAM -AGC ACC CAG TCC GCC CTG AGC A- BHQ1	703–724	1.25	
Internal controls	IC_e	ACT-1005-F	CAG CAC AAT GAA GAT CAA GAT CAT C	1005–1029	2.5	
		ACT-1135-R	CGG ACT CAT CGT ACT CCT GCT T	1135–1114	2.5	10
		ACT-1081-HEX	HEX -TCG CTG TCC ACC TTC CAG CAG ATG T- BHQ1	1081–1105	1.25	

Table 7.3 Thermal Program

Process	Temp.	Time	
Reverse transcription	50°C	30 min	
Inactivation of RT /Activation of Taq	95°C	15 min	
Denaturation	95°C	30 sec	45 Cycles
Annealing	55°C	30 sec	45 Cycles
Elongation	72°C	30 sec	45 Cycles

Table 7.4 Summary of Outcomes and Overall Result

Test Result	Internal Control	Overall Result
Negative	Negative	Invalid. Repeat testing.
Negative	Positive	Negative result reported.
Positive	Positive	Positive result reported.
Positive	Negative	Internal control failure! Repeat and/or run on a gel. Follow up positive result.

7.2.2.4 Thermal Program

The thermal program is illustrated in Table 7.3.

7.2.2.5 Read Out

For each RT-PCR, a quantification cycle (Cq) is determined according to the PCR cycle number at which the fluorescence of the reaction crosses a value that is statistically higher than the background, which is determined by the respective software associated with each system. Positive samples will have a Cq value, whereas negative samples will have no Cq value. The lower the Cq, the more amplicon produced, and therefore the more viral template added to the RT-PCR. Melt curve analysis should be performed (1 min 95°C, 1 min 55°C, increase 0.5°C per cycle for 10 s; 55–95°C) for verification of positive results (Table 7.4).

7.2.3 Real-Time Polymerase Chain Reaction for Lyssavirus Species using Specific Taqman Probes

7.2.3.1 Materials/Equipment

For lyssavirus species-specific detection, the AgPath-ID™ One-Step RT-PCR Kit (Applied Biosystems, Carlsbad, USA) was used by combining the assay developed by Wakeley et al. (2005) (R13 MP)[5] with another (R14 MP)[11] assay, both targeting the N gene of lyssaviruses. Furthermore, the published assay by Wakeley et al. (2005)[5] was extended

Table 7.5 Mastermix Per Sample and Per Assay Added in this Order

Item	Components	Volume
1.	RNase-free water	1.75 μL
2.	2× Reaction Mix (AgPath-ID™ One-Step RT-PCR Kit)	6.25 μL
3.	Primer/probe mix (see Table 7.2 for details)	1.00 μL
4.	Internal control (see Table 7.2 for details)	1.00 μL
5.	RNA template or RNase free water for no template control (NTC)	2.50 μL
	Total	12.50 μL

Table 7.6 Thermal Program

Process	Temp.	Time	
Reverse transcription	45°C	10 min	
Inactivation of RT/Activation of Taq	95°C	15 min	
Denaturation	95°C	15 sec	
Annealing	56°C	20 sec	45 Cycles
Elongation	72°C	30 sec	

by an additional BBLV probe (Table 7.2). Further probes for the detection of other lyssavirus species are currently being developed/validated. Both assays were optimized using a total reaction volume of 12.5 μL.

7.2.3.2 Preparation of Mastermix

Prior to the preparation of a mastermix, a primer and probe mix needs to be prepared (Table 7.5). For each assay, 20 μL of forward and reverse primers (100 pmol/μL) plus 2.5 μL of the respective probe is added to 0.1×TE buffer (pH 8.0) for a total volume of 200 μL (for details, see Table 7.2). Additionally, a probe specific to an artificial positive control could be added.[11]

7.2.3.3 Thermal Program

The thermal program is illustrated in Table 7.6.

7.2.3.4 Read Out

Fluorescence emitted following the Fluorescence Resonance Energy Transfer (FRET) is measured per cycle, and a Cq is determined in the same way as described above. Results are only valid if all controls have expected values.

7.3 DISCUSSION

Real-time RT-PCR has become a standard technique for the diagnosis of infectious diseases, and will remain the main molecular technique at least for the near future. Initial constraints including cross-contamination and false positive results seem to have been largely resolved by various preventive measures and quality-control systems. Although not recommended as a primary diagnostic technique, in fact, in a recent ring-trial in Europe, the performance of (real-time) PCR gave more concordant results than virus isolation in cell culture.[12]

Numerous diagnostic real-time RT-PCRs have been developed for lyssavirus detection. Considering the diversity of this genus, they all have certain limitations as regards their diagnostic range.[2–3,11] The choice of an intercalating dye approach versus a more species-specific approach will depend on the intended application. For a laboratory conducting surveillance on brain material and where a considerable number of negative samples would be expected, the use of the broader intercalating dye real-time PCR would be recommended. The more specific and sensitive probe-based assays may be preferable in a laboratory conducting lyssavirus diagnosis on samples which are expected to have a restricted lyssavirus species range or contain low viral loads.

Here, we present a molecular diagnostic that uses a two-level cascade protocol. On a first level, two independent pan-lyssavirus RT-PCR systems target (1) the N gene based on the protocol by Hayman et al. (2011)[6] using one universal primer pair,[5] or (2) highly conserved regions in the L-gene.[7] In a double-check strategy, one assay is able to balance the impairment of the corresponding assay, thereby increasing diagnostic sensitivity and safety. In the case of a doubtful result, the melting curve analysis can be consulted to avoid a false positive result due to the formation of primer dimers and/or sequencing of the product can give further confirmation.

In a second level, lyssavirus species-specific assays using TaqMan probes can be used to further confirm and specify the results. However, it has been shown that the diversity within the rabies virus (RABV) also hampers a universal probe-based assay, thus a combination of two assays can improve the diagnostic utility.[11] One approach is to use one universal primer pair,[5] which appears user friendly. However, this leads to a limitation of the assay as it confines the sequence region for probe positioning, which adversely affects the specificity and averts an arbitrary expansion of the system.

Here we have presented assays for limited sample material using a total reaction volume of 12.5 μL. This reduction was made without losing performance of the assays. With the smaller volumes needed, PCRs become more cost-efficient. Also, as one application for these pan-lyssa assays is screening of bat samples, including oral swabs, the lower amount of RNA needed also allows for several other pathogens to be tested at the same time.

The diagnostic real-time RT-PCRs shown here are examples for lyssavirus detection and offer possibilities for laboratories to introduce this technique. Surely, whenever novel virus variants or even novel lyssavirus species are discovered, there may be a need to modify existing assays. However, with the tools shown here, we are confident that laboratories can establish this technique for multiple own purposes.

When established, laboratories should regularly check the performance of their assays by using internal controls and by taking part in national and or international ring-trials. Because all assays should be fit-for-purpose in the individual laboratory, it is not advisable to recommend one specific assay. Rather, if laboratories want to use RT-PCR as a confirmatory test additional to FAT, they need to show congruent results when testing a panel of RABV or other lyssaviruses that represents the global and local diversity.

Whilst positive PCR results should result in prompt action concerning the provision of post-exposure prophylaxis, they should be followed up using classical virological techniques, i.e. virus isolation to allow for future characterization and use of the virus and enable case notification to OIE.

ACKNOWLEDGMENTS

This work was funded by the German Ministry for Education and Research (BMBF, Grant 01KI1016A), by the Department for Environment, Food and Rural Affairs (Defra), UK (Grant ROAME SE0427), and by the EU FP7-funded Research Infrastructure Grant European Virus Archive (No. 19 228292).

REFERENCES

1. Nadin-Davis SA, Sheen M, Wandeler AI. Development of real-time reverse transcriptase polymerase chain reaction methods for human rabies diagnosis. *J Med Virol* 2009;**84**:1484–97.
2. Coertse J, Weyer J, Nel LH, Markotter W. Improved PCR methods for detection of African rabies and rabies-related lyssaviruses. *J Clin Microbiol* 2010;**48**:3949–55.
3. Fooks AR, Johnson N, Freuling CM, Wakeley PR, Banyard AC, McElhinney LM, et al. Emerging technologies for the detection of rabies virus: challenges and hopes in the 21st century. *PLoS Negl Trop Dis* 2009;**3**:e530.
4. Panning M, Baumgarte S, Pfefferle S, Maier T, Martens A, Drosten C. Comparative analysis of rabies virus reverse transcription-PCR and virus isolation using samples from a patient infected with rabies virus. *J Clin Microbiol* 2010;**48**:2960–2.

5. Wakeley PR, Johnson N, McElhinney LM, Marston D, Sawyer J, Fooks AR. Development of a real-time, TaqMan reverse transcription-PCR assay for detection and differentiation of lyssavirus genotypes 1, 5, and 6. *J Clin Microbiol* 2005;**43**:2786–92.
6. Hayman DTS, Banyard AC, Wakeley PR, Harkess G, Marston D, Wood JL, et al. A universal real-time assay for the detection of Lyssaviruses. *J Virol Methods* 2011;**177**:87–93.
7. Fischer M, Freuling C, Müller T, Wegelt A, Kooi EA, Rasmussen TB, et al. Molecular double check strategy for the identification and characterization of European Lyssaviruses. . *J Virol Methods.* 2014;**203**:23–32.
8. Marston DA, Horton DL, Ngeleja C, Hampson K, McElhinney LM, Banyard AC, et al. Ikoma lyssavirus, highly divergent novel lyssavirus in an African civet. *Emerg Infect Dis* 2012;**18**:664–7.
9. Hoffmann B, Depner K, Schirrmeier H, Beer M. A universal heterologous internal control system for duplex real-time RT-PCR assays used in a detection system for pestiviruses. *J Virol Methods* 2006;**136**:200–9.
10. Toussaint JF, Sailleau C, Breard E, Zientara S, De Clercq K. Bluetongue virus detection by two real-time RT-qPCRs targeting two different genomic segments. *J Virol Methods* 2007;**140**:115–23.
11. Hoffmann B, Freuling CM, Wakeley PR, Rasmussen TB, Leech S, Fooks AR, et al. Improved Safety for Molecular Diagnosis of Classical Rabies Viruses by Use of a TaqMan Real-Time Reverse Transcription-PCR "Double Check" Strategy. *J Clin Microbiol* 2010;**48**:3970–8.
12. Robardet E, Picard-Meyer E, Andrieu S, Servat A, Cliquet F. International interlaboratory trials on rabies diagnosis: An overview of results and variation in reference diagnosis techniques (fluorescent antibody test, rabies tissue culture infection test, mouse inoculation test) and molecular biology techniques. *J Virol Methods* 2011;**177**:15–25.
13. Schatz J, Ohlendorf B, Busse P, Pelz G, Dolch D, Teubner J, et al. Twenty years of active bat rabies surveillance in Germany: a detailed analysis and future perspectives. *Epidemiol Infect* 2013;**6**:1–12.
14. Freuling CM, Abendroth B, Beer M, Fischer M, Hanke D, Hoffmann B, et al. Molecular diagnostics for the detection of Bokeloh bat lyssavirus in a bat from Bavaria, Germany. *Virus Res* 2013;**177**:201–4.

CHAPTER EIGHT

Reverse Transcription-Loop-Mediated Isothermal Amplification System for the Detection of Rabies Virus

Takuya Itou[1], Wanda Markotter[2], and Louis H. Nel[2]
[1]Nihon University Veterinary Research Center, Fujisawa, Kanagawa, Japan
[2]Department of Microbiology and Plant Pathology, University of Pretoria, Pretoria, South Africa

Chapter Contents

8.1 Introduction 85
8.2 Materials 86
  8.2.1 Reagents 86
  8.2.2 Primer Set for Reverse Transcription-Loop-Mediated Isothermal Amplification 86
  8.2.3 Equipment 89
  8.2.4 Biological Materials and Sample Preparation 90
8.3 Methods 90
  8.3.1 Preparation of Mastermix 90
  8.3.2 Mixing of Mastermix and Sample Solution 91
  8.3.3 Amplification Reaction 91
  8.3.4 Detection and Interpretation of Results 91
    8.3.4.1 Turbidity Detection *91*
    8.3.4.2 Fluorescence Detection *91*
    8.3.4.3 Electrophoresis *92*
    8.3.4.4 Validation of the Specificity of Loop-Mediated Isothermal Amplification Reaction *93*
8.4 Discussion 93
Acknowledgments 94
References 94

8.1 INTRODUCTION

The innovative loop-mediated isothermal amplification (LAMP) technique, providing a rapid, simple, sensitive, and inexpensive method for specific DNA sequence amplification, was first demonstrated in 2000.[1]

C. Rupprecht & T. Nagarajan (Eds): Current Laboratory Techniques in Rabies Diagnosis, Research, and Prevention, Volume 1.
Doi: http://dx.doi.org/10.1016/B978-0-12-800014-4.00008-1

LAMP depends on auto-cycling strand-displacement DNA synthesis conducted by *Bst* DNA polymerase. This reaction proceeds without denaturation of DNA templates,[2] and thus can be performed at an isothermal temperature. It provides high amplification efficiency with DNA being amplified 10^9–10^{10} times in 15–60 minutes. The amplified products consist of a series of stem-loop DNA in various lengths.[3] The results can be determined by visual inspection of the turbidity due to precipitation of white magnesium pyrophosphate, a byproduct of DNA synthesis,[4] or by visual and UV inspection of DNA amplification with fluorescent dye.[5]

For the amplification of target RNA, the reverse transcription (RT)-LAMP method can synthesize cDNA from target RNA and apply LAMP technology to specifically amplify the resultant cDNA. After mixing and incubating at a constant temperature between 60–65°C, amplification and detection can be carried out in a single step. The RT-LAMP method has potential application for pathogen detection, and has been developed for diagnosis of many RNA viral diseases. For rabies virus (RABV), several instances of the application of RT-LAMP have been reported in recent years.[6–10]

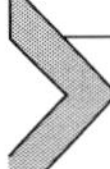

8.2 MATERIALS

8.2.1 Reagents

- AMV reverse transcriptase (New England BioLabs)
- *Bst* DNA Polymerase Lg Frag (New England Biolabs)
- Betaine (Sigma)
- 10×ThermoPol II (Mg-free) reaction buffer (New England Biolabs)
- dNTPs (100 mM)
- $MgSO_4$
- Purified primers (HPLC or Cartridge purification grade)

These reagents, as a mixture that excludes the specific primers to be used, are commercially available as "Loopamp RNA Amplification Kit (RT-LAMP)" from Eiken Chemical.[3]

8.2.2 Primer Set for Reverse Transcription-Loop-Mediated Isothermal Amplification

To assess the applicability of RT-LAMP, a primer set can be designed by using PrimerExplorer V4 software (Eiken Chemical). The primer sets for RABV detection used in the previous studies are shown in Table 8.1. The concept of primer design for effective gene amplification and elongation

Table 8.1 Details of Oligonucleotide Primers Designed to RT-LAMP of Rabies Virus

Primer	Sequence (5′- 3′)	Position	Target Strain	Ref.
C-F3	ACATGTCCGGAAGACT	250–265[b]	CVS-11	6
C-B3	CAGACTCAGGAGAAGACC	424–441[b]		
C-BLP	GGCATGGAATTGACAAGGGACC	373–394[b]		
C-FIP	ACTAGAGAGTTTGGGGTGA-GGACCAGCTATGGAATCC	308–326 + 266–283[b]		
C-BIP	ACGGGAATTGGGCTCTGAC-CTAAAGATGCATGTTCAG	350–368 + 403–421[b]		
P-F3	ACATGCCCTGAAGATT	250–265[b]	Philippines dog	6
P-B3	AAGACTCAGGAGAAGACC	424–441[b]		
P-BLP	GGTAGGAGCTGACAAGGGACC	373–394[b]		
P-FIP	ACAAGGGAATCAGGGGTGA-GGACTAGCTATGGGATCT	308–326 + 266–283[b]		
P-BIP	AAGGAAATTGGGCTCTGAC-CTAAAGACGCATGTTCTG	350–368 + 403–421[b]		
F3	GCCCCCGACTTAAACAAAGC	118–137[c]	Cosmopolitan lineage and vampire bat lineage	7
B3	TTCCCCTCTACATCAGTACG	319–338[c]		
FIP	ACTGCATTGCTGCTGCCAAGTA-GCATGAACGCCGCCAAAC	199–220 + 158–175[c]		
BIP	TGTCCGGAAGACTGGACCAGCT-ACAAGAGAATCTGGGGTGAT	235–256 + 289–308[c]		
FLoop	GGAGCATACATCATCAGGATCNA	176–198[c]		
BLoop	ATGGAATCCTGATTGCACGAMA	257–278[c]		
Rab1F3	AGCCCCCGACTTAAACAAAG[a]	–	Cosmopolitan lineage	8
Rab1B3	CTGTCAGAGCCCAATTTCCT[a]	–		
Rab1FIP	GCATTGCTGCTGCCAAGTAGGATTTT CAGGCATGAATGCAGCCA[a]	–		
Rab1BIP	CGTGTCCAGAAGACTGGACCAGTTTTA TTTCCACCAGAGAATCC[a]	–		
Rab1FLOOP	ACATACATCATCAGGATCAAGT[a]	–		

(Continued)

Table 8.1 (Continued)

Primer	Sequence (5′- 3′)	Position	Target Strain	Ref.
Rab1BLOOP	CTATGGAATCTTGATCGCACG[a]	–		
Rab4F3	GCCCCCGATTTGAACAA[a]	–	Arctic lineage	8
Rab4B3	GGGAATTGGGCTTTGACG[a]	–		
Rab4FIP	ACTGCATCGCAGCTGCTAAGTAGGA TTTTCAGGCTTGAATGCTGCCAA[a]	–		
Rab4BIP	CATGTCCTGAAGACTGGACCAGTTTTA TCTCCACAAGAGAATCTGGGGT[a]	–		
Rab4FLOOP	ACATACATCAGGATCAAGC[a]	–		
Rab4BLOOP	CTATGGGATCTTGATTGCAAG[a]	–		
F3	GAAAAGGAGACAAGATCACC	363–382[d]	Africa 1b lineage	9
B3	CCGGTGTTTTGTCCTGAT	528–545[d]		
FIP	CCTTGTCAGCTCCATGCCTCCCGGACTCTCTAGTGGAAAT	383–460[d]		
BIP	ACCCCACTGTCTCTGAGCATTGCTCAACCTATACAGACTCA	461–524[d]		
CVSF3	AGCCCCCGACTTGAACAAAG	–	CVS	10
CVSB3	CTGTCAGAGCCCAATTCCCG	–		
CVSFIP	GCATTGCTGCTGCCAAGTAGGATTTTC AGGCATGAATGCCGCCAA	–		
CVSBIP	CATGTCCGGAAGACTGGACCAGTTTTAT CTCCACTAGAGAGTTTGG	–		
CVSFLOOP	GCATACATCCGGATCAAGT	–		
CVSBLOOP	CTATGGAATCCTGATTGCACG	–		

[a]No information (-).
[b]Based on the nucleotide sequence of the CVS-11 strain (GenBank accession no AB069973).
[c]Based on the nucleotide sequence of the Pasteur virus strain (GenBank accession no M13215).
[d]Based on the nucleotide sequence of the Pasteur virus strain (GenBank accession no NC_001542).

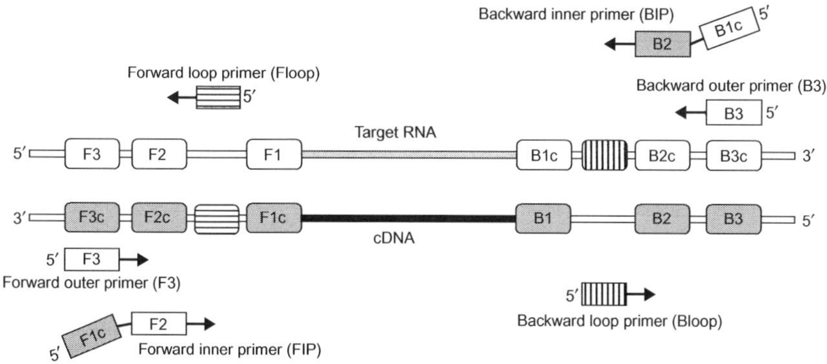

Figure 8.1 Schematic diagram of RT-LAMP primers showing the position of the six primers spanning the target gene. This diagram was partially adopted from the Eiken GENOME SITE provided by Eiken Chemical. The inner primers FIP (BIP) are composed of F2 (B2) and F1c (B1c). The outer primers are designed at the region of F3 and B3. The loop primers Floop and Bloop are designed between F1c (B1c) and F2c (B2c). The U (Uracil) on the target RNA sequence will be transcribed into T (Thymine) for primer designing.

reactions can be explored by accessible animation in Eiken GENOME SITE.[3] The standard set of four primers (F3, B3, FIP, BIP) consisting of two outer primers and two inner primers is basically used for the gene amplification in RT-LAMP (Figure 8.1). The outer primers are referred to as the forward outer primer (F3) and the backward outer primer (B3). The two inner primers are referred to as the forward inner primer (FIP: F1c + F2) and the backward inner primer (BIP: B1c + B2). The use of the primers through OPC or HPLC purification is advisable. Two further loop-binding primers (Floop and Bloop) have been optionally added to increase the rate of strand displacement and synthesis.

8.2.3 Equipment

1. For real-time turbidity detection:
   - Loopamp Realtime Turbidimeter (Eiken Chemical)
2. For visual and real-time fluorescence detections:
   - Incubator with hot bonnet (temperature accuracy within ±0.5°C) and fluorescence scanner (e.g. ordinary real-time PCR detection system or ESEQuant Tube Scanner, QIAGEN)
   - Heat block (for termination of the LAMP reaction)
   - UV transilluminator (wavelength at 240–260 nm or 350–370 nm)

8.2.4 Biological Materials and Sample Preparation

For detection, brain samples are used. Viral RNA extraction of RABV from brain tissue is conducted according to the method of reverse transcription – polymerase chain reaction (RT-PCR) (see Chapters 6 and 7). Commercial extraction kits such as QIAamp Viral RNA Kit (QIAGEN) and TRIzol (Life Technologies) are commonly available.

8.3 METHODS

8.3.1 Preparation of Mastermix

1. After frozen reagents are thawed at room temperature, prepare the following Mastermix reagent on ice (Table 8.2). Once the reagents are thawed, keep them on ice. For real-time fluorescence detection, add fluorescent dye such as 1 μL of Fluorescent Detection Reagent (Eiken Chemical) or 0.4 μM of SYTO-9 green fluorescent dye (Life Technologies) and maintain the total mixture amount at 20 μL.
2. After dispensing, gently tap the tubes a few times, or mix the solution by repeatedly inversing the tube, or mix thrice by vortexing for 1 sec. After mixing well, centrifuge the tubes for a few seconds. Avoid too

Table 8.2 Composition of Mastermix[a]

Reagents		Amount/Final Conc.
10×ThermoPol II (Mg-free) reaction buffer		2.5 μL
DNTPs		0.5–1.0 mM each
$MgSO_4$		8 mM
Betaine (Sigma)		1 M
Bst DNA Polymerase Lg Frag		8–16 units
AMV reverse transcriptase (or equivalent enzyme)		0.2–1.0 U
Primer:	FIP	40 pmol
	BIP	40 pmol
	Floop[b]	20 pmol
	Bloop[b]	20 pmol
	F3	5 pmol
	B3	5 pmol
Distilled Water (DW)		X μL (ad q.s.)
		Total 20.0 μL/tube

[a]Mastermix reagents excluding primers are replaced by Reaction Mix and Enzyme Mix in Loopamp RNA Amplification Kit (RT-LAMP).
[b]These loop primers are optional and their use is merely to accelerate the LAMP reaction.

much mixing by vortexing because of the enzyme inactivation. The Mastermix should be prepared immediately before using.

8.3.2 Mixing of Mastermix and Sample Solution

1. Dispense 20 μL of the Mastermix into a micro tube on ice.
2. Add 5 μL of extracted sample RNA to the Mastermix, and the volume should be 25 μL in total. For the negative control reaction, use 5 μL of DW instead of sample RNA. Thoroughly mix the solution by pipetting or tapping the tube with the cap closed, and then spin down. Take care not to make bubbles when mixing.

8.3.3 Amplification Reaction

1. Set the reaction tubes in a Turbidimeter, fluorescence scanner, or the incubator, and incubate at 60–65°C for 30–60 minutes. (The condition in the reaction should be optimized by preliminary examinations with specific primers.)
2. Inactivate the enzyme and terminate the reaction by incubating the mixture for 2 minutes at 95°C.

8.3.4 Detection and Interpretation of Results

8.3.4.1 Turbidity Detection

The turbidity of magnesium pyrophosphate, a byproduct of the LAMP reaction, is formed in proportion to the amount of amplified products. Because LAMP can yield extremely large amounts of amplified products, white turbidity can be visually observed. Real-time turbidity detection can be conducted with the turbidimeter (e.g. Loopamp Realtime Turbidimeter). A representative result of this method is shown in Figure 8.2.

8.3.4.2 Fluorescence Detection

Visual fluorescence detection can be achieved by the addition of 2 μL of 10–100-fold diluted SYBR green I (Lonza) to the reaction tube after the amplification reaction (Figure 8.3). A UV transilluminator and protective goggles are required. The fluorescence of samples should be evaluated by comparison with that of positive and negative controls. The incubation can be done in commercially available incubators with a hot bonnet or in the Loopamp Turbidimeter. Real-time fluorescence detection can be conducted with the fluorescence scanner (ordinary real-time PCR detection system or ESEQuant Tube Scanner).

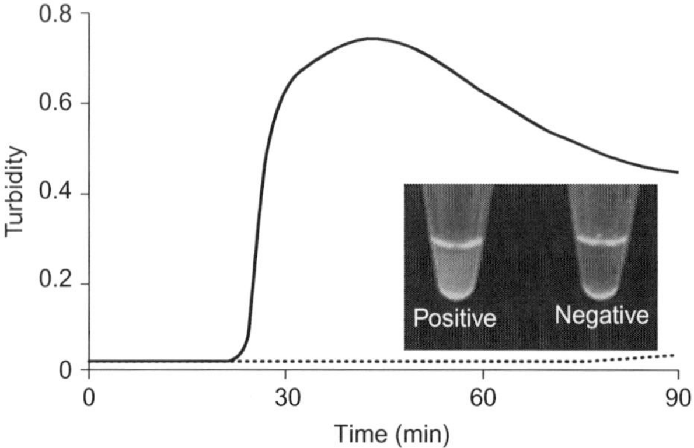

Figure 8.2 Real-time monitoring of the representative results of the RT-LAMP assay. Solid line shows the RT-LAMP reaction of a RABV-positive sample. The dotted line shows the reaction of a negative sample. The inset shows representative tubes after the RT-LAMP reaction. A positive reaction results from the formation of magnesium pyrophosphate, a white precipitate byproduct.

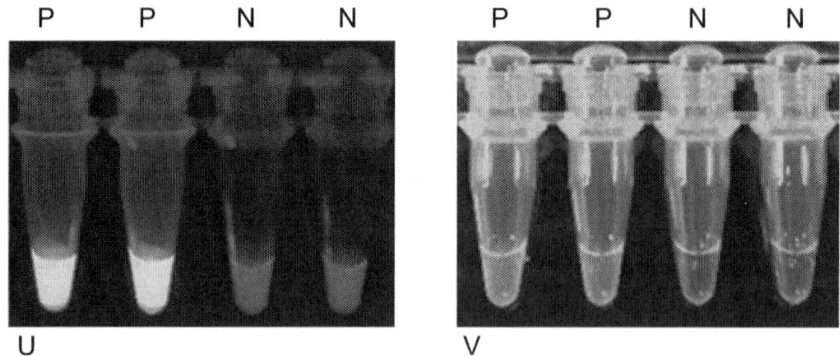

Figure 8.3 The visual inspection with SYBR green I after RT-LAMP. The panel U and V indicate the results under UV and visible lights, respectively. P, RABV positive sample; N, negative control.

8.3.4.3 Electrophoresis

To avoid contamination, extra care should be taken when handling the amplification products during the electrophoresis process.

- 1–2 μL of the reaction solution is applied for electrophoresis with the 2% agarose gel.
- A typical ladder pattern can be observed in electrophoresis (Figure 8.4) because the amplified products consist of various sizes of inverted repeats of the target sequence on the same strands.

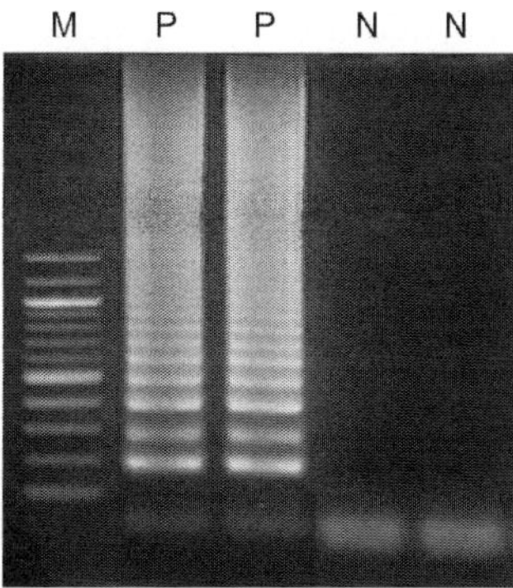

Figure 8.4 Agarose gel electrophoresis of RT-LAMP product. M, 100 bp ladder marker; P, rabies virus-positive sample; N, negative control.

8.3.4.4 Validation of the Specificity of Loop-Mediated Isothermal Amplification Reaction

To check whether the amplified product is derived from the target region, sequencing or restriction enzyme digestion of the amplicon can be performed.

8.4 DISCUSSION

The RT-LAMP technology, which has remarkably high amplification efficiency, realizes highly sensitive detection of specific nucleotide sequences in about 1 hour. The sensitivity of RT-LAMP is 10–1,000 times higher than that of conventional gel-based RT-PCR,[6,7] and equivalent to real-time RT-PCR.[11] Thus, the RT-LAMP is a promising new technology for simple and rapid genetic detection for RABV. Also, LAMP reactions have high specificity and can discriminate slight differences in the sequences of target genes because the length of binding sites complimentary to LAMP primers are longer than that of PCR primers. Hence, the design of LAMP primers with high sensitivity and specificity is key to processing adequate LAMP reactions. In previous reports of RT-LAMP for RABV detection, all LAMP primer sets were targeted at the nucleoprotein gene, which is relatively conserved in RABV genomes. Saito et al. have reported risk for false negative results in RT-LAMP applied to RABV isolates from different parts of the world, depending on the LAMP primers selected. Too much specificity in LAMP primers can lead to binding failure, if there is much sequence variation in the RABVs to be detected.[7] In the area where multiple genetic lineages of RABV exist (e.g. The Americas), the development of primer sets based on phylogenetic surveillance is recommended. Ultimately, universal RT-LAMP primer sets that can detect all of the Lyssavirus isolates around the world should be developed – unlike developments in the real time RT-PCR assays. Hayman et al. demonstrated that RT-LAMP successfully detects multiple lineages of African RABV by using a combination of two sets of LAMP primers for a total of 12 primers.[8] Additional trials using these primer sets or further improved primers are necessary for the construction of universal primer sets.

The RT-LAMP is prone to some of the same problems as other molecular detection techniques, such as unexpected cross-contamination during the opening of the tube cap, due to the high amplification efficiency. To avoid false positive results by contamination, the sample and reagent preparation should be performed in different clean areas, and the use of multiple negative controls between samples is recommended.

The RT-LAMP technique is carried out inexpensively without a thermal cycler, and the result is detectable by a portable scanner (e.g. ESEQuant Tube Scanner). Furthermore, an easy-to-use lateral flow device (LFD) format can be adapted for visual detection of LAMP products.[8] Therefore, the RT-LAMP or the modified technology is practicable in rural areas where rabies occurs frequently. The LAMP is an innovative technology that may well contribute more extensively to the future molecular diagnosis of rabies.

ACKNOWLEDGMENTS

The authors acknowledge the following individuals for constructive comments and review of the manuscript: Dr. Yuki Kobayashi and Dr. Takao Segawa, Nihon University Veterinary Research Center.

REFERENCES

1. Notomi T, Okayama H, Masubuchi H, Yonekawa T, Watanabe K, Amino N, et al. Loop-mediated isothermal amplification of DNA. *Nucleic Acids Res* 2000;**28**:E63.
2. Nagamine K, Watanabe K, Ohtsuka K, Hase T, Notomi T. Loop-mediated isothermal amplification reaction using a nondenatured template. *Clin Chem* 2001;**47**:1742–3.
3. Eiken GENOME SITE: http://loopamp.eiken.co.jp/e/ [accessed 26.02.2014].
4. Mori Y, Nagamine K, Tomita N, Notomi T. Detection of loop-mediated isothermal amplification reaction by turbidity derived from magnesium pyrophosphate formation. *Biochem Biophys Res Commun* 2001;**289**:150–4.
5. Tomita N, Mori Y, Kanda H, Notomi T. Loop-mediated isothermal amplification (LAMP) of gene sequences and simple visual detection of products. *Nat Protoc* 2008;**3**(5):877–82.

6. Boldbaatar B, Inoue S, Sugiura N, Noguchi A, Orbina JR, Demetria C, et al. Rapid detection of rabies virus by reverse transcription loop-mediated isothermal amplification. *Jpn J Infect Dis* 2009;**62**(3):187–91.
7. Saitou Y, Kobayashi Y, Hirano S, Mochizuki N, Itou T, Ito FH, et al. A method for simultaneous detection and identification of Brazilian dog- and vampire bat-related rabies virus by reverse transcription loop-mediated isothermal amplification assay. *J Virol Methods* 2010;**168**(1–2):13–17.
8. Hayman DT, Johnson N, Horton DL, Hedge J, Wakeley PR, Banyard AC, et al. Evolutionary history of rabies in Ghana. *PLoS Negl Trop Dis* 2011;**5**(4):e1001.
9. Muleya W, Namangala B, Mweene A, Zulu L, Fandamu P, Banda D, et al. Molecular epidemiology and a loop-mediated isothermal amplification method for diagnosis of infection with rabies virus in Zambia. *Virus Res* 2012;**163**(1):160–8.
10. Fooks AR, Johnson N, Freuling CM, Wakeley PR, Banyard AC, McElhinney LM, et al. Emerging technologies for the detection of rabies virus: challenges and hopes in the 21st century. *PLoS Negl Trop Dis* 2009;**3**(9):e530.
11. Chen Z, Liao Y, Ke X, Zhou J, Chen Y, Gao L, et al. Comparison of reverse transcription loop-mediated isothermal amplification, conventional PCR and real-time PCR assays for Japanese encephalitis virus. *Mol Biol Rep* 2011;**38**(6):4063–70.

CHAPTER NINE

Detection of Viral Nucleic Acids by *In Situ* Hybridization

Susan A. Nadin-Davis
Centre of Expertise for Rabies, Canadian Food Inspection Agency, Ottawa Laboratory Fallowfield, Ottawa, Canada

Chapter Contents

9.1 Introduction 97
9.2 Materials 98
  9.2.1 Reagents 98
  9.2.2 Equipment 100
  9.2.3 Biological Materials 100
  9.2.4 Laboratory Animals 100
9.3 Methods 100
  9.3.1 Generation of Control Tissues 100
  9.3.2 Section Preparation 100
  9.3.3 Production and Evaluation of Probe 101
  9.3.4 *In Situ* Hybridization 102
    9.3.4.1 Slide Preparation 102
    9.3.4.2 Probe Hybridization 102
    9.3.4.3 Probe Detection 102
  9.3.5 Interpretation of Results 103
9.4 Discussion 103
  9.4.1 Experimental Tips 103
  9.4.2 Critical Parameters, Troubleshooting, and Precautions 104
  9.4.3 Alternative Materials and/or Methods 104
  9.4.4 Time Considerations 104
  9.4.5 Limitations 105
  9.4.6 Future Considerations 105
Acknowledgments 105
References 106

9.1 INTRODUCTION

Detection of rabies virus (RABV) in fresh brain tissue can be readily achieved by a variety of methods, including the gold standard direct fluorescent antibody test (DFAT) for antigen detection, the reverse

C. Rupprecht & T. Nagarajan (Eds): Current Laboratory Techniques in Rabies Diagnosis, Research, and Prevention, Volume 1.
Doi: http://dx.doi.org/10.1016/B978-0-12-800014-4.00009-3

transcription-polymerase chain reaction (RT-PCR), or other nucleic acid amplification methods for viral RNA detection.[1] However, the success of these methods is often much less than optimal when applied to fixed tissues because the fixation process can destroy the conformational epitopes detected by many antibodies employed for the DFAT, and damages the RNA molecules targeted by RT-PCR. While these challenges can sometimes be overcome by optimal application of anti-rabies conjugates,[2,3] or by amplification of short regions of viral RNA,[3,4] success is not always guaranteed. Accordingly, alternative methods for RABV detection in formalin-fixed tissues were developed, including immunohistochemical methods for antigen detection[5] and *in situ* hybridization (ISH) techniques for viral RNA detection.[2,5,6] Such methods permit examination of the cellular and subcellular distribution of RABV at various stages of infection, generating knowledge that could provide important insights into viral pathogenesis. While the earliest methods of ISH employed radiolabeled probes,[5,6] later protocols used probes labeled with the commonly employed steroid hapten, digoxigenin (DIG), which can be visualized colorimetrically.[2,7]

The ISH method described here built upon these earlier methods by developing a protocol that permitted both detection and typing of RABV variants that circulate across Canada. As described previously,[8] it employs a DIG-labeled single-stranded RNA probe targeting a conserved N gene sequence for detection of a wide range of RABV and a panel of variant-specific probes for viral typing. Use of the N gene probe for RABV detection is detailed below, while the use of variant-specific probes for viral typing is described in Chapter 18. Each probe is generated by RNA polymerase-driven transcription of a plasmid into which the target sequence has been cloned. To prevent run-through into the vector, the plasmid DNA is cleaved by a selected restriction endonuclease prior to transcription. All probes are of negative sense orientation to allow detection of positive sense sequence, as infected tissue usually contains larger quantities of viral mRNA than genomic RNA.[5]

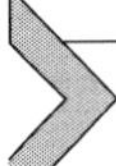

9.2 MATERIALS

9.2.1 Reagents

- 1× phosphate buffered saline (PBS).
- 1× TE buffer.
- 3% paraformaldehyde.

- Acetic anhydride.
- Alkaline phosphatase (AP) developing buffer.
- BCIP-NBT substrate-chromogen solution.
- DIG Northern starter kit for RNA labeling (Roche-Applied Science). Contents:
  - Labeling Mix, 5× concentrated.
  - Transcription Buffer, 5× concentrated.
  - SP6 RNA Polymerase, 20 U/μL.
  - T7 RNA Polymerase, 20 U/μL.
  - T3 RNA Polymerase, 20 U/μL.
  - Anti-DIG AP conjugate, Fab fragments, 750 U/mL.
  - DNaseI (RNase free) 10 U/μL.
  - CDP-*Star*, ready-to-use (optional, for chemiluminescent analysis of probe labeling).
  - Actin RNA Probe, DIG-labeled Antisense Probe, length 588 bases, 10 ng/μL.
  - DIG Easy Hyb Granules.
  - Blocking Solution, 10× concentrated.
- Ethanol (molecular biology grade).
- Formamide.
- Glass coverslips (sterile).
- Hybridization solution.
- Paraffin.
- Phenol-chloroform.
- Positively charged nylon membrane.
- Proteinase K.
- Proteinase K buffer.
- Restriction endonuclease BglII (provided with appropriate buffer).
- RNase inhibitor (e.g. Protector RNase inhibitor available from Roche Applied Science).
- Rubber cement.
- SSC buffer at various concentrations (5×, 2×, 1×, 0.5×, 0.1×).
- Superfrost plus charged glass slides (Fisher Scientific).
- TBS buffer.
- Triethanolamine.
- Washing buffer.
- Water (nuclease-free).
- Xylene.

9.2.2 Equipment

- Items for tissue embedding/processing.
- Microtome for tissue sectioning.
- Hybridization chamber.
- Incubator (range from ambient temperature up to 80°C or higher).
- Olympus BX50 microscope or equivalent.
- Digital camera compatible for use with microscope (for documentation of results).
- Microcentrifuge.

9.2.3 Biological Materials

DNA of plasmid pGEM7zf(+)-AFX-N containing a 388 bp insert corresponding to a 3′ terminal coding sequence of the RABV N gene of isolate 90N5850SK (Genbank Accession # L20671). For further details on this clone refer to Nadin-Davis et al. (2003).[8]

9.2.4 Laboratory Animals

Mice, for generation of positive and negative control brain tissue sections.

9.3 METHODS

9.3.1 Generation of Control Tissues

To generate positive control material, mice are inoculated intra-cerebrally with 10% suspensions of RABV-infected brain tissue. Once an animal displays clinical signs it is euthanized, the brain is removed and divided into two halves. One half is reserved for confirmation of RABV infection using a standard method (e.g. DFAT), and the other half is fixed in 3% paraformaldehyde at room temperature for up to 48 hrs. The tissue is then processed and paraffin-embedded at a temperature not exceeding 60°C. Brains from un-inoculated mice are similarly treated to generate material for use as negative controls.

9.3.2 Section Preparation

1. Cut embedded brain tissue into 5 μm sections and mount onto Superfrost plus charged slides.
2. Bake the slides at 60°C for 2 hr.
3. Use three 5 min changes in xylene to de-paraffinize sections and then clear with three 5 min changes in 100% ethanol. Allow to dry.

9.3.3 Production and Evaluation of Probe

The 388 base RNA probe is generated from the plasmid pGEM7Zf(+)-AFX-N by the following steps:

1. Digest 10 μg plasmid DNA with Bgl II. An aliquot can be electrophoresed through an agarose gel to confirm complete digestion.
2. Add 1 μL proteinase K (20 μg/mL) and continue incubation for 15 min to remove any nuclease activity. Add 1 volume phenol-chloroform, mix periodically over 2–3 mins and spin at 12,000 *g* for 3 mins. Recover the aqueous phase.
3. Precipitate the restricted DNA by addition of 0.1 volume 3 M sodium acetate and 2 volumes ethanol. Incubate at −70°C for 30 mins or −20°C overnight. Spin at 12,000 *g* for 15 mins, wash the pellet with 70% ethanol, and dry. Dissolve in sterile water to a final concentration of 1 μg/μL and store at −20°C.
4. Prepare a transcription reaction using the DIG Northern starter kit as follows: Mix 1 μg linearized plasmid DNA, 4 μL 5× labeling mix, 4 μL 5× transcription buffer, 2 μL SP6 RNA polymerase (20 U/μL), 1 μL RNase inhibitor, and bring the final volume to 20 μL with nuclease-free water. Incubate at 42°C for 1 hr. Add 2 U RNase-free DNase 1 and incubate at 37°C for 15 min to digest template. Add 2 μL 0.2M EDTA (pH 8.0) to terminate the reaction. Recover the labeled transcript by ethanol precipitation as detailed in Section 9.3.3 Step 3 and re-suspend in 20 μL sterile nuclease-free water.
5. To evaluate labeling efficiency, spot 1 μL aliquots of 1 in 10 serial dilutions of the probe (1/100 to 1/100,000) onto a positively charged nylon membrane. Dilutions of the actin control probe provided with the kit can be applied in parallel for comparison. Mark the membrane with the location of each application. Fix the RNA to the membrane by UV crosslinking or baking at either 80°C for 2 hr or 120°C for 30 min. Wash the membrane in 20 mL washing buffer with shaking for 2 min. Discard the buffer and incubate the membrane in 10 mL 1× blocking solution for 30 min. Discard the buffer and incubate the membrane in a freshly prepared 10 mL antibody solution (1× blocking solution containing anti-DIG AP conjugate diluted 1/10,000) at room temperature for 30 min. Wash the membrane 2 × 15 min with the 10 mL washing buffer and then equilibrate up to 5 min in the 10 mL detection buffer. Apply the BCIP-NBT substrate chromogen solution for 90 mins. Probes generating a colorimetric signal at a dilution ≥1/1,000 can be used for ISH. Bring the stock probe to a volume of

50 µL using RNase-free 10 mM Tris-HCl and dispense into 5 × 10 µL aliquots for storage at −80°C.

9.3.4 *In Situ* Hybridization

The ISH procedure is divided into three main steps: slide preparation, probe hybridization, and probe detection. For each of the steps below, removal of a solution prior to addition of the next solution is assumed.

9.3.4.1 Slide Preparation

1. Hydrate sections using one change of 70% ethanol, two changes of sterile water and then immerse in 0.2 N HCl at room temperature for 30 min.
2. Following rinses in water and 1× PBS, incubate in 1× PBS at 37°C for 20 min. Incubate in proteinase K buffer containing 20 µg/mL proteinase K at 37°C for 30 min.
3. Wash in three changes of water. Incubate in 0.1 M triethanolamine (TEA) for 5 min followed by 0.1 M TEA containing 0.25% acetic anhydride for 10 min. This treatment blocks positively charged amino groups and destroys residual proteinase K activity.
4. Rinse in three changes of water, incubate in 2× SSC at 70°C for 30 min, dehydrate in absolute ethanol, and allow to dry.

9.3.4.2 Probe Hybridization

1. Add 25 to 40 µL of hybridization buffer (depending on size of section) containing the probe diluted 200× to each section. [Note: if the probe is highly labeled and gives a signal at a dilution of 1/100,000 (see Section 9.3.3 Step 5), it may be diluted 400×.] Cover with a sterile siliconized glass coverslip that is sealed using rubber cement.
2. Seal slides within a hybridization chamber humidified with 75% formamide. Incubate the slides at 80°C for 30 min, then reduce temperature to 70°C and incubate overnight.

9.3.4.3 Probe Detection

1. Remove the coverslips and wash the slides sequentially in 2×, 1×, and 0.5× SSC for 15 min each, and then in 0.1× SSC at 80°C for 1 hr.
2. Incubate in blocking solution for 1 hr at room temperature.
3. Apply 50 µL of a 1/500 dilution of the anti-DIG AP conjugate and incubate at room temperature for at least 3 hr or at 4°C overnight.
4. Wash in three 5 min changes of TBS buffer to remove excess conjugate, and then equilibrate in AP developing buffer for 10 min.

5. Apply BCIP-NBT chromogen substrate and monitor color development periodically under the microscope over several hours. The reaction should be terminated by monitoring the signal-to-background ratio, as weak signals may be visible many hours after application of the substrate. Development of clear positive signals in the positive control sections may be used as a guide to judge how long to allow the reaction to develop. Terminate the reaction by removal of the substrate and washing in 1× TE buffer.
6. Staining patterns may be photographed for long-term record keeping.

9.3.5 Interpretation of Results

Each ISH assay must be evaluated independently. Negative controls must present minimal staining of all tissue types. Positive controls must exhibit purple-stained intra-cytoplasmic inclusions, varying in size from fine particles to large inclusion bodies, in the perikarya of a proportion of all neurons. Test samples are scored as positive if they exhibit this staining pattern, while staining of all other tissue elements is regarded as being nonspecific in nature. There can be significant variations in the extent of staining depending on the viral load within the section. Positive sections may be scored as 1+ to 3+ depending on staining intensity, similar to the scoring of DFAT results. The reader is referred to the photomicrographs published in Nadin-Davis et al. (2003) for examples of the expected staining morphology.[8]

9.4 DISCUSSION

9.4.1 Experimental Tips

The biggest challenge to the successful performance of this procedure is ensuring that the probe has access to its target sequence. This is achieved by careful optimization of the proteinase K treatment of the tissue by altering enzyme concentration (a range of 20 μg/mL to 200 μg/mL can be evaluated) and incubation time. This enzymatic treatment is the most crucial step of the procedure and has a large impact on final signal intensity. While the conditions described here were optimal for the samples examined in this laboratory, modification of these parameters may improve results obtained by others. Proteinase K breaks the extensive protein lattice crosslinks generated during fixation and thereby increases probe access to the RNA target, but excessive digestion results in loss of tissue architecture and complicates interpretation of observed staining. Thus, the ideal balance between probe accessibility and tissue integrity is required.

9.4.2 Critical Parameters, Troubleshooting, and Precautions

Because this protocol employs an RNA probe, minimizing the possibility of introducing RNases that degrade this probe at any point in the procedure is critical to its success. Measures that will help to avoid this problem include baking of all glassware (e.g. slides and coverslips) and use of gloves at all times during the procedure. Repeated freezing and thawing of probes should be avoided as this can result in probe degradation. Aliquoting of probes into small volumes for storage minimizes this problem. If handled carefully, the probe can be stored at −80°C for several months. Lack of a signal in positive tissue sections may indicate probe degradation at some point in the procedure. If high background staining is observed, raising the hybridization temperature may reduce this problem.

9.4.3 Alternative Materials and/or Methods

The 388-base RNA probe employed here is long enough to be reasonably specific for viral RNA, but sufficiently small enough to be able to diffuse into the tissue as needed for binding to its target. This N gene region is sufficiently conserved so as to permit detection of a wide range of RABV variants known to circulate in the Americas. While it is anticipated that this probe, or others developed from the corresponding N gene region of other RABV variants, will detect most known RABV variants, the applicability of a probe to detect lyssaviruses from parts of the world other than Canada should be investigated prior to use. *In silico* comparison of the probe's sequence with those of target RABVs may indicate its likely utility, although empirical study, in which its binding to southern or dot blots of RT-PCR products of the RABV variant(s) in question is performed, would establish its suitability. Probe size can impact signal intensity, although other investigators have used smaller[2] and larger[7] probes successfully.

9.4.4 Time Considerations

The ISH method is labor intensive. Restriction and recovery of plasmid DNA requires 4–5 hr. Probe production and evaluation takes 7–8 hr. Slide preparation requires about 4 hr. Hybridization is performed overnight, and probe detection is best performed with an overnight incubation with the anti-DIG AP conjugate so as to allow monitoring of color development the following day. Thus, the entire procedure as described here requires approximately 5 days for completion.

9.4.5 Limitations

The procedures described here have been validated using 3% paraformaldehyde as fixative. Application of these methods to materials prepared in other fixatives would require careful evaluation. The length of time that tissue is exposed to formalin has a significant impact on the success of this procedure. Fixation periods of greater than 1 week prior to embedding will significantly reduce the signal that can be achieved, and can even result in complete loss of staining.

9.4.6 Future Considerations

A few investigators have reported successful application of *in situ* RT-PCR for RABV detection in fixed tissues.[9,10] While this methodology may prove valuable in the future, it is the experience of this laboratory that such a procedure requires extensive optimization to achieve robust results.

ACKNOWLEDGMENTS

I thank M. Sheen and Drs. G. Mitchell and C. Fehlner-Gardiner for critical review of this work.

Additional Information

Composition of Buffers and Solutions

- AP developing buffer – 10 mM Tris-HCl, pH 9.5, 10 mM NaCl, 5 mM $MgCl_2$.
- 1× Blocking solution – 1:10 dilution of 10× blocking solution (Roche) in Maleic acid buffer (0.1 M Maleic acid, 0.15 M NaCl, pH 7.5).
- Detection buffer – 0.1 M Tris-HCl, 0.1 M NaCl, pH 9.5.
- Hybridization solution –75% formamide, 10% dextran sulphate, 3× SSC, 1× Denhardt's solution, 50 mM sodium phosphate, pH 7.5, 200 μg/mL yeast tRNA carrier.
- Proteinase K buffer – 50 mM Tris-HCl, pH 8.0, 2 mM $CaCl_2$.
- SSC buffer 10× –1.5 M NaCl, 0.15 M sodium citrate.
- TBS buffer –10 mM Tris-HCl, pH 8.0, 15 mM NaCl.

Examples of Major Sources of Supplies/Equipment

Roche Applied Sciences markets a full line of products for the generation and detection of DIG-labeled probes. The DIG Northern starter kit is suggested for those applying the DIG system for the first time. Subsequent purchases of the necessary consumables either separately or in alternate packs may be more cost effective.

Disclaimer: While sources of some materials are suggested, this does not constitute official endorsement of specific products.

REFERENCES

1. Hanlon CA, Nadin-Davis SA. Laboratory diagnosis of rabies. In: Jackson AC, editor. *Rabies* 3rd edition. Oxford: Academic Press; 2013. p. 409–59.
2. Warner CK, Whitfield SG, Fekadu M, Ho H. Procedures for the reproducible detection of rabies virus antigen mRNA and genome in situ in formalin-fixed tissues. *J Virol Methods* 1997;**67**:5–12.
3. Kulonen K, Fekadu M, Whitfield S, Warner CK. An evaluation of immunofluorescence and PCR methods for detection of rabies in archival carnoy-fixed, paraffin-embedded brain tissue. *J Vet Med B* 1999;**46**:151–5.
4. Wacharapluesadee S, Ruangvejvorachai P, Hemachudha T. A simple method for detection of rabies viral sequences in 16-year old archival brain specimens with one-week fixation in formalin. *J Virol Methods* 2006;**134**:267–71.
5. Jackson AC, Reimer DL, Wunner WH. Detection of rabies virus RNA in the central nervous system of experimentally infected mice using in situ hybridization with RNA probes. *J Virol Methods* 1989;**25**:1–11.
6. Jackson AC, Wunner WH. Detection of rabies virus genomic RNA and mRNA in mouse and human brains by using in situ hybridization. *J Virol* 1991;**65**:2839–44.
7. Jackson AC. Detection of rabies virus mRNA in mouse brain by using in situ hybridization with digoxigenin-labelled RNA probes. *Mol Cell Probes* 1992;**6**:131–6.
8. Nadin-Davis SA, Sheen M, Wandeler AI. Use of discriminatory probes for strain typing of formalin-fixed, rabies virus-infected tissues by in situ hybridization. *J Clin Microbiol* 2003;**41**:4343–52.
9. Nuovo GJ, DeFaria DL, Chanona-Vilchi JG, Zhang Y. Molecular detection of rabies encephalitis and correlation with cytokine expression. *Mod Pathol* 2005,**18**.62–7.
10. Praveena PE, Jayakumar R, Balachandran C, Thirumurugan G, Raj GD, Manohar BM. Detection of rabies virus genes by in-situ polymerase chain reaction. *Vet Res Communs* 2007;**31**:775–81.

CHAPTER TEN

Genetic Characterization via Pyrosequencing

Paola De Benedictis and Cristian De Battisti
FAO Reference Centre for Rabies, Istituto Zooprofilattico Sperimentale delle Venezie, Legnaro (Padova), Italy

Chapter Contents

10.1 Introduction 107
10.2 Materials 108
  10.2.1 Reagents and Equipment 108
  10.2.2 Diagnostic Samples 109
10.3 Methods 109
  10.3.1 Preliminary Phase – Population of the Local Database (IdentiFire Library) 109
  10.3.2 Phase 1 – RNA Extraction, One-Step RT-PCR, Gel Electrophoresis Analysis 109
  10.3.3 Phase 2 – Pyrosequencing Reaction and Result Analysis 110
10.4 Discussion 111
  10.4.1 Experimental Tips 111
  10.4.2 Critical Parameters and Troubleshooting 111
  10.4.3 Precautions 113
  10.4.4 Alternative Methods 113
  10.4.5 Time Consideration 114
  10.4.6 Limitations 114
  10.4.7 Future Considerations 115
Acknowledgments 115
References 115

10.1 INTRODUCTION

The pyrosequencing technique is based on the sequencing-by-synthesis principle[1] and on the real-time luminometric detection of pyrophosphate (PPi), which is released during nucleotide incorporation. The reader is referred to Chapter 20 for further details about the basic technical principles and test characteristics.[1–3]

C. Rupprecht & T. Nagarajan (Eds): Current Laboratory Techniques in Rabies Diagnosis, Research, and Prevention, Volume 1.
Doi: http://dx.doi.org/10.1016/B978-0-12-800014-4.00010-X

To date, pyrosequencing has been applied successfully to lyssavirus detection and typing by targeting the 3′ terminus of the N gene,[4] but also to the rapid identification of vaccine-associated cases, which may occur within the context of oral rabies vaccination (ORV) campaigns in wildlife.[5] Both methods could be applied as complementary to each other for viral typing, with the former able to detect and distinguish a viral infection caused by any lyssavirus species, and the latter developed for the rapid assessment of a potential vaccine-associated rabies virus infection case in the field. The protocol described in Chapter 20, and the one presented herein, are based on the sequence analysis (SQA) principle.

For lyssavirus species typing, refer to Chapter 20. In the following paragraphs, a method that is able to differentiate animal rabies cases resulting from a vaccine-associated infection versus a field strain will be presented. The protocol is based on the amplification of the G gene harboring previously reported polymorphisms of replication-competent vaccine strains.[6] In particular, a polymorphism at codon 333 was found in SAG2 (R333E; due to nucleotide substitutions C4373G, G4374A, and G4375A) and SAD VA1 (R333G; due to nucleotide substitutions C4373G and G4375A), with sequence positions in reference to the CVS-11 strain (GenBank GQ918139).[6]

10.2 MATERIALS

10.2.1 Reagents and Equipment

Information on reagents and equipment necessary to perform the protocol are available in the Additional Information section of Chapter 20.

The only differences consist in the reagents necessary for One-Step reverse transcription – polymerase chain reaction (RT-PCR) and for further sequencing, as reported below:

- One-Step RT-PCR: The SuperScript® III One-Step RT-PCR System with Platinum® Taq DNA Polymerase (Life Technologies, Carlsbad, U.S.), RNasin® Plus RNase Inhibitor (Promega, Madison, USA), and primers (Eurofins MWG Operon, Ebersberg, Germany).
    - RV_G_PyroF AACTTGTCCCYGGGTTYGG Position* 4293–4311.
    - RV_G_PyroR CTGRAGGAGGGATGAYTGC Position* 4522–4504.
- Sequencing primer (Eurofins MWG Operon, Ebersberg, Germany).
- RV_G_Pyro Seq YGATGCYCACTACAAGTCAG Position* 4351–4370.

Position* referred to the challenge virus strain (CVS) of rabies virus (RABV) (GenBank accession number GQ918139).

10.2.2 Diagnostic Samples

Diagnostic samples include animal brain tissue, although the protocol has been successfully applied to cell culture supernatants as well.

10.3 METHODS

10.3.1 Preliminary Phase – Population of the Local Database (IdentiFire Library)

The final identification of the sequences obtained is based on their alignment on a local database specifically populated with the sequences of interest (the IdentiFire library).

- Download reference sequences from the public databases (i.e. GenBank) as txt file format. Select further reference sequences available in the laboratory.
- Align the sequences using Mega4 software.[7]
- Reduce the sequences to a short fragment, including the 40 nucleotides target of the protocol (corresponding to position between 4371 and 4411 of the GenBank accession number sequence GQ918139).
- Create the local database by saving all sequences available in a single file FASTA format.

10.3.2 Phase 1 – RNA Extraction, One-Step RT-PCR, Gel Electrophoresis Analysis

Manual or automated (using TissueLyser II, Qiagen, Hilden, Germany) sample preparation, viral RNA extraction, and storage are performed according to the instructions reported in the correspondent paragraph of Chapter 20.

For One-Step RT-PCR, use the SuperScript® III One-Step RT-PCR System with Platinum Taq DNA Polymerase according to the manufacturer's instructions (Life Technologies, Carlsbad, U.S.), with the following final concentration:

- 200 nM of each primer (1 μL), 1× of PCR Buffer 2× (25 μL), 2 μL of SuperScript III Platinum Taq.
- 20 U of RNAse inhibitor 40U/μL (0.5 μL).
- 5 μL of isolated RNA.
- Water (15.5 μL) to reach a final volume of 50 μL.

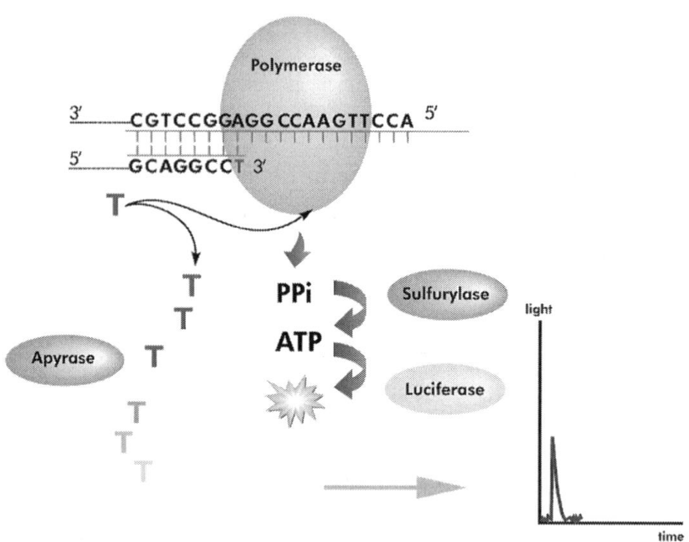

Figure 10.1 Partial alignment (from residue 4371 of GenBank accession number GQ918139) of rabies virus G gene sequences targeted for vaccine versus field strain typing. The well-known polymorphisms at codon 333 (residues 4373–4375) and the two further identified nucleotide substitutions (residues 4391 and 4396) are clearly distinguishable in all vaccine strains.

Three different cycles of amplification have to be applied using scalable annealing temperature: 30 min at 55°C, 2 min at 94°C, then 45 cycles at 94°C for 30 sec, 60/58/57°C (15 cycles at each temperature) for 30 sec, and 68°C for 30 sec, followed by 68°C for 5 min. Biotinylated amplicons are detected by 7% polyacrylamide gel electrophoresis followed by silver staining.

The protocol is based on the amplification of the G gene encoding for codon 333 (showing either nucleotide or amino acid substitutions in vaccine strains) and for two additional nucleotide polymorphisms, as identified in all vaccine strains analyzed, namely A4391C and C4396T (see Figure 10.1). The following pyrosequencing analysis will provide evidence on the possible occurrence of the above-mentioned polymorphisms.

10.3.3 Phase 2 – Pyrosequencing Reaction and Result Analysis

Pyrosequencing reactions are performed using the Pyromark ID platform (Biotage, Uppsala, Sweden) as described in Chapter 20.

Single-stranded biotinylated DNAs are transferred to a 40 μL annealing buffer containing pyrosequencing primers at a final concentration of 0.5 μM, and then placed into the PyroMark ID instrument. Volumes of the PyroMark Gold reagents are provided by the software and are pipetted into the dispensing cartridge that is further placed in the PyroMark platform.[8]

Result analysis is performed as described in Chapter 20. Results analyzed in real time by the software are reported as an IdentiFire report (IdentiFire Software, Biotage, Uppsala, Sweden) (see Figure 10.2). According to the results obtained on field samples using the described protocol, score values greater or at least equal to 90 are considered as satisfactory of a correct classification. Such a high score value is based on a good quality of the pyrogram (highlighted as a blue colorimetric indication on the IdentiFire report; refer to Chapter 20) and on a high homology between query and reference sequences found on the local alignment.

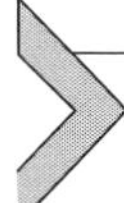

10.4 DISCUSSION

10.4.1 Experimental Tips

For experimental tips, refer to the corresponding paragraph in Chapter 20.

10.4.2 Critical Parameters and Troubleshooting

Several challenges are known when developing a pyrosequencing protocol,[1,3] and have been partially mentioned in Chapter 20.

The protocol described herein has been applied directly to an ORV survey for vaccine-associated cases in Italy, thus demonstrating the feasibility of applying pyrosequencing to a large-scale survey. However, in the case of application of the protocol to different epidemiological and geographical scenarios/context, *in silico* testing and possible adjustment of the primer set (both for vaccine and field strains) are strongly recommended. Such a critical parameter is strictly dependent on the target region amplified by the One-Step RT-PCR (portion of the G gene encoding for the antigenic site III). As for a new primer set design, homopolymeric stretches of DNA, GC-rich sequences, and conserved regions should be avoided because of possible negative interactions on the efficiency of the amplification reaction and, consequently, on the pyrosequences obtained.

As mentioned earlier (refer to Chapter 20), the correct choice and identification of the sequences that will populate the IdentiFire library, as well as their continuous update, is a crucial and preliminary activity that will affect the success of the pyrosequencing protocol.

A score value greater than 90 is recommended to ensure a correct classification of the viral strain infecting the sample of origin (see Figure 10.2).

(A)

2013-11-26

IdentiFire
Detailed Report

Sample ID: **3-2665rs08**

Well:	A3
Notes:	20dna+5beads
PSQ run:	20100312 Rabbia POS DIAGN
Entry ID:	SQA 15
Sequence library:	Library RV N gene (2013-11-26, 11.33.23)
Query sequence:	TGCCGACAAGATTGTATTCAAGGTCAATAA

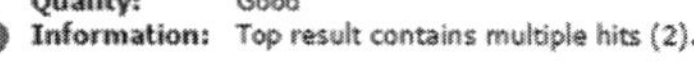
Result: RV 1 , EU293113 9001/RV/FRENCH GUYANA/BAT **Score: 100**

Quality: Good
Information: Top result contains multiple hits (2).

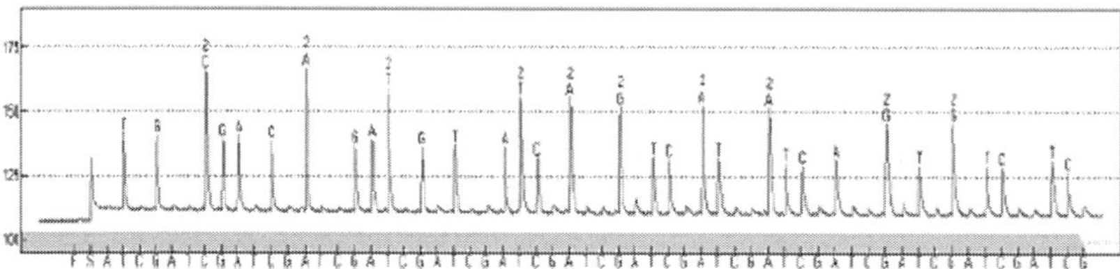

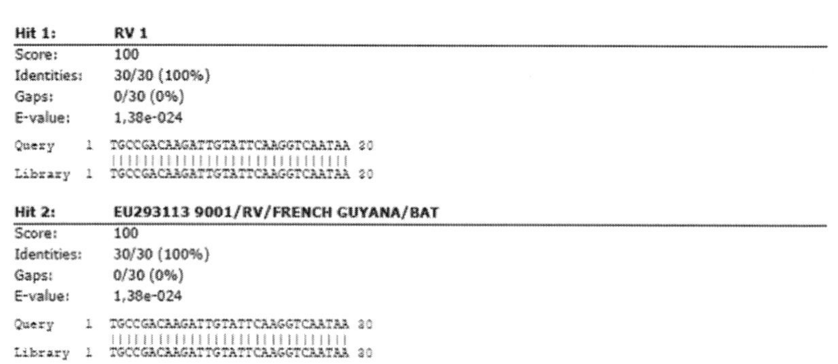
Hit 1: **RV 1**

Score:	100
Identities:	30/30 (100%)
Gaps:	0/30 (0%)
E-value:	1,38e-024

```
Query   1  TGCCGACAAGATTGTATTCAAGGTCAATAA 30
           ||||||||||||||||||||||||||||||
Library 1  TGCCGACAAGATTGTATTCAAGGTCAATAA 30
```

Hit 2: **EU293113 9001/RV/FRENCH GUYANA/BAT**

Score:	100
Identities:	30/30 (100%)
Gaps:	0/30 (0%)
E-value:	1,38e-024

```
Query   1  TGCCGACAAGATTGTATTCAAGGTCAATAA 30
           ||||||||||||||||||||||||||||||
Library 1  TGCCGACAAGATTGTATTCAAGGTCAATAA 30
```

Figure 10.2 IdentiFire detailed reports as generated by the IdentiFire Software (Biotage, Uppsala, Sweden). Examples of a correct classification of a sample as a field rabies virus (a) and as a vaccine virus SAG2 strain (b). The quality of the pyrogram has been recognized as Good and is clearly mentioned in the reports. A score value of 100 has been assigned to both samples. Information available on the report: sample ID and corresponding query sequence, protocol adopted (*Notes*), run ID (*PSQ run*), number and type of enzymatic reactions setup (*Entry ID*), the IdentiFire library used for the analysis.

(B)

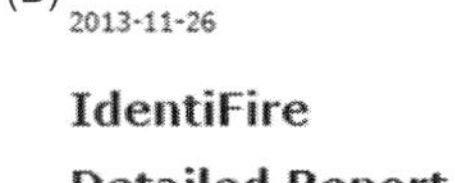
2013-11-26

IdentiFire
Detailed Report

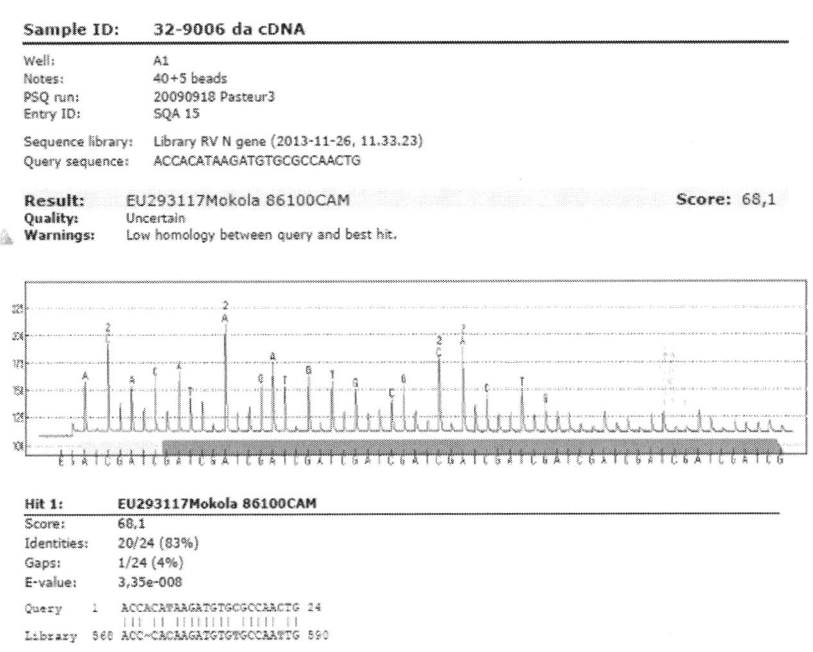

Sample ID: 32-9006 da cDNA

Well: A1
Notes: 40+5 beads
PSQ run: 20090918 Pasteur3
Entry ID: SQA 15

Sequence library: Library RV N gene (2013-11-26, 11.33.23)
Query sequence: ACCACATAAGATGTGCGCCAACTG

Result: EU293117Mokola 86100CAM **Score:** 68,1
Quality: Uncertain
Warnings: Low homology between query and best hit.

Hit 1: EU293117Mokola 86100CAM
Score: 68,1
Identities: 20/24 (83%)
Gaps: 1/24 (4%)
E-value: 3,35e-008

```
Query     1   ACCACATAAGATGTGCGCCAACTG 24
              ||| || ||||||| ||||| ||
Library 568   ACC-CACAAGATGTGTGCCAATTG 590
```

Figure 10.2 (Continued)

In the case of doubtful IdentiFire reports, a conventional Sanger sequencing of the entire 230 bp targeted by the One-Step RT-PCR protocol is recommended.

For further specifications on technical troubleshooting (related to the hardware of the instrument or to the pyrogram quality), refer to the trouble shooting section which can be found on the Qiagen website.[9]

10.4.3 Precautions

Refer to Chapter 20 for additional information on the precautions recommended when applying this method.

10.4.4 Alternative Methods

Available alternative methods, such as an improved pyrosequencing technology or Sanger sequencing of the amplicons obtained by OneStep

RT-PCR, have been fully mentioned in the corresponding section of Chapter 20.

As for pyrosequencing typing of lyssaviruses described in Chapter 20, the SQA method for viral classification was also applied to the protocol developed for revealing substitutions at the G gene. The Single Nucleotide Polymorphism (SNP) genotyping could be also applied for differentiating vaccine associated from field cases. However, a prerequisite for applying SNP is that the regions flanking the target polymorphism have to be extremely conserved. According to our experience, particular attention must be taken when considering the target region of this protocol, which presents several substitution sites, thus not allowing a correct classification through the SNP method.

Among other existing methods already developed to identify the occurrence of vaccine-associated cases in the field, MAb typing and molecular techniques could be applied.[10] However, as with any technique, MAb typing alone can fail, leading to an underestimation of the real burden of vaccine-associated cases.[11] Similarly, the utility of probe-based, real-time RT-PCR is influenced by mutations arising during viral replication or by the occurrence of new strains.[12] Thus, it can be assumed that no 100%-sensitive protocols exist and that the performances of each new development should be assessed carefully and continuously while they are applied under field conditions.

10.4.5 Time Consideration

Using the method described, a maximum time span of about 5–6 hours (depending on the number of samples to run) from the submission to the final typing is needed to analyze a suspect sample under laboratory conditions. The amplified product is obtained in about 3 hours and analyzed by gel electrophoresis in one additional hour (Phase 1). Starting from the amplified products, a pyrosequencing analysis takes less than 2 hours to obtain the final result (Phase 2).

10.4.6 Limitations

Despite the continued progress in use of the pyrosequencing technique (flexibility, automation, accuracy, etc.), some limitations have reduced its application to specific fields.[3] This protocol has been directly applied to an ORV survey for vaccine-associated cases in Italy, thus indicating the feasibility of applying pyrosequencing to a large-scale survey. However, a prior assessment of primer set efficiency and, possibly, a new set design, are recommended when applying the protocol in a different epidemiological context. One of the main limitations can be given by the misinterpretations of homopolymeric stretches of DNA (three or more identical nucleotides) due to the nonlinear chemiluminescent intensity, which may not be detected properly.[1,2] This limitation can affect the efficiency of the test, especially if there are no reference sequences in the IdentiFire library to be aligned with the query sequence.[1]

It should also be mentioned that the application of the pyrosequencing method described is limited to viral typing (as in the case of a vaccine Vs non-vaccine strain), and the short sequence fragments obtained in Phase 2 could not be exploited for phylogenetic or phylogeographic analysis.

10.4.7 Future Considerations

The performance of the protocol described could be affected by the hypervariability of the region targeted by the protocol (refer to Sections 10.4.2 and 10.4.6), rather than by the limited read length obtainable applying this technique. However, it should also be acknowledged that the currently available improved technology (Section 10.4.4) may extend the read length of the pyrosequenced fragments (to 140 nucleotides and more), possibly making primer set design easier on a relatively more conserved region of the G gene.

ACKNOWLEDGMENTS

Laurent Dacheux and Hervé Bourhy, Pasteur Institute (Paris, France), are gratefully acknowledged for provision of some strains used for protocol development. All the staff at the FAO Reference Centre for Rabies, as well as those involved in the management and rollout of ORV campaigns, IZSVe-Italy, are acknowledged for technical help and advice. The authors especially acknowledge Francesca Ellero, who has revised the manuscript.

REFERENCES

1. Ronaghi M. Pyrosequencing sheds light on DNA sequencing. *Genome Res* 2001;**11**(1):3–11.
2. Nyrén P. The history of pyrosequencing. *Methods Mol Biol* 2007;**373**:1–14.
3. Frakuddin MD, Chowdhury A, Nur Hossain MD, Mannan KSB, Mazumdar RM. Pyrosequencing-principles and applications. *Int Journ Life Science & Pharma Res* 2012;**2**(2):65–76.
4. De Benedictis P, De Battisti C, Dacheux L, Marciano S, Ormelli S, Salomoni A, et al. Lyssavirus detection and typing using pyrosequencing. *J Clin Microbiol* 2011;**49**(5):1932–8.
5. De Benedictis P, De Battisti C, Marciano S, Mutinelli F, Capua I, Cattoli G. Pyrosequencing of the rabies virus glycoprotein gene to demonstrate the absence of vaccine-associated rabies cases following oral vaccination. *Vet J* 2013;**195**(3):388–90.

6. Geue L, Schares S, Schnick C, Kliemt J, Beckert A, Freuling C, et al. Genetic characterisation of attenuated SAD rabies virus strains used for oral vaccination of wildlife. *Vaccine* 2008;**26**:3227–35.
7. Tamura K, Dudley J, Nei M, Kumar S. MEGA4: Molecular Evolutionary Genetics Analysis (MEGA) software version 4.0. *Mol Biol Evol* 2007;**24**(8):1596–9.
8. Cummings PJ, Ahmed R, Durocher JA, Jessen A, Vardi T, Obom KM. Pyrosequencing for microbial identification and characterization. *J Vis Exp* 2013;**2**(78):e50405.
9. Qiagen. Homepage. <http://www.qiagen.com> [accessed 1.03.14].
10. World Health Organization, 2013. WHO Expert Consultation on rabies. Second Report, Geneva, WHO technical report series; no. 982.<http://apps.who.int/iris/bitstream/10665/85346/1/9789240690943_eng.pdf?ua=1> [accessed 28.03.14].
11. Mueller T, Batza HJ, Beckert A, Bunzenthal C, Cox JH, Freuling CM, et al. Analysis of vaccine-virus-associated rabies cases in red foxes (Vulpes vulpes) after oral rabies vaccination campaigns in Germany and Austria. *Arch Virol* 2009;**154**:1081–91.
12. Hoffmann B, Freuling CM, Wakeley PR, Rasmussen TB, Leech S, Fooks AR, et al. Improved safety for the molecular diagnosis of classical Rabies viruses using a TaqMan real-time RT-PCR "double check" strategy. *J Clin Microbiol* 2010;**48**:3970–8.

Demonstration of Viral Antibodies and Immune Complexes

CHAPTER ELEVEN

Demonstration of Immune Complexes by Capture Enzyme-Linked Immunosorbent Assay

Shampur N. Madhusudana[1], Kader Muhamuda[2], and Mani Reeta[1]
[1]Department of Neurovirology, WHO Collaborating Centre for Reference and Research on Rabies, National Institute of Mental Health and Neurosciences, Bangalore, India
[2]Department of Microbiology and Molecular Genetics, University of Pittsburgh School of Medicine, Pittsburgh, PA, USA

Chapter Contents

11.1 Introduction 119
11.2 Materials 121
    11.2.1 Reagents 121
    11.2.2 Equipment and Other Consumables 121
    11.2.3 Biological Materials 121
11.3 Methods 122
11.4 Discussion 122
    11.4.1 Experimental Tips 122
    11.4.2 Precautions 123
    11.4.3 Alternative Materials and/or Methods 123
    11.4.4 Limitations and Future Considerations 123
Acknowledgments 123
References 124

11.1 INTRODUCTION

Rabies, an acute progressive, fatal viral encephalitis is responsible for tens of thousands of human deaths worldwide.[1] Two distinct forms of rabies, furious and paralytic, are recognized in humans. The classical furious (encephalitic) form, which constitutes about 80% of human rabies cases, rarely poses diagnostic difficulties due to the distinctive clinical signs and symptoms. In clinical practice, the paralytic or atypical form, which constitutes about 20% of human rabies cases, poses a diagnostic dilemma because it is often clinically indistinguishable from Guillain-Barre

C. Rupprecht & T. Nagarajan (Eds): Current Laboratory Techniques in Rabies Diagnosis, Research, and Prevention, Volume 1.
Doi: http://dx.doi.org/10.1016/B978-0-12-800014-4.00011-1

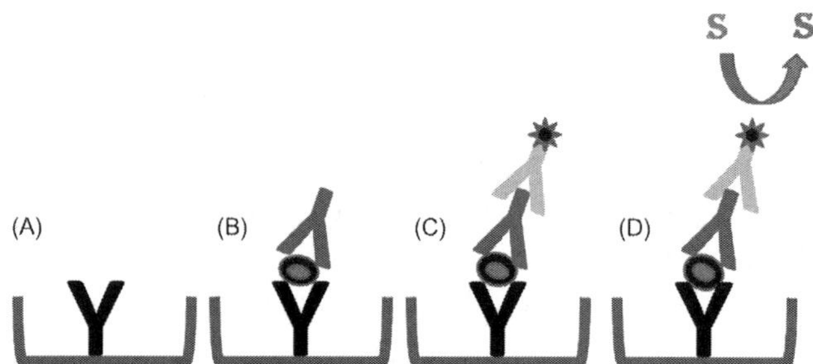

Figure 11.1 Rabies-Specific Immune Complex Capture ELISA. (A) Solid-phase coated with murine N-MAbs or G-MAbs. (B) MAbs capture the antigen component of rabies specific immune complexes. (C) Antibody component of immune complexes revealed by addition of biotinylated goat anti-human IgG. (D) Addition of Streptavidin – HRP, followed by substrate addition and development of color.

syndrome (GBS). The situation is further compounded by a lack of history of animal bite, psychiatric, or other atypical clinical manifestations, the unavailability of a definitive diagnostic test for GBS and the limited availability of tests for the antemortem diagnosis of human rabies.[2]

Immune complexes, which are three-dimensional structures forming a lattice-type of sequence between antigens and antibodies, have been shown to be present in several viral and bacterial human infections, and often their presence has been utilized as a diagnostic marker.[3–6] Paralytic rabies occurs more frequently in partially vaccinated individuals and is considered to have an immuno-pathological basis for its clinical manifestations. Because rabies viral RNA, antigen, and antibodies have been demonstrated in cerebrospinal fluid (CSF) samples of patients with rabies in various studies,[7,8] an assay to detect the presence of specific immune complexes in patients that are clinically suspect for paralytic rabies was developed and evaluated.[9]

For demonstration of immune complexes, a capture Enzyme-Linked Immunosorbent Assay (ELISA) based on monoclonal antibodies (MAbs) to rabies nucleoprotein (N) and glycoprotein (G) proteins is employed (Figure 11.1). Murine MAbs to rabies N (N-MAbs) and G (G-MAbs) are used separately to capture the respective antigenic component of immune complexes. These MAbs have been developed and characterized[10] using established procedures.[11–13] Model immune complexes containing rabies viral antigens were prepared by incubating 1 μL aliquots of serum samples from individuals vaccinated against rabies, with increasing amounts of purified rabies virus. Samples were then analyzed by the assay for detection of immune complexes against both N and G.[4] The study included CSF samples collected antemortem from 30 paralytic rabies-suspect patients in whom the diagnosis was later confirmed by autopsy, 30 CSF samples from individuals undergoing spinal anesthesia as negative controls, and 30 CSF samples from patients with other viral encephalitides as disease controls. Twenty-three out of 30 CSF samples (76.6%) showed the presence of immune complexes to both rabies N and G. None of the CSF samples from negative controls and disease controls were positive, confirming a 100% specificity of the assay. The results of this study indicate that detection of specific immune complexes can be utilized as one of the diagnostic tests for antemortem diagnosis of human rabies. As immune complexes to both N and G were present in all positive cases in equal proportion, detection of immune complexes to any one of the antigens can be employed preferably to N, as this protein is more conserved among different strains.[9] The test is a simple, rapid procedure capable of providing results within a few hours.

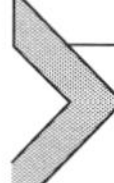

11.2 MATERIALS

11.2.1 Reagents

- Biotinylated goat antihuman IgG.
- N-MAbs, G-MAbs.
- Streptavidin-horse radish peroxidase (Streptavidin-HRP).
- Substrate – Hydrogen peroxide (H_2O_2) and O-Phenylenediamine (OPD).
- Stop solution (4N H_2SO_4).
- Phosphate buffered saline (PBS) with Tween 20 (PBST-1X).

11.2.2 Equipment and Other Consumables

- ELISA reader.
- Incubator.
- Refrigerator.
- Polystyrene ELISA strips.
- Micropipettes and tips.

11.2.3 Biological Materials

- Human clinical sample – CSF.
- Positive control – CSF sample positive for immune complexes.
- Negative control – CSF sample negative for immune complexes.

11.3 METHODS

- Coat the polystyrene ELISA microwells with murine N-MAbs or G-MAbs in carbonate–bicarbonate buffer at a concentration of 0.1 µg per well. (Carbonate–bicarbonate buffer, pH 9.6, 100 µL/well).
- Incubate overnight at 4°C.
- Quench the strips with 1% skimmed milk powder in PBST (150 µL/well) at room temperature for 1 hour. Wash the plates 3 times with PBST.
- Dilute the CSF samples (and positive and negative controls) 1:10 in PBST and add 100 µL/well. Run each sample in duplicate wells.
- Incubate at room temperature for 2 hours.
- Wash the plates 3 times with PBST.
- Add 100 µL/well of biotinylated goat antihuman IgG (1:10,000 dilution in PBST).
- Incubate at 37°C for 2 hours. Wash the plates 3 times with PBST.
- Add 100 µL of Streptavidin-HRP and incubate for 15 minutes at room temperature.
- Wash the plates 3 times with PBST.
- Add 100 µL of substrate (H_2O_2 and OPD) and incubate in the dark for 20 minutes.
- Stop the reaction by adding 50 µL of 4N H_2SO_4 and read the optical density (OD) using an ELISA reader at 492 nm wave length within 30 minutes of stopping the reaction.
- Interpretation of results:
  - Calculate the Mean OD value of negative samples.
  - Cut-off: The mean+3 SD (standard deviation) of OD values of the negative samples is taken as the cut-off.
  - CSF samples showing an OD value greater than the cut-off is considered positive for the presence of immune complexes.

11.4 DISCUSSION

11.4.1 Experimental Tips

- Allow all the reagents to reach room temperature before starting the procedure.
- Washing is a critical step in this procedure – respect the recommended number of washing cycles. Make sure that all wells are completely filled and then completely emptied. Incorrect washing may lead to inaccurate results.

- Do not allow the microwells to dry between the end of the washing operation and the reagent distribution.
- Check the pipettes and other equipment for accuracy and correct operation.
- Use a new tip for each clinical sample.

11.4.2 Precautions

- OPD is a suspect carcinogen. Avoid direct contact. Use gloves and mask when handling tablets.

11.4.3 Alternative Materials and/or Methods

- Because OPD is a suspect carcinogen, it can be substituted with 3,3′,5,5′-tetramethylbenzidine (TMB).
- Presently no other method is known or has been evaluated for detection of rabies virus-specific immune complexes.

11.4.4 Limitations and Future Considerations

Though the presence of immune complexes in the CSF is suggestive of rabies, a negative test does not rule out rabies. Therefore, other tests for the antemortem diagnosis of rabies should also be performed. Paralytic rabies cases pose a diagnostic dilemma as they are clinically indistinguishable from GBS, and hence a lumbar puncture is routinely performed on such patients to differentiate these two conditions. Therefore, this assay has been evaluated in our laboratory on CSF samples from patients with paralytic rabies only.[9] Though the furious (encephalitic) form of human rabies can be diagnosed clinically based on the classical symptoms, it will be of interest to know if such immune complexes are present in the CSF of these patients as well. This may validate the clinical diagnosis of hydrophobia patients and assist in laboratory confirmation of rabies in patients presenting with atypical manifestations, as well as in epidemiological surveys.

The commercial non-availability of murine N-MAbs and G-MAbs may be a limitation for the use of this assay in laboratories that do not have either access to or facilities for development and characterization of these murine MAbs. The WHO or OIE reference laboratories should be contacted for the availability of such reagents.

ACKNOWLEDGMENTS

We thank Dr. V. Ravi and Dr. Anita Desai for their technical advice during the development and standardization of the assay.

Additional Information

Composition of Media, Buffers, Solutions, Etc.

- **Coating Buffer: 0.05 M Carbonate–Bicarbonate Buffer, pH 9.6.**
  - Solution A: Dissolve 0.53 g of Na_2CO_3 (anhydrous) in 100 mL of distilled water (DW).
  - Solution B: Dissolve 0.42 g of $NaHCO_3$ in 100 mL of DW.
  - To 32 mL of Solution A, add Solution B slowly until the pH is 9.6.
- **PBS, pH 7.2 (1×).**
  - NaCl: 8.0 g
  - KCl: 0.2 g
  - KH_2PO_4: 0.2 g
  - $Na_2HPO_4 \cdot 2H_2O$: 1.5 g
  - DW: 100 mL
  - Adjust the pH to 7.2 with 0.1 N NaOH or 0.1 N NaCl.
- **Wash Buffer: Phosphate Buffered Saline with Tween 20 (0.05%).**
  - Add 0.5 mL of Tween 20 (Polyoxyethylene sorbitan monolaurate) to 1 L of 1× PBS to make PBST.
- **Phosphate Citrate Buffer, pH 5.0.**
  - Solution A: Dissolve 11.9 g of Na2HPO4 • 2H2O in 1,000 mL of DW.
  - Solution B: Dissolve 7.0 g of citric acid in 1,000 mL of DW.
  - Mix 55 mL of Solution A with 45 mL of Solution B. Adjust the pH to 5.0 with solution A or B.
- **Substrate for ELISA**
- O-Phenylenediamine (OPD): 4 mg.
- Phosphate citrate buffer: 10.0 mL.
- H_2O_2 (30% w/v), added just before use: 10.0 μL.

Examples of Major Sources of Supplies

- ELISA strips: Maxisorp, Nunc, USA; Greiner Bio One, USA.
- Biotinylated goat antihuman IgG: Bangalore Genei, India; Millipore, USA.
- Streptavidin-HRP: Millipore,USA; Dako, Denmark.
- OPD: Sigma Aldrich,USA; Thermo Scientific, USA.

REFERENCES

1. World Health Organization WHO Expert Consultation on Rabies. Second report. *World Health Organ Tech Rep Ser* 2013;**982**:1–139.
2. Gadre G, Satishchandra P, Mahadevan A, Suja MS, Madhusudana SN, Sundaram C, et al. Rabies viral encephalitis: clinical determinants in diagnosis with special reference to paralytic form. *J Neurol Neurosurg Psychiatry* 2010;**81**:812–20.
3. Patil SA, Gourie-Devi M, Anand AR, Vijaya AN, Pratima N, Neelam K, et al. Significance of mycobacterial immune complexes (IgG) in the diagnosis of tuberculous meningitis. *Tuber Lung Dis* 1996;**77**:164–7.

4. Salmi A, Lund G, Ziola B, Ruenamen M. Circulating immune complexes in patients with sub-acute scelorosing panencephalitis. *Clin Immunol Immunopathol* 1986;**41**:16–25.
5. Desai A, Ravi V, Chandramukhi A, Gouri-Devi M. Detection of immune complexes in the CSF of Japanese encephalitis patients: correlation of findings with outcome. *Intervirology* 1994;**37**:352–5.
6. Coyle PK, Schutzer SE, Belman AL, Krupp LB, Golightly MG. Cerebrospinal fluid immune complexes in patients exposed to *Borrelia burgdorferi*: detection of Borrelia-specific and -nonspecific complexes. *Ann Neurol.* 1990;**28**:739–44.
7. Vasanth JP, Madhusudana SN, Abhilash KV, Suja MS, Muhamuda K. Development and evaluation of an enzyme immunoassay for rapid diagnosis of rabies in humans and animals. *Indian J Pathol Microbiol* 2004;**47**:574–8.
8. Mani RS, Madhusudana SN, Mahadevan A, Reddy V, Belludi AY, Shankar SK. Utility of real-time TaqMan PCR for antemortem and postmortem diagnosis of human rabies. *J Med Virol* 2013 doi: http://dx.doi.org/10.1002/jmv.23814. [Epub ahead of print].
9. Muhamuda K, Madhusudana SN, Ravi V, Desai A. Presence of rabies specific immune complexes in cerebrospinal fluid can help in ante-mortem diagnosis of human paralytic rabies. *J Clin Virol* 2006;**37**:162–7.
10. Muhamuda K, Madhusudana SN, Ravi V. Use of neutralizing murine monoclonal antibodies to rabies glycoprotein in passive immunotherapy against rabies. *Hum Vaccin* 2007;**3**:192–5.
11. Lafon M. Techniques for the production screening and characterization of monoclonal antibodies. In: Meslin FX, Kaplan MM, Kopwrosky H, editors. *Laboratory techniques in rabies* 4th ed. Geneva: WHO; 1996. p. 133–44.
12. Wiktor TJ, Koprowski H. Monoclonal antibodies against rabies virus produced by somatic cell hybridization: detection of antigenic variants. *Proc Natl Acad Sci USA* 1978;**75**:3938–42.
13. Müller T, Dietzschold B, Ertl H, Fooks AR, Freuling C, Fehlner-Gardiner C, et al. Development of a mouse monoclonal antibody cocktail for post-exposure rabies prophylaxis in humans. *PLoS Negl Trop Dis* 2009;**3**:e542.

CHAPTER TWELVE

Demonstration of Viral Antibodies by an Immunochromatographic Strip Test

Akira Nishizono
Department of Microbiology, Faculty of Medicine, Oita University, Japan

Chapter Contents

12.1 Introduction 127
12.2 Materials 128
  12.2.1 Reagents 128
  12.2.2 Equipment 129
12.3 Methods 129
12.4 Discussion 130
Acknowledgments 131
References 131

12.1 INTRODUCTION

Pre-exposure prophylaxis (PrEP) against rabies confers acquired immunity by vaccination, which is an effective strategy for people who may be at high risk of infection because of their occupation, as well as those who travel to, or live in, rabies-endemic areas. The mass vaccination of dogs is the most cost-effective means of achieving a significant and lasting reduction in the number of human deaths from rabies, but these preventive efforts are often not prioritized by national governments even in the worst affected rabies-endemic countries. Reducing the number of rabies victims by animal vaccination requires the production of an adequate level of antibody in vaccinated animals, particularly dogs and humans.

Two virus neutralization tests, namely the rapid fluorescent focus inhibition test (RFFIT) and fluorescent antibody virus neutralization (FAVN) test, are generally used to measure the level of virus-neutralizing

C. Rupprecht & T. Nagarajan (Eds): Current Laboratory Techniques in Rabies Diagnosis, Research, and Prevention, Volume 1.
Doi: http://dx.doi.org/10.1016/B978-0-12-800014-4.00012-3

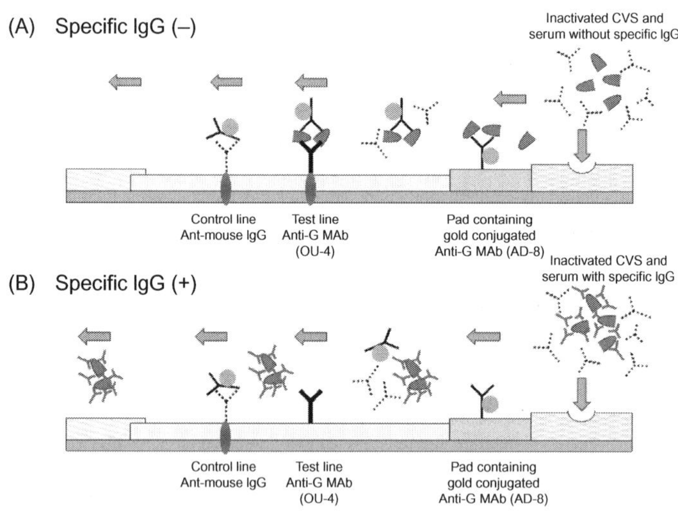

Figure 12.1 *In vitro* binding using inactivated RABV.

antibodies (VNA). However, these tests are time consuming and expensive, and they require live rabies vaccine (RABV), which means that they can only be performed in restricted reference laboratories.

We developed a novel, fast, and simple assay called the "Rapid Neutralizing Antibody" (RAPINA) test, which is an alternative to the usual techniques for antibody determination.[1] This test is based on the principle of immunochromatography combined with *in vitro* binding using inactivated RABV (Figure 12.1). The RAPINA test has good positive and negative predictive value, i.e. an approximate corresponding VNA value of ~0.5 IU/mL, which is the level recommended by the World Health Organization (WHO) and the World Organization for Animal Health (OIE).[2]

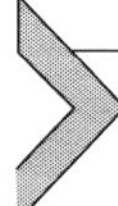

12.2 MATERIALS

12.2.1 Reagents

The RAPINA test has two components: an immunochromatographic strip for detecting the RABV G and inactivated RABV.

1. The immunochromatographic strip test can detect RABV G using two types of anti-G monoclonal antibodies (MAbs): one MAb labeled with a gold particle as the detection antibody, and another MAb as the capture antibody. Both MAbs possess broad activities against fixed and street RABV strains.

2. The inactivated RABV is prepared by inactivating the Challenge Virus Standard (CVS-11) RABV strain with formalin, where the inactivation process is confirmed by infecting mouse neuroblastoma cells and inoculating into the brain of suckling mice.

12.2.2 Equipment

Tubes for serum dilution, pipette tips, and a water bath.

12.3 METHODS

1. Step 1: Serum or plasma (either with ethylenediaminetetraacetic acid (EDTA)-2Na or Heparin-Na) can be used in the RAPINA test. Before performing the RAPINA test, sera or plasma should be heat-inactivated at 56°C (135°F) for 30 min.
2. Step 2: For qualitative estimates of the antibody level, the heat-inactivated and undiluted serum samples were mixed with an equal volume of inactivated CVS-11 RABV and incubated at 37°C (98.6°F) for 30 min. Then, 100 μL of the mixture is placed in the sample port in the immunochromatographic strip and allowed to react for 15 min at room temperature (not to be read after 20 min) to detect the G.
3. Interpretation of the results: If the band was visible in the control and the test lines (Figure 12.2A), the result of the RAPINA test was defined as <0.5 corresponding international units (CIU/mL), which may correspond in theory to a VNA level of <0.5 IU/mL in a viral neutralization tests (RFFIT or FAVN). By contrast, if the band was visible only in the control line (Figure 12.2B), the RAPINA value was ≥0.5 CIU/mL, which corresponded to a VNA level of ≥0.5 IU/mL. The test was considered invalid if there was no visible band on the control line (Figure 12.2C).
4. To estimate the antibody titer semi-quantitatively using the RAPINA test, heat-inactivated serum samples were diluted eight-fold with phosphate-buffered saline (PBS) (−) and incubated with inactivated CVS-11 RABV. The test was then performed in the manner described in Step 3. If the test line was visible (Figure 12.2A), the RAPINA value was interpreted as <4.0 CIU/mL; if the test line was not visible (Figure 12.2B), the RAPINA value was interpreted as ≥4.0 CIU/mL. Therefore, the antibody level was divided semi-quantitatively into categories of <0.5 CIU/mL, 0.5–4.0 CIU/mL, and >4.0 CIU/mL using undiluted and eight-fold-diluted serum or plasma.

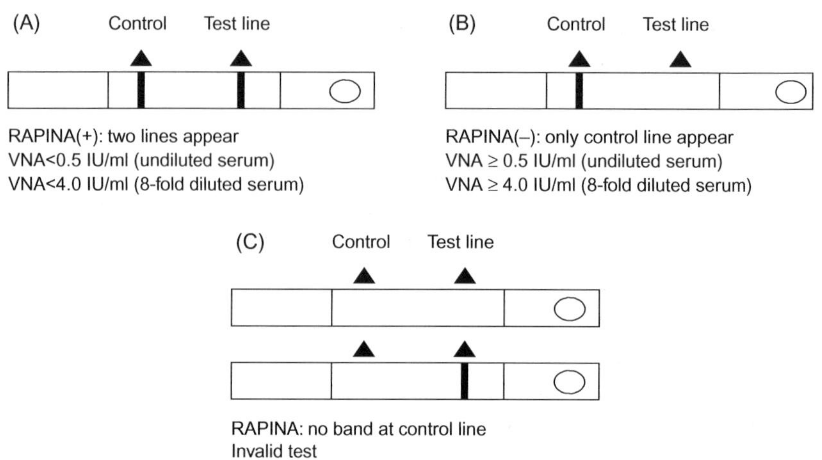

Figure 12.2 RAPINA test results: (A) Band visible in test and control lines. (B) Band visible in control line. (C) Band not visible in control line.

12.4 DISCUSSION

In this study, we developed a novel method for estimating the antibody titer; the positive predictive values for subjects with antibody were 99.5% in humans and 99.7% in dogs ($\kappa = 0.98$). The negative predictive values for subjects without antibody were 98.6% in humans and 95.6% in dogs ($\kappa = 0.96$). The RAPINA test results were closely correlated with the RFFIT results.[2] The advantages of the RAPINA test compared with the RFFIT are that it is simple, requires minimal equipment, and is less time consuming. It can also be used with large or small numbers of serum samples, which means that antibody predictions can be performed at any time and for any purpose. The RAPINA test is a versatile and practical method for measuring the antibody level in humans and other animals.

The RAPINA test could be potentially considered for evaluations after vaccination in humans, particularly PrEP assessments. PrEP can reduce the risk of rabies infection, and determination of the antibody titer may avoid or reduce the need for unnecessary booster vaccinations. PrEP is also an efficacious anti-rabies measure before exposure to risky activities in rabies-endemic areas. However, there are individual differences in the duration of the immune memory, or the remaining antibodies, and the immune response after vaccination. A low-cost or affordable PrEP regimen might be facilitated by serological assessments using the RAPINA test. A preventive measure used in countries where rabies is endemic is the immunization of children against rabies, which is the most vulnerable age group.[3] The RAPINA test could also be used in vaccination clinics to evaluate the effects of the vaccine in immune-compromised hosts, including HIV-positive patients.

For animals, the RAPINA test could be used to monitor the success of mass vaccination programs in potentially endemic and rabies-free countries. Some dog serum samples exhibit cytotoxicity in BHK-21 or neuroblastoma cells, which are used in the neutralization test. To avoid this problem, these serum samples must be diluted before measuring the VNA level. However, a low-level VNA might not be detected by the RFFIT or FAVN test, whereas it could be evaluated using the RAPINA test. Unlike an Enzyme-Linked Immunosorbent Assay (ELISA), the RAPINA assay can predict the antibody level in the serum of any type of animal (except the mouse) because it does not require a secondary antibody.

There are no special experimental tips or precautions. However, it is important that the sera or plasma samples are heat-inactivated before the assay. The detection power is markedly reduced without heat inactivation.

The RAPINA test might be considered as an intermediate type of binding assay between the viral neutralization tests (RFFIT or FAVN) and a solid-phase immuno-binding assay (ELISA).

ACKNOWLEDGMENTS

We greatly appreciate Dr. Pakamatz Khawplod, Queen Saovabha Memorial Institute, Thai Red Cross Society, for helpful discussion and advice.

REFERENCES

1. Shiota S, Mannen K, Matsumoto T, Yamada K, Yasui T, Takayama K, et al. Development and evaluation of a rapid neutralizing antibody test for rabies. *J Virol Methods.* 2009;**161**(1):58–62.
2. Nishizono A, Yamada K, Khawplod P, Shiota S, Perera D, Matsumoto T, et al. Evaluation of an improved rapid neutralizing antibody detection test (RAPINA) for qualitative and semiquantitative detection of rabies neutralizing antibody in humans and dogs. *Vaccine* 2012;**30**(26):3891–6.
3. Dodet B, The Asian Rabies Expert Bureau (AREB). The Rabies in Asia conference (RIACON), Conference Proceedings. Hanoi, 2009;**13**.

CHAPTER THIRTEEN

Rabies Diagnosis: Demonstration of Viral Antibodies by Flow Cytometry

Juliano Bordignon[1] and Carlos Roberto Zanetti[2]
[1]Department of Virology, Instituto Carlos Chagas, ICC/Fiocruz, Curitiba, Paraná, Brazil
[2]Department of Microbiology, Immunology and Parasitology, Federal University of Santa Catarina, Florianópolis, Santa Catarina, Brazil

Chapter Contents

13.1 Introduction 133
13.2 Materials 134
13.2.1 Reagents 134
13.2.2 Equipment 135
13.2.3 Biological Materials 135
13.3 Methods 135
13.3.1 Major Step 1 135
13.3.2 Major Step 2 135
13.3.3 Interpretation of Results and Determination of Rabies Virus-Neutralizing Antibodies Levels 136
13.4 Discussion 136
13.4.1 Experimental Tips 138
13.4.2 Critical Parameters and Troubleshooting 138
13.4.3 Precautions 138
13.4.4 Alternative Materials and/or Methods 138
13.4.5 Time Considerations 139
13.4.6 Limitations 139
13.4.7 Future Considerations 139
Acknowledgments 139
References 140

13.1 INTRODUCTION

Virus neutralization is not the only mechanism by which antibodies protect against rabies,[1] and several authors have demonstrated the important role played by cellular immunity.[2–4] Nevertheless, rabies virus-neutralizing antibodies (RVNA) are key immunological parameters

C. Rupprecht & T. Nagarajan (Eds): Current Laboratory Techniques in Rabies Diagnosis, Research, and Prevention, Volume 1.
Doi: http://dx.doi.org/10.1016/B978-0-12-800014-4.00013-5

that are routinely tested.[2] RVNA are essential for protection against rabies, and their presence at a level of at least 0.5 IU/mL is considered to indicate successful immunization in humans.[5,6] The gold-standard technique for RVNA quantification is based on the ability of antibodies against the rabies virus (RABV) glycoprotein to neutralize the infection of a cell line with a known concentration of virus, assessed by observing the inhibition of fluorescent focus development. Rapid fluorescent focus inhibition (RFFIT) and simplified focus inhibition (SFIMT) are suitable assays for this purpose.[7,8] In addition to fluorescent focus inhibition (FFI) tests, several Enzyme-Linked Immunosorbent Assays (ELISAs) have been developed as an alternative to cell culture-based tests.[9–11] The FFI tests are effective, but time-consuming and dependent on human intervention for the microscopic analysis of fluorescence. ELISA techniques are also of limited value because they generally assess the levels of both neutralizing and non-neutralizing antibodies, thus providing limited inferences about protection. Flow cytometry (FC) can be used to analyze cell structure and phenotype on the basis of light scattering and fluorescence staining. Several authors have demonstrated the use of FC for the detection and quantification of antibodies against a range of viruses, including cytomegalovirus, herpes simplex virus (HSV), and human immunodeficiency virus (HIV).[12,13] We describe here the use of FC as a sensitive and reproducible technique for the demonstration of neutralizing antibodies by detection of residual RABV.

13.2 MATERIALS

13.2.1 Reagents

- 4% *p*-Formaldehyde.
- Fluorescent anti-rabies conjugate.
- 24-Well cell culture dishes.
- DMEM-F12 and MEM cell culture media.
- Fetal calf serum (FCS).
- 1× phosphate buffer pH 7.4.
- 1× phosphate buffer pH 7.4 plus 0.05% Tween 20 – wash buffer.
- Coulter IsotonII® diluter.
- FACS lysissolution (Becton, Dickinson, and company).
- 100 mM glutamine solution.
- Penicillin/streptomycin (10 IU/μg/mL) and amphotericin (25 μg/mL).
- Trypsin–EDTA solution (Sigma Cat #4049).

13.2.2 Equipment

- Flow cytometer.
- Cell culture facility with CO_2 incubator, freezers and refrigerators, a centrifuge, laminar flow cabinets, and other devices commonly used in cell culture laboratories.

13.2.3 Biological Materials

- BHK-21 (C-13; ATCC CCL10) and mouse neuroblastoma cells (MNA) have been used for the detection of non-neutralized virus by a serum sample in an FFI test.[14,15] The use of other cell lines might require minor modifications to the protocol.
- Standard and titrated RABV strains, such as Pasteur virus (PV) or challenge standard virus (CVS), should be used as positive controls for RABV infection.
- An international reference serum (IRS) sample with a known titer of neutralizing antibodies.
- Clinical or field serum samples.

13.3 METHODS

13.3.1 Major Step 1

For virus neutralization, 100 µL of a 1:25 dilution of each serum sample (See Section 13.4.2, *Critical Parameters and Troubleshooting*) is first incubated in a 24-well plate with 1 $CCID_{50}$ of the PV strain (100 µL) for 45 minutes at 37°C, under an atmosphere containing 5% CO_2. Concomitantly, a serial dilution of an IRS (pre-diluted to 20 IU/mL) is prepared for the establishment of a standard curve for virus neutralization. An additional well containing a serum sample known to not contain RVNA is also set up for determination of the maximum percentage of infection. After incubation, 5.0×10^5 BHK-21 cells are added to each well and the plates are incubated for an additional 24 h to allow the infection of the cells with any non-neutralized virus present.

13.3.2 Major Step 2

After incubation for 24 h, the cells are washed once (1× PBS [phosphate buffered saline]) and detached by the addition of 200 µL of trypsin-EDTA to each well, and the incubation of the plate for 5 min at 37°C. The cells are recovered in MEM supplemented with 10% FCS. BHK-21 cells are washed twice with 1 mL of wash buffer and centrifuged for 5 min at about 1,900×*g* at room temperature. The cells are washed and then fixed/permeabilized by incubation for 10 minutes at room temperature with a 1:1 (v/v) mixture of 4% *p*-formaldehyde and a 1:10 dilution (in water) of FACS lysis fluid (Becton, Dickinson & Company) (500 μL/well). The cells are then washed twice, as previously described. BHK-21 cells are recovered in 100 μL of an optimal predetermined dilution (1:20) of polyclonal rabies virus nucleoprotein FITC-conjugate (Sanofi Pasteur) in 1 × PBS supplemented with 0.2% Evans blue. The cells are incubated for 45 minutes at 37°C and washed twice, as described above. Finally, BHK-21 cells are recovered in 1 mL of IsotonII® (Beckman Coulter). They may be stored at 4°C for up to 24 h before flow cytometric analysis.

13.3.3 Interpretation of Results and Determination of Rabies Virus-Neutralizing Antibodies Levels

As previously described in Chapter 4, *Rabies Diagnosis: Demonstration of Viral Antigens by Flow Cytometry*, a gate based on size (FSC) and complexity (SSC) and excluding cell debris and doublets, must be set up before flow cytometric analysis. The equipment settings used depend on the type of equipment (analog or digital), and the cell line used. An isotype control (an unrelated antibody conjugated to the same fluorophore as the specific antibody) should be used to assess nonspecific binding. The numbers of positive cells in controls or samples are determined with a gate established with uninfected controls, with a maximum of 1 to 2% nonspecific staining. Finally, once the system has been set up, a percentage of infected cells can be determined for each serum sample tested and for IRS. The assistance of a trained technician is recommended, to ensure that the flow cytometer is set up correctly.

Based on the infection rates obtained with serial dilutions of the IRS and with the negative control serum sample, a standard curve is plotted, correlating the percentage infection inhibition and RVNA content in IU/mL (Figure 13.1). The amount of RVNA in each experimental serum sample is determined by extrapolation, by plotting the percentage infection inhibition obtained for each sample on the standard curve constructed with the IRS dilutions.

13.4 DISCUSSION

Antibodies against viruses can be detected directly by flow cytometry, by conjugating specific antigens to microspheres and incubating them

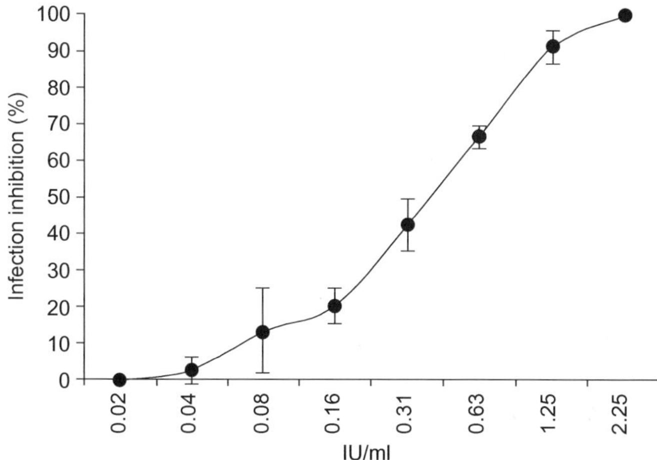

Figure 13.1 Standard curve of infection inhibition. Serial two-fold dilutions of an IRS (20 IU/mL; Copenhagen, Denmark) were mixed with 1 $CCID_{50}$ of rabies virus (PV) and then added to BHK-21 cells. The cells were incubated for 24 h, and the cell monolayers were then disaggregated by trypsinization, fixed with *p*-formaldehyde, and stained with a rabies virus nucleocapsid protein-specific fluorophore-conjugated antibody. Flow cytometry analysis was carried out on a FACSCalibur® flow cytometer. The percentage inhibition of infection achieved with each dilution of the IRS was determined by comparison with infected positive control cells (incubated with a control serum known to contain no RVNA). The results shown are the means ±SD of four independent experiments. For details, see Bordignon et al. (2002).[14]

with biological samples. McHugh et al. (1988) demonstrated the simultaneous detection of antibodies directed against cytomegalovirus and herpes simplex virus in a microsphere fluorescence assay.[12] Antibodies against HIV, *Helicobacter pylori* antigens, and *Toxoplasma gondii* have also been detected by flow cytometry.[13,16,17] Moreover, the amount of each antibody can be quantified by determining mean fluorescence intensity (MFI) by flow cytometry. As rabies laboratories are generally interested in the detection and quantification of neutralizing antibodies in large numbers of serum samples, the protocol described here is designed for the automated indirect detection of RVNA based on the neutralization of BHK-21 cell infection with a standard RABV strain. The determination of RVNA levels by flow cytometry has been shown to be reproducible (as indicated by the low standard deviation values obtained in four independent experiments, as shown in Figure 13.1). The results of this technique were correlated with those of a regular infection neutralization assay ($r = 0.773$) in tests on 104 dog serum samples.[14]

Vengatesan et al. (2006) compared flow cytometry and RFFIT, and reported a similar correlation ($r = 0.74$) between the results obtained with these two techniques.[15] They used MNA cells in a similar experimental design and expressed their results as equivalent units (EU/mL) of RVNA determined by a regression method.[15]

However, it should be stressed that large multicenter studies comparing classical neutralizing techniques with flow cytometry are required to establish definitively the suitability of automated flow cytometry methods for RVNA quantification. Nevertheless, the protocols already available make it possible to envisage the use of flow cytometry as an automated and sensitive technique for RVNA determination, particularly in situations in which it is necessary to analyze large numbers of samples.[14,15]

13.4.1 Experimental Tips

- Before RABV neutralization, serum samples should be incubated for 30 minutes at 56°C in a water bath to inhibit the activity of the complement system.
- If necessary, cells could be analyzed by flow cytometry three to four days after staining. However, if such a delay is envisaged, IsotonII should be replaced with sterile 3% *p*-formaldehyde (in PBS). Cells should be stored in the dark at 4°C.

  See also Chapter 4 for additional information.

13.4.2 Critical Parameters and Troubleshooting

Experimental serum samples should always be analyzed at the same, fixed dilution (1:25). This dilution makes it possible to eliminate the nonspecific neutralization displayed by some negative samples without interfering with the neutralizing activity of positive serum samples, even if their antibody titers are low. This parameter should be checked if the method is to be applied under different conditions.

13.4.3 Precautions

See Chapter 4 for further information.

13.4.4 Alternative Materials and/or Methods

RABV antigens (e.g. glycoprotein) can be coupled to microspheres and used for the quantification of antibodies against RABV. Some companies produce and sell bead kits for conjugation to a protein of interest (see

Functional Bead Conjugation on the Becton, Dickinson & Company website, for example).

13.4.5 Time Considerations

The analysis of RVNA by flow cytometry is no more time-consuming than traditional methods. The times required to neutralize RABV and to infect and stain cells are similar for the two techniques. The major advantage of flow cytometry is the automation of cell fluorescence reading, which is much more rapid than for microscopy-based techniques.

13.4.6 Limitations

See Chapter 4 for further information.

13.4.7 Future Considerations

Flow cytometry is increasingly widely used due to the recent release onto the market of new, smaller, less expensive machines. These changes are making it possible to envisage the routine use of this technique in rabies laboratories in the near future. Flow cytometry can provide faster, automated, sensitive, and reproducible RVNA titer determinations, and this is of the utmost importance for public health purposes.

ACKNOWLEDGMENTS

We thank Prof. André Báfica from the Laboratory of Immunobiology of the Federal University of Santa Catarina for critical reading and helpful discussions on this book chapter.

Additional Information

Examples of Major Sources of Supplies/Equipment (including websites):

- Becton, Dickinson and company (www.bdbiosciences.com).
- Millipore (www.millipore.com).
- Miltenyi Biotec (www.miltenyibiotec.com).
- Beckman Coulter (www.beckmancoulter.com).
- Bio-Rad (www.bio-rad.com).

REFERENCES

1. Moore SM, Hanlon CA. Rabies-specific antibodies: measuring surrogates of protection against a fatal disease. *PLoS Negl Trop Dis* 2010;**4**(3):e595.
2. Wiktor TJ. Cell-mediated immunity and post-exposure protection from rabies by inactivated vaccines of tissue culture origin. *Dev Biol Stand* 1978;**40**:255–64.
3. Kawano H, Mifune K, Ohuchi M, Mannen K, Cho S, Hiramatsu K, et al. Protection against rabies in mice by a cytotoxic T cell clone recognizing the glycoprotein of rabies virus. *J Gen Virol* 1990;**71**(pt.2):281–7.
4. Perry LL, Lodmell DL. Role of $CD4^+$ and $CD8^+$ T cells in murine resistance to street rabies virus. *J Virol* 1991;**65**(7):3429–34.
5. Wiktor TJ, Gyorgy E, Schlumberger D, Sokol F, Koprowski H. Antigenic properties of rabies components. *J Immunol* 1973;**110**:269–76.
6. World Health Organization. 1992. World Health Organization Expert Committee on Rabies: Eighth Report. 824.
7. Smith JS, Yager PA, Baer GM. A rapid reproducible test for determining rabies-virus neutralizing antibody. *Bull Wld Hlth Org* 1973;**48**:535–41.
8. Favoretto SR, Carrieri ML, Tino MS, Zanetti CR. Pereira OAC. Simplified fluorescent inhibition microtest for the titration of rabies neutralizing antibodies. *Rev Inst Med Trop S Paulo* 1993;**35**:171–5.
9. Atanasiu P, Savy V, Perrin P. Épuvreimmunoenzymatique pour la detection rapide des anticorps antirabiques. *Ann Inst Pasteur Microbiol* 1977;**128A**:489–98.
10. Perrin P, Vermisse P, Delagneau JF, Lucas G, Rollin PE, Sureau P. The influence of the type of immunosorbent on rabies antibody EIA: advantages of purified glycoprotein over whole virus. *J Biol Stand* 1986;**14**:95–102.
11. Piza AS, Santos JL, Chaves LB, Zanetti CR. An ELISA suitable for the detection of rabies virus or rabies virus antibodies in serum samples from humans vaccinated with either cell-culture vaccine or suckling mouse-brain vaccine. *Rev Inst Med Trop São Paulo* 1999;**41**(1):39–43.
12. McHugh TM, Miner RC, Logan LH, Stites DP. Simultaneous detection of antibodies to cytomegalovirus and herpes simplex virus by using flow cytometry and a microsphere-based fluorescence immunoassay. *J Clin Microbiol* 1988;**26**:1957–61.
13. Sligh JM, Roodman ST, Tsai CC. Flow cytometric indirect immunofluorescence assay with high sensitivity and specificity for detection of antibodies to human immunodeficiency virus (HIV). *Am J Clin Pathol* 1989;**91**:210–4.
14. Bordignon J, Comin F, Ferreira SC, Caporale GMM, Lima Filho JHC, Zanetti CR. Calculating rabies virusneutralizing antibody titres by flow cytometry. *Rev Inst Med Trop Sao Paulo* 2002;**44**(3):151–4.
15. Vengatesan D, Raj GD, Raja A, Ramadass P, Gunaseelan L. Detection of rabies virus antigen or antibody using flow cytometry. *Cytometry* 2006;**70B**:335–43.
16. Best LM, van Zanten SJOV, Bezanson GS, Haldane DJM, Malatjalian DA. Serological detection of Helicobacter pylori by a flow microsphere immunofluorescence assay. *J Clin Microbiol* 1992;**30**:2311–7.
17. Cozon G, Roure C, Lizard G, Greenland T, Larget-Piet D, Ganilho F, et al. An improved assay for the detection of Toxoplasma gondii antibodies in human serum by flow cytometry. *Cytometry* 1993;**14**:569–75.

CHAPTER FOURTEEN

Demonstration of Rabies Antibody by a Latex Agglutination Test

Songsri Kasempimolporn, Narongsak Chaiyabutr, and Visith Sitprija
Queen Saovabha Memorial Institute, Thai Red Cross Society, Bangkok, Thailand

Chapter Contents

14.1 Introduction 141
14.2 Materials 142
  14.2.1 Reagents 142
  14.2.2 Biological Materials 142
  14.2.3 Equipment 142
14.3 Methods 142
  14.3.1 Latex Coating 142
  14.3.2 Slide Agglutination 143
14.4 Discussion 143
  14.4.1 Experimental Tips 143
  14.4.2 Technical Notes 144
References 145

14.1 INTRODUCTION

Several serological tests have been developed for measuring neutralizing antibodies to rabies virus (RABV) in serum. In all routine assays, procedures are conducted in well-equipped rabies laboratories. The most widely used method is the rapid fluorescent focus inhibition test, or RFFIT.[1] The RFFIT, being a bioassay, requires relatively sophisticated laboratory facilities, including the production of RABV and cell cultures, which are laborious, time consuming, and costly. To overcome this disadvantage, a number of enzyme-linked immunosorbent assays (ELISAs) have been described to quantify the level of RABV antibodies.[2–4] The ELISA is relatively less time consuming and less expensive than the RFFIT, but even so it still cannot be carried out in remote areas where facilities or resources to perform such tests are not available.

C. Rupprecht & T. Nagarajan (Eds): Current Laboratory Techniques in Rabies Diagnosis, Research, and Prevention, Volume 1.
Doi: http://dx.doi.org/10.1016/B978-0-12-800014-4.00014-7

The utility of a Latex Agglutination Test (LAT) for the determination of RABV antibodies in the sera of humans and animals, especially equines, has been reported.[5–7] Latex particles are coated with whole RABV, glycoprotein or vaccine. The antigen-coated particles are specifically agglutinated by RABV antibodies, and the agglutination is evident macroscopically within minutes. The LAT is simple to perform, uses an inactivated viral antigen, and needs minimum laboratory equipment. The performance of LAT is comparable to that of the standard serologic technique. A rapid screening test, such as the LAT, is desirable for the production of therapeutic equine rabies immune globulin (ERIG) to evaluate the level of antibodies simply and quickly, and to determine whether additional (booster) doses of vaccine are required. Although the LAT does not directly measure virus-neutralizing activity, it can be used as an inexpensive alternative for process controls in the preparation of ERIG products and as a testing tool for evidence of seroconversion following rabies vaccination.

14.2 MATERIALS

14.2.1 Reagents

- Polystyrene latex beads (0.8 μm in diameter; 10% suspension [w/v]).
- 0.05 M Carbonate buffer, pH 9.6 (see Additional Information).
- Phosphate buffered saline (PBS), pH 7.4.
- Blocking buffer (carbonate–bicarbonate buffer with 5% trehalose and 0.3% bovine serum albumin [BSA]).

14.2.2 Biological Materials

- Inactivated RABV, purified glycoprotein, or cell culture rabies vaccine.
- Test sera.

14.2.3 Equipment

- Double-concave slides or clear pressed glass with 10 depressed cavities.

14.3 METHODS

14.3.1 Latex Coating

1. Wash 1 mL of 10% latex beads with carbonate buffer by centrifugation at 12,500× *g* for 15 min at 4°C.
2. Re-suspend the latex beads in 3 mL of carbonate buffer and add 1.75 mL of either a 1 mg/mL solution of RABV antigen or a 1:10 dilution of inactivated rabies vaccine.

3. Gently shake at room temperature for 3 h and at 4°C overnight.
4. Wash the latex beads twice with PBS by centrifugation at 12,500×*g* for 15 min at 4°C.
5. Block the unsaturated sites on the latex bead surface with 0.3% BSA and 5% trehalose in carbonate buffer at room temperature for 30 min.
6. Repeat the washing step.
7. Re-suspend the sensitized latex beads in PBS at a concentration of 3% and store at 4°C.

14.3.2 Slide Agglutination

1. Add 25 μL aliquot of sensitized latex beads onto a double-concave slide and 25 μL aliquot of the serum to be tested.
2. Rock gently for 5 to 15 min at room temperature and read agglutination visually against a dark background. A positive reaction for antibodies to RABV is indicated by the formation of fine granular particles. The degree of agglutination is scored as 4+, 3+, 2+, or 1+ depending on the size of agglutinated particles. A negative reaction leaves the serum and latex suspension with a milky, smooth consistency.

14.4 DISCUSSION

14.4.1 Experimental Tips

1. With the LAT developed for determining antibodies, large-sized latex particles (0.8–1.0 μm) are preferable.
2. The use of commercial inactivated cell-culture rabies vaccine for human use as a coating antigen in the LAT simplifies the experimental technique.
3. Reference immune sera should be tested initially with the LAT and with the neutralization test. A score of agglutination corresponds to a range of neutralizing antibody titers. The degree of agglutination scored in the LAT with positive sera is proportional to RFFIT titers. Indicating the criteria for the scoring system for LAT is important, as the scoring is based on "classification (1+ to 4+)" by an investigator. Therefore, the investigators have to determine LAT scores of the serum samples and have to statistically compare the LAT score with neutralizing antibody titers. However, interpretation of positive LAT corresponding to titers might differ among the experimental systems. The discrepancies could reside with the coating antigen used. Different sources of serum might also be a cause.

4. The LAT can be performed for routine use under field conditions if sensitized beads are supplied from a central laboratory. Sensitized latex beads are stable for 2–3 months at 4°C.
5. Sera contaminated by bacteria cannot be used with the LAT because a strong nonspecific agglutination may occur.

14.4.2 Technical Notes

1. Immunoassays based on LAT can be employed with either antibody- or antigen-coating of the latex beads. However, due to the existence of a vast variety of antigens with different physical-chemical characteristics, an empirical optimization is required for each assay if antigen (e.g. whole virus, RABV glycoprotein, cell-culture rabies vaccine) is used as immobilized molecules on the latex bead surface. A preliminary study could be necessary. In contrast, it is widely known that most of the antibody molecules have a similar physical-chemical pattern.
2. There are also other factors to optimize, for example, particle size, reaction time, adsorption conditions which determine the amount and mode of attachment of the antigen, agglutination conditions, etc.[8]
3. LAT is not proposed for antibody titration, but rather for the rapid screening of RABV antibodies, and its sensitivity seems to be adequate. If the results of LAT are directly correlated with RFFIT, and if the degree of agglutination correlates with the level of RABV-neutralizing antibody, then the presence of moderate levels of agglutinating antibodies should equate with acceptable neutralizing antibody titers.
4. Some serum samples might show high antibody titers but exhibit a low degree of agglutination, and vice versa. These discrepant results may be due to a different degree of accuracy and/or diversity between the two test systems. In addition, there exist components in serum that may affect the specific agglutination of sensitized latex.
5. The LAT is subjective in interpretation, and performing the assay requires some degree of skill.

Additional Information

Sources of Supplies

- Polystyrene latex beads (Sigma, St Louis, MO, USA).
- Purified Vero cell rabies vaccine (Verorab®, Sanofi Pasteur, France).
- Double-concave glass slides (Various sources).
- Clear pressed glass with 10 depressed cavities (Marienfeld, Germany).

Preparation of Buffer

- Carbonate-bicarbonate buffer, pH 9.6, 0.05 M.
- Dissolve 2.93 g of sodium hydrogen carbonate ($NaHCO_3$) and 1.59 g of sodium carbonate (Na_2CO_3) in 1 liter of distilled water.

REFERENCES

1. Smith JS, Yager PA, Baer GM. A rapid reproducible test for determining rabies neutralizing antibody. *Bull WHO* 1973;**48**:535–41.
2. Nicholson KG, Prestage H. Enzyme-linked immunosorbent assay: a rapid reproducible test for the measurement of rabies antibody. *J Med Virol* 1982;**9**:43–5.
3. Mannen K, Mifune K, Reid-Sanden FL, Smith JS, Yager PA, Sumner JW, et al. Microneutralization test for rabies virus based on an enzyme immunoassay. *J Clin Microbiol* 1987;**25**:2440–2.
4. Grassi M, Wandeler AI, Peterhans E. Enzyme-linked immunosorbent assay for determination of antibodies to the envelope glycoprotein of rabies virus. *J Clin Microbiol* 1989;**27**:899–902.
5. Perrin P, Versmisse P, Sureau P. A rabies agglutination test (RAT) for rabies antibody detection. *J Biol Stand* 1988;**16**:281–6.
6. Madhusudana SN, Saraswati S. Development and evaluation of a latex agglutination test for rabies antibodies. *J Clin Virol* 2003;**27**:129–35.
7. Saengseesom W, Kasempimolporn S, Akesowan S, Ouisuwan S, Sitprija V. Use of latex agglutination test to determine rabies antibodies in production of rabies antisera in horse. *SoutheastAsian J Trop Med Public Health* 2010;**41**:1387–92.
8. Ortega-Vinuesa JL, Bastos-Gonzalez D. A review of factors affecting the performances of latex agglutination tests. *J Biomater Sci Polymer Edn* 2001;**12**:379–408.

CHAPTER FIFTEEN

Demonstration of Viral Antibodies by Pseudotype Virus Neutralization

Edward Wright
Viral Pseudotype Unit, Faculty of Science and Technology, University of Westminster, London, United Kingdom

Chapter Contents

15.1 Introduction 147
15.2 Materials 148
15.2.1 Reagents 148
15.2.2 Equipment 149
15.2.3 Biological Materials 149
15.3 Methods 149
15.3.1 Generation of Pseudotyped Virus Stock 149
15.3.2 Titration of Pseudotyped Virus Stock 151
15.3.3 Setting Up a PVNA 151
15.3.4 Reporter Gene Assay 151
15.3.5 Interpretation of Results 152
15.4 Discussion 153
15.4.1 Experimental Tips and Critical Parameters 153
15.4.2 Limitations 154
15.4.3 Future Considerations 154
Acknowledgments 154
References 155

15.1 INTRODUCTION

The study of highly pathogenic Advisory Committee on Dangerous Pathogens category 3 viruses such as the rabies virus (RABV) is restricted to institutes that have access to high-containment biosafety level (BSL) 2 or 3 laboratories. While it is not possible to circumvent this requirement for many studies, serological investigations can be undertaken in low-containment BSL 1 and 2 laboratories by utilizing pseudotyped virus (PV) as an alternative source of viral antigen. Pseudotyping of viruses has been widely used for gene

C. Rupprecht & T. Nagarajan (Eds): Current Laboratory Techniques in Rabies Diagnosis, Research, and Prevention, Volume 1.
Doi: http://dx.doi.org/10.1016/B978-0-12-800014-4.00015-9

therapy applications and to elucidate virus–host interactions since it was first undertaken in the 1960s.[1] Refinement of the system through different generations of PV over the subsequent decades has resulted in a highly flexible and safe system that can be used in any laboratory that has access to basic tissue culture facilities.[2] Consequently, the uses of PV have expanded, and they are now routinely employed for much broader purposes, e.g. neutralization assays,[3] vaccine immunogens,[4–7] and enzyme-linked immunosorbent assay (ELISA) antigens.[8]

These recombinant viruses comprise a reporter gene, maintained as the viral genome, within a core or capsid of one virus, typically a retrovirus or vesicular stomatitis virus (VSV). Surrounding this is the plasma membrane-originating viral envelope that has the surface envelope glycoprotein from the virus of interest, expressed in its native conformation, studded on the surface. PV are generated through the transfection of producer cells with plasmids that carry the genes for the components mentioned above. As the only genetic information contained within the PV is the reporter (or transfer) gene, these viruses are only able to undergo one round of infection, resulting in stable integration of the gene within the cell chromosome. Expression of this reporter gene can be determined by using specific substrate-enzymatic assays. Recently, PV-based neutralization assays (PVNA) covering many different families of viruses have been shown to be highly sensitive and specific when compared to existing gold standard assays,[9–14] not observed for all other assays used to monitor antibody levels. PVNA also have the advantage that they only measure the level of virus neutralizing antibodies (VNAb), thereby providing a more accurate insight of immune-protection. RABV-based PV studies initially were focused on their use as gene therapy vectors for targeting neuronal cells,[15,16] however, there is a growing body of evidence to suggest RABV PVNA are an attractive alternative when undertaking RABV serological studies.[17–20]

15.2 MATERIALS

The following lists comprise the major items that would need to be sourced by a laboratory for the production of PV and conducting a PVNA.

15.2.1 Reagents

- Plasmids encoding the PV reporter, core and envelope G genes.
- Tissue culture media to maintain mammalian cell lines, e.g. Dulbecco's Modified Eagle's Medium (DMEM) supplemented with 10% fetal bovine serum and antibiotics/antimycotics as required.

- Transfection reagent, e.g. Fugene 6 (Promega) or polyethylenimine (PEI).
- Reporter gene substrate, which depends on the reporter used.

15.2.2 Equipment

- A humidified, CO_2 incubator for the mammalian cell lines.
- Facilities to measure the level of expression of the reporter gene, e.g. plate reader, light/fluorescent microscope, flow cytometer, etc.

15.2.3 Biological Materials

- A PV producer cell line, typically HEK 293T-17 cells (ATCC CRL-11268).
- A permissive target cell line, typically BHK-21 cells (ATCC CCL-10).
- RABV PV stocks.
- Serum samples, antibodies, or antiviral drugs to be tested.

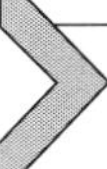

15.3 METHODS

15.3.1 Generation of Pseudotyped Virus Stock

Day 1: Seed HEK 293T-17 cells into the desired sized dish or plate; see Table 15.1 for quantities.

Day 2: At least 2 hours (h) prior to transfection, the cell culture media needs to be changed (Table 15.1). Following this, in separate 1.5 mL Eppendorf tubes, prepare the plasmid DNA and transfection reagent mixes (Table 15.1). It is important that the DNA and reagent are added directly to the serum free medium (SFM) and not to the side of the tube. Incubate these for 5 minutes (mins) at room temperature (RT). Then add the DNA/dH_2O (Fugene) or DNA/SFM (PEI) mix to the transfection reagent/SFM mix, incubate for a further 20 mins at RT, and mix frequently. Following this, swirl dishes/plates simultaneously while the transfection mix is added drop wise. Incubate overnight in a 5% CO_2, humidified 37°C incubator.

Day 3: Aspirate off the medium on the cells and replace with fresh media.

Day 4: Harvest the supernatant containing the PV (the first of two collections) and pass through a 0.45 µm cellulose-based filter or centrifuge at 3,000 rpm for 10 mins (4°C). Replace with fresh media.

Day 5: Harvest the supernatant containing the PV (the final of two collections) and pass through a 0.45 µm cellulose-based filter or centrifuge at 3,000 rpm for 10 mins (4°C). PV stocks are ideally stored at −80°C, where they remain stable for many years, otherwise storage at −20°C, +4°C, and RT can be undertaken for months, weeks or days respectively.

Table 15.1 Transfection Reagents[a]

Plate/Dish size	293T-17 Cells to Seed per Well	Vol Media (ml)	Vol SFM for T^n Reagent (μl)[b]	Vol Fugene 6/PEI (μl)	Vol dH_2O/SFM for DNA (μl)[c]	Plasmid (μg)		
						Core	Reporter	Envelope
10 cm dish	5×10^6	7	200	18/35	20	1	1.5	1
6 cm dish	1.5×10^6	3	120	12/20	15	0.7	1.1	0.7
6-well plate	8×10^5	1.5	100	9/15	15	0.6	0.9	0.6

[a]The quantities given should be used as a guide as it is likely each laboratory will have to optimize these values.
[b]For both transfection (T^n) reagents (Fugene 6 and PEI) dilute the transfection reagent in the stated volume of SFM.
[c]For Fugene 6 or PEI make your DNA mixes up to the stated volumes with dH_2O or SFM respectively.

15.3.2 Titration of Pseudotyped Virus Stock

PV that carries a fluorescent or lacZ reporter gene can be serially diluted, and foci of infection counted to determine infectious units per mL (IFU/mL). For luciferase-based PV, an end-point dilution method can be used to determine $TCID_{50}$ values.[21] Briefly, the PV is serially diluted in quadruplicate, and the dilution of virus that resulted in infection of 50% of replicates can then be calculated. Typically, 2×10^4 BHK-21 cells are added to each well of a 96-well plate after the virus has been diluted.

15.3.3 Setting Up a PVNA

In a 96-well plate make doubling dilutions of serum in duplicate or triplicate as required. Add 90µL of cell culture media and 10µL of serum to the first well, and 50µL of cell culture media to all the other wells. Mix and serially dilute 50µL from the first well across the other wells in the dilution series, disposing of the "spare" 50µL from the last well. The total volume of media/serum in each well following this step should be 50µL. It is important to include "cells alone," "virus alone," and "cells and virus" controls that need to be set up with 100µL, 150µL, and 50µL of media in them, respectively, at this point. Add 50µL of diluted PV, containing 30-90 $TCID_{50}$ (luciferase-based PV) or 200 IFU/mL (fluorescent- and lacZ-based PV), to each well apart from the "cells alone" control wells. The total volume in each well is now 100µL, except the "virus alone," which has 200µL. Spin the plate at ~500rpm for 5 second (s) to bring down any liquid on the side of the wells, and then incubate for 1h at 37°C (5% CO_2). Following this, add 100µL of DMEM containing 2×10^4 BHK cells to each well (apart from the "virus alone" control). The total volume in all wells is now 200µL. Spin plates for 5s at 500rpm and incubate for 48h (luciferase and lacZ) to 72h (fluorescence) at 37°C (5% CO_2).

15.3.4 Reporter Gene Assay

Detect fluorescence using an appropriate microscope or flow cytometer, and if using microscopy, replace media with phosphate buffered saline (PBS). For firefly and/or renilla luciferase-based PV, use the required substrates as directed by the manufacturer, such as Bright-Glo or Dual-Glo (Promega), and read the plate in a luminometer. If lacZ PV were used, stain cells with substrate buffer and X-gal as follows. Replace the liquid in the wells with 0.5% glutaraldehyde (in PBS) and incubate at 4°C for 10mins. Discard and incubate in PBS for a further 5mins at 4°C. Discard and wash the wells twice with PBS followed by incubating each well with 50µL X-gal buffer containing 1mg/mL X-gal. Wrap in foil and incubate for 1h at 37°C. After this, the wells are observed under a microscope, where infected cells will be stained blue. For long-term storage (i.e. months), tip off the X-gal buffer, wash the cells once with PBS, then half-fill the wells with PBS and store at 4°C. The standard result for each reporter gene is shown in Figure 15.1.

15.3.5 Interpretation of Results

Similar to a fluorescent antibody virus neutralization assay (such as the RFFIT or FAVN), the reported result corresponds to the concentration

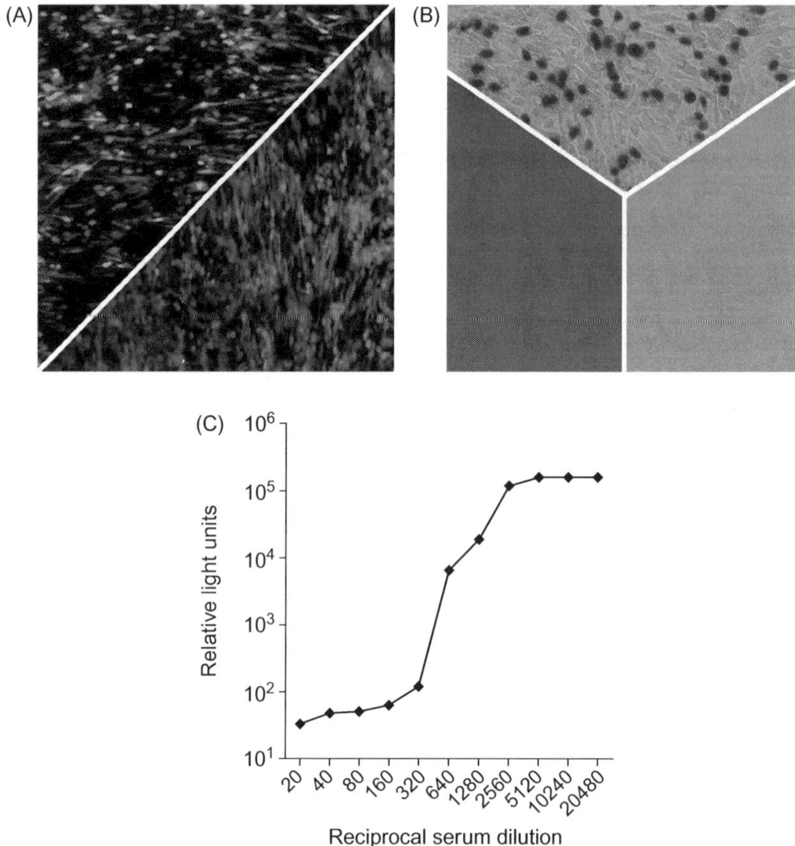

Figure 15.1 Pseudotyped virus reporter gene repertoire. Typical observations for PV-infected BHK-21 cells when using (A) green (top left; cytoplasm appears green) or cherry-red (bottom right, cytoplasm appears red) fluorescent reporter proteins; (B) lacZ reporter gene with X-gal (top, using a light microscope the nuclei will appear dark blue), ONPG (right, by eye the supernatant will appear yellow), CPRG (left, by eye the supernatant will appear dark red) substrates; and (C) firefly or renilla luciferase reporter gene with expression reported as relative light units, which inversely correlates with the level of PV neutralization.

of antibody at which 100% of input virus is neutralized (IC_{100}). However, for the PVNA, the figure is the greatest reciprocal dilution of the sample at which IC_{100} is achieved. Background levels of reporter expression are established using the controls described above.

15.4 DISCUSSION

15.4.1 Experimental Tips and Critical Parameters

For luciferase-based assays it is important to spin the neutralization assay plates before the serum/PV incubation step as the reporter is very sensitive so even tiny volumes of PV that remain on the side of the well will produce a strong signal in the presence of target cells. It is highly recommended that white plates (Nunc) be used for the luciferase assays, as this eliminates leakage of signal between wells when the plate is read. However, when using completely opaque plates, it is not possible to observe cell monolayer growth, so serum cytotoxicity needs to be checked in transparent plates or the PVNA run in a transparent plate with the cell lysate being transferred to an opaque plate for the reporter gene assay if cytotoxicity is expected. If using fluorescent proteins, the cell culture media must be replaced with PBS prior to microscopy, as components within it can mask the fluorescent signal. To enhance the signal further, sodium butyrate can be added to the PVNA culture media, which leads to hyperacetylation of histones and consequently increased expression of the reporter gene.

As with all neutralization assays, it is critical that the level of virus input is the same across all the wells. Variation in virus input can greatly influence the observed antibody titers. When establishing what virus input to use, those detailed within this chapter should be a starting point, and users should aim for a concentration where the control positive serum neutralizes 100% of the RABV PV at a serum dilution of 1:40. This enables a negative sample to be confidently recorded while allowing as wide a range of serum dilutions to be tested.

At least one isolate from each lyssavirus species has been successfully pseudotyped, including newly putative species. However, when G from the Arctic-like lineage of RABV isolates were used, very low or no titer was achieved. This was overcome by creating chimeric forms of the RABV G swapping the cytoplasmic tail for that of VSV.[22]

15.4.2 Limitations

The current gold standard RABV neutralization assays, such as rapid fluorescent focus inhibition test (RFFIT) or FAVN, report antibody titers as international units/mL (IU/mL). The PVNA currently only reports antibody titers in reciprocal dilutions. Establishing a procedure for converting these figures to IU/mL would enable greater standardization.

Innate host restriction factors, such as TRIM proteins, block retrovirus infection at early stages.[23] As retroviral cores are used to generate PV, it is important to check whether any observed refractivity to infection is not a consequence of these factors. BHK-21 cells are fully permissive to RABV PV infection.[17]

15.4.3 Future Considerations

Each reporter gene has its own advantages: luciferase is rapid to read but expensive, fluorescent proteins require no substrates but complex equipment, while lacZ is inexpensive but time consuming. Work is ongoing to further expand the panel of reporter genes to include secreted alkaline phosphatase and other luciferase isoforms. These have the potential to be simultaneously inexpensive and rapid to screen. The large range of reporters also leads to a greater potential to multiplex the PVNA, assaying for the presence of antibodies to more than one virus in the same well. Encouraging data on a dual-PVNA has previously been reported,[9,24–26] but this could be expanded to three or four PV, each carrying a different reporter gene.

To enable laboratories in resource-limited countries to use PV, their ability to be lyophilized is under investigation. Preliminary data suggests PV can be efficiently freeze-dried and re-suspended with minimal loss of viral titer.[27]

ACKNOWLEDGMENTS

I thank my colleagues and collaborators at the Viral Pseudotype Unit and Animal Health and Veterinary Laboratories Agency for providing essential information and feedback on the PVNA. I am especially grateful to Nigel Temperton and Emma Bentley for critically reviewing this chapter.

Additional Information

Composition of Reagents

- DMEM – high glucose (4,500 mg/L glucose), L-glutamine, sodium pyruvate, and sodium bicarbonate. Liquid, sterile-filtered, and suitable for cell culture. Supplemented with 10% FBS, 100 units penicillin/mL, and 100 μg streptomycin/mL.
- X-gal substrate buffer – 10 mM deoxycholric acid, 5 mM potassium ferrocynate, 5 mM potassium ferricynate, 4.3 mM magnesium chloride, and 0.02% (v/v) NP40. X-gal is dissolved in dimethylformamide.
- PEI (Polysciences, #23966, linear, MW 25,000) – made up at 1 mg/mL in distilled water, sterilized using a 0.22 μm filter, and stored at – 20°C.

REFERENCES

1. Rubin H. Genetic control of cellular susceptibility to pseudotypes of rous sarcoma virus. *Virology* 1965;**26**:270–6.
2. Bouard D, Alazard-Dany D. Cosset F-L. Viral vectors: from virology to transgene expression. *Br J Pharmacol* 2009;**157**:153–65.
3. Mather S, Scott S, Temperton N, Wright E, King B, Daly J. Current progress with serological assays for exotic emerging/re-emerging viruses. *Future Virol* 2013;**8**:745–55.
4. Szecsi J, Gabriel G, Edfeldt G, Michelet M, Klenk HD. Cosset FL. DNA vaccination with a single-plasmid construct coding for viruslike particles protects mice against infection with a highly pathogenic avian influenza A virus. *J Infect Dis* 2009; **200**:181–90.
5. Szecsi J, Boson B, Johnsson P, Dupeyrot-Lacas P, Matrosovich M, Klenk HD, et al. Induction of neutralising antibodies by virus-like particles harbouring surface proteins from highly pathogenic H5N1 and H7N1 influenza viruses. *Virol J* 2006;**3**:70.
6. Yang ZY, Wei CJ, Kong WP, Wu L, Xu L, Smith DF, et al. Immunization by avian H5 influenza hemagglutinin mutants with altered receptor binding specificity. *Science* 2007;**317**:825–8.
7. Yu L, Bai W, Wu X, Zhang L, Zhang L, Li P, et al. A recombinant pseudotyped lentivirus expressing the envelope glycoprotein of Hantaan virus induced protective immunity in mice. *Virol J* 2013;**10**:301.
8. Wang W, Xie H, Ye Z, Vassell R, Weiss CD. Characterization of lentiviral pseudotypes with influenza H5N1 hemagglutinin and their performance in neutralization assays. *J Virol Methods* 2010;**165**:305–10.
9. Wright E, Hayman DT, Vaughan A, Temperton NJ, Wood JL, Cunningham AA, et al. Virus neutralising activity of African fruit bat (Eidolon helvum) sera against emerging lyssaviruses. *Virology* 2010;**408**:183–9.
10. Temperton NJ, Chan PK, Simmons G, Zambon MC, Tedder RS, Takeuchi Y, et al. Longitudinally profiling neutralizing antibody response to SARS coronavirus with pseudotypes. *Emerg Infect Dis* 2005;**11**:411–6.

11. Gierer S, Bertram S, Kaup F, Wrensch F, Heurich A, Krämer-Kühl A, et al. The spike protein of the emerging betacoronavirus EMC uses a novel coronavirus receptor for entry, can be activated by TMPRSS2, and is targeted by neutralizing antibodies. *J Virol* 2013;**87**:5502–11.
12. Molesti E, Milani A, Terregino C, Cattoli G, Temperton NJ. Comparative serological assays for the study of h5 and h7 avian influenza viruses. *Influenza Res Treat* 2013;**2013**:286158.
13. Khetawat D, Broder CC. A functional henipavirus envelope glycoprotein pseudotyped lentivirus assay system. *Virol J* 2010;**7**:312.
14. Salvador B, Zhou Y, Michault A, Muench MO, Simmons G. Characterization of Chikungunya pseudotyped viruses: Identification of refractory cell lines and demonstration of cellular tropism differences mediated by mutations in E1 glycoprotein. *Virology* 2009;**393**:33–41.
15. Desmaris N, Bosch A, Salaun C, Petit C, Prevost MC, Tordo N, et al. Production and neurotropism of lentivirus vectors pseudotyped with lyssavirus envelope glycoproteins. *Mol Ther* 2001;**4**:149–56.
16. Mazarakis ND, Azzouz M, Rohll JB, Ellard FM, Wilkes FJ, Olsen AL, et al. Rabies virus glycoprotein pseudotyping of lentiviral vectors enables retrograde axonal transport and access to the nervous system after peripheral delivery. *Hum Mol Genet* 2001;**10**:2109–21.
17. Wright E, Temperton NJ, Marston DA, McElhinney LM, Fooks AR, Weiss RA. Investigating antibody neutralization of lyssaviruses using lentiviral pseudotypes: a cross-species comparison. *J Gen Virol* 2008;**89**:2204–13.
18. Wright E, McNabb S, Goddard T, Horton DL, Lembo T, Nel LH, et al. A robust lentiviral pseudotype neutralisation assay for in-field serosurveillance of rabies and lyssaviruses in Africa. *Vaccine* 2009;**27**:7178–86.
19. Hultberg A, Temperton NJ, Rosseels V, Koenders M, Gonzalez-Pajuelo M, Schepens B, et al. Llama-derived single domain antibodies to build multivalent, superpotent and broadened neutralizing anti-viral molecules. *PLoS One* 2011;**6**:e17665.
20. Both L, van Dolleweerd C, Wright E, Banyard AC, Bulmer-Thomas B, Selden D, et al. Production, characterization, and antigen specificity of recombinant 62-71-3, a candidate monoclonal antibody for rabies prophylaxis in humans. *FASEB J* 2013;**27**:2055–65.
21. Condit RC. Principles ofVirology. *Fields Virol* 2001;**1**:19–51.
22. Bentley E, Ali R, Horton D, Banyard AC, Wright E. The construction of chimeric rabies virus glycoproteins rescues Arctic-like rabies pseudovirus production. In: Pseudotype viruses: applications and troubleshooting. Conference Paper. October 2013, London, UK.
23. Fletcher AJ, Towers GJ. Inhibition of retroviral replication by members of the TRIM protein family. *Curr Top Microbiol Immunol* 2013;**371**:29–66.
24. Peel AJ, Baker KS, Crameri G, Barr JA, Hayman DTS, Wright E, et al. Henipavirus neutralising antibodies in an isolated island population of African fruit bats. *PLoS One* 2012;**7**:e30346.
25. Zhou Y, Agudelo J, Lu K, Goetz DH, Hansell E, Chen YT, et al. Inhibitors of SARS-CoV entry–identification using an internally-controlled dual envelope pseudovirion assay. *Antiviral Res* 2011;**92**:187–94.
26. Temperton N, Cattoli G, Wright E. A multiplex assay for the evaluation of avian influenza H5 and H7 neutralizing antibodies directed against haemagglutinin. *Soc Gen Microbiol Spring Conf* 2011.
27. Mather S, Temperton N, Wright E, Scott S. Development of a pseudotype neutralisation assay-based diagnostic kit for in-field vaccine evaluation and serosurveillance for highly pathogenic viruses. *Soc Gen Microbiol Spring Meet* 2013.

SECTION D

Typing/Differentiation of Lyssaviruses

CHAPTER SIXTEEN

Sanger Sequencing of Lyssaviruses

Lorraine M. McElhinney[1,2], Denise A. Marston[1], Richard J. Ellis[3], Conrad M. Freuling[4], Thomas F. Müller[4], and Anthony R. Fooks[1,2]

[1]Wildlife Zoonosis and Vector-Borne Disease Research Group, Animal Health and Veterinary Laboratories Agency, Addlestone, Surrey, United Kingdom
[2]Institute of Infection and Global Health, University of Liverpool, Liverpool, United Kingdom
[3]Central Sequencing Unit, Animal Health and Veterinary Laboratories Agency, Addlestone, Surrey, United Kingdom
[4]Friedrich-Loeffler-Institut, Federal Research Institute for Animal Health, Institute of Molecular Virology and Cell Biology, Greifswald-Insel Riems, Germany

Chapter Contents

16.1 Background 159
16.2 Methodology 161
16.2.1 Preparing Polymerase Chain Reaction Amplicons for the ABI Genetic Analyzers 161
16.2.2 Purifying the Polymerase Chain Reaction Product 162
16.2.3 Big Dye Terminator Cycle Sequencing Reactions 162
16.2.4 Ethanol Precipitation of Sequencing Reactions 163
16.2.5 Preparing 96-Well Sample Plate(s) for the Genetic Analyzers 166
16.2.6 Loading 96-Well Sample Plate(s) 166
16.2.7 Checking Run Reagents 166
16.2.8 Starting the Run on the 3130xl 167
16.2.9 Starting the Run on the 3730 167
16.2.10 Analyzing Output Data 168
16.3 Discussion 169
Acknowledgments 170
References 170

16.1 BACKGROUND

Sanger dideoxy terminator sequencing was developed by Fred Sanger and colleagues in 1977.[1] Based on the selective incorporation of chain-terminating dideoxynucleotides by DNA polymerase during *in vitro* DNA replication, it was the most widely used sequencing method for over 30 years. More recently, Sanger sequencing has been supplanted by "Next-Generation"

C. Rupprecht & T. Nagarajan (Eds): Current Laboratory Techniques in Rabies Diagnosis, Research, and Prevention, Volume 1.
Doi: http://dx.doi.org/10.1016/B978-0-12-800014-4.00016-0

sequencing methods, principally for large-scale, automated genome analyses. However, the Sanger method remains in wide use, particularly when partial genome characterization is sufficient for virus typing or for smaller-scale genome projects. The Sanger method is also the approach best suited to obtaining long contiguous DNA sequence reads (>500 nucleotides).

Rabies diagnostic laboratories introduced reverse transcription (RT)-polymerase chain reaction (PCR) methods to amplify lyssavirus RNA in the early 1990s.[2] Since then, a wide array of molecular-based diagnostic assays have been established.[3] The amplified lyssavirus-specific nucleic acid (PCR product or amplicon), generated from infected material by these molecular assays, facilitates the subsequent typing of the virus. Sanger sequencing enabled the rapid, reliable, and relatively inexpensive typing of lyssavirus PCR amplicons compared to conventional monoclonal antibody typing.[4] Within just a few years, the genetic characterization of lyssaviruses was to become routine, quickly followed by an explosion of published molecular epidemiological and phylogenetic studies. A specific genomic region has not been prescribed or standardized for lyssavirus molecular typing, however, a considerable amount of data has been published for partial and complete nucleoprotein gene sequences, and this region has been shown to be sufficiently discriminatory for virus typing and evolutionary studies.[5–8]

Sanger sequencing of overlapping PCR products[9,10] or long distance PCR products,[11,12] sometimes referred to as "walking the genome," has enabled the generation of complete genome sequences for a number of representative lyssaviruses. The multiple primer sets employed are designed specifically for the target viral species or strains, and are often available from the authors on request, or supplied as supplementary data.[12]

Sanger sequencing, also referred to as dideoxy sequencing or chain termination, is based on the use of fluorescently labeled dideoxynucleoside triphosphates (ddNTPs), in addition to the normal deoxynucleoside triphosphates (dNTPs) found in DNA. Modern day automated sequencing employs an approach called "dye-terminator sequencing." Each of the chain terminator ddNTPs, is labeled with a different fluorescent dye, each of which emit light at different wavelengths. The four labeled ddNTPs (ddATP, ddGTP, ddCTP, or ddTTP) are added with the four normal dNTPs and the DNA polymerase. Before the DNA can be sequenced, it is heat denatured into single strands. Next a primer is annealed to one of the template strands. The primer is specifically constructed so that its 3′ end is located next to the DNA sequence of interest. Following rounds of template DNA extension from the bound primer (sequencing reaction), the resulting DNA fragments all start from the same primer site, but are terminated by the incorporation of the ddNTP at different sites along the template sequence, thus generating fragments of different sizes that can be separated by gel electrophoresis. Unincorporated dye terminators must be completely removed before the samples can be analyzed by electrophoresis. Excess dye terminators in sequencing reactions obscure data in the early part of the sequence, and can interfere with the base calling software. Hence, prior to loading the sequence reactions onto the automated sequencers, they are purified to remove excess ddNTPS. Automated DNA sequencers perform capillary electrophoresis for size separation and detection. They record the incorporation of the labeled ddNTPs as fluorescent peak trace chromatograms.

16.2 METHODOLOGY

The following describes the protocols employed when using the ABI 3130xl, 16 capillary and ABI3730, 48 capillary Genetic Analyzer instruments. Other machines and reagents are available for Sanger sequencing.

The ABI3730 and ABI3130 Genetic Analyzers are fully automated capillary sequencers. They are capable of determining the nucleotide sequence of any given DNA samples, and can also be used to estimate the sizes of DNA fragments that have been prepared with ABI dye-labeled reagents. The ABI 3130xl, 16 capillary Genetic Analyzer can run 2 × 96 well plates, while the ABI 3730, 48 capillary Genetic Analyzer can run up to 16 96-well plates. Both machines can utilize capillary arrays of different lengths depending on the samples processed. Each set of samples, containing labeled DNA fragments, is automatically denatured and then separated by capillary electrophoresis. The replaceable medium (polymer) is automatically replaced in the capillaries after each separation. Detection is facilitated by laser-induced fluorescence in four spectral channels. The four-channel raw data sets generated by each of the capillaries are processed automatically to produce high-quality base sequences or fragment lists after separation.

16.2.1 Preparing Polymerase Chain Reaction Amplicons for the ABI Genetic Analyzers

After a PCR product is amplified from a lyssavirus-positive sample, it needs to be purified and then sequenced using florescence-based terminator cycle sequencing. The following describes the protocol to be followed when using the ABI Big Dye Terminator ready reaction kit. The kit works by providing a ready mix that contains the enzyme, dNTPs, magnesium chloride, buffer and dye terminators (ddNTPs), all at the appropriate quantities. Purified amplicon DNA and one primer are added to the mix and cycled to obtain dye-terminated products, which can be precipitated and analyzed on the ABI genetic analyzers.

- Sequencing must be performed in both the reverse and forward direction (i.e. using 3′ and 5′ directed primers) so consensus (complementary) sequence can be obtained.
- Sequencing in each direction should be replicated, at least in duplicate, prior to publication of the sequence to reduce the likelihood of errors.
- If cloned PCR products are sequenced, ensure at least five independent clones are sequenced, and use the consensus sequence for publication to ensure minor variants are not used as the representative sequence.

16.2.2 Purifying the Polymerase Chain Reaction Product

After visualising the PCR product by agarose gel electrophoresis, the remaining PCR product (amplicon) should be purified prior to sequencing. A number of commercial kits are available for the purification of PCR products (e.g. QiaQuick PCR Purification kit; Qiagen). If the PCR product band appears weak on the gel, it can be loaded and extracted directly from the gel using a commercially available kit (e.g. MinElute gel extraction kit (Qiagen)), which will enable the product to be eluted in as little as 10 μL Elution Buffer, thereby increasing the concentration. For large-scale purification (96-well plates), automated purification may be preferable (e.g. Millipure Multiscreen HTS PCR 96-well purification system or Beckman Coulter Agencourt AMPure clean up system).

- Ensure fresh tips are used between samples to avoid cross-contamination.
- Elute purified PCR product in 10–50 μL Elution Buffer depending on amount of product and purification kit used.
- Quantify the DNA concentration using a spectrometer (e.g. the Nanodrop or Polarstar Galaxy plate reader).

16.2.3 Big Dye Terminator Cycle Sequencing Reactions

On ice, add the purified amplicon DNA (approximately 50–100 ng – usually 2–5 μL) to the appropriate wells on a 96-well plate. If plasmid DNA is being used, pre-heat the DNA at 96°C for 1 minute and keep on ice. If many primers are being used for sequencing (forward, reverse, etc.), make a mastermix without the primer and add primers to appropriate wells at this stage. See below as an example for calculating amount of DNA to be added.

- PCR products: (PCR product length/100) × 2)/DNA conc. (ng/μL).
- Plasmids: 100/DNA concentration (ng/μL).

Make a mastermix using the volumes shown below. The amount of water can be amended to allow for variation in the amount of DNA product added, ensuring final volume is 20 μL. The primer employed will depend on the template sequence to be analyzed (examples given in Tables 16.1 and 16.2).

Sequencing Mix volume added per reaction:

- DNA, 5.0 μL
- Molecular grade water, 6.0 μL
- Big Dye Sequencing mix, 4.0 μL
- 5× sequencing buffer, 4.0 μL
- Primer (3.2 pmol/μL), 1.0 μL
- Total volume, 20.0 μL

Examples of primers are shown in Table 16.1. See Table 16.2 for details of primers used for long-distance PCR to provide overlapping amplicons for European bat lyssaviruses (EBLV) genome sequencing.

Cover the plate with foil, sealing the lid and centrifuge briefly, to collect liquid at the bottom of the wells if necessary.

Transfer the plate to a PCR machine and run on the suitable ABI sequencing program. These can be modified to suit primers if necessary (see below).

Cycling conditions for Big Dye Sequencing reactions:

- 96°C, 10 sec ⎫
- 50°C, 5 sec ⎬ ×25
- 60°C, 4 mins ⎭
- 4°C, hold

16.2.4 Ethanol Precipitation of Sequencing Reactions

The purification of the sequencing reactions facilitates the removal of unincorporated labeled ddNTPs, which would interfere with base calling. Once the cycling is complete, remove the plate from the PCR machine and clean up the sequencing products. The Ethanol/EDTA/NaOAc clean-up method is recommended. While ethanol/EDTA can be used, the smallest PCR fragments may not be precipitated.

Note: Isopropanol precipitation is NOT recommended.

- Add 2 μL 125 mM EDTA to each well.
- Add 2 μL 3M Sodium Acetate to each well.
- Add 50 μL of 100% Ethanol to each well.

Table 16.1 Examples of Primers used for Sequencing PCR Amplicons

Gene	Forward Primer		Reverse Primer	
Partial Nucleoprotein 1st round	JW12	ATGTAACACCYCTACAATG	JW6UNI	CARTTVGCRCACATYTTRTG
Partial Nucleoprotein 2nd round	JW12	ATGTAACACCYCTACAATG	JW10UNI	GTCATYARWGTRTGRTGYTC
Glycoprotein	GP01 (GT1)	GGAAAGATGGTTCCTCAGG	GP02 (GT1)	AACTTGAGTTTGCATGCTCC
Plasmid pCR II	T7	TAATACGACTCACTATAGG	SP6	ATTTAGGAGACTATAGAA

Table 16.2 Oligonucleotides used in the Long-Distance PCR and RLM-RACE Strategies for EBLV-1 and -2 Sanger Genome Sequencing

Primer	Orientation	Position EBLV-1	Position EBLV-2	Sequence (5′-3′)	Used for	EBLV
L FOR 1	M	10264–10285 6	10228–10250	GTDGATCAAGAGGTVCGCCATGC	L LD primer	1/2
L REV 2	G	11694–11671	11658–11635	CTTGTAAATCAACCTKATCCAGTG	L LD primer	1/2
JW12 long SacI	M		44–78	GCAGAGAGCTCATGTAACACCTCTACAATGGATGC	LD PCR	2
GG2 long XbaI	G		4946–4913	CTAAGTCTAGAGGTATTTGGTCCATTAGACAAGC	LD PCR	2
GG2 FOR	M		4913–4935	GCTTGTCTAATGGACCAAATACC	LD PCR	2
L REV long dist	G		10626–10605	TGAGGCCATGAAGTCATTCACC	LD PCR	2
JW12 EBLV LD	M	52–77	52–77	GAAATGTAACACCCCTACAATGGATG	LD PCR	1/2
EBLV G REV LD	G	4584–4562	4596–4574	GTGGAWGGGTCAGCCAAGGGATG	LD PCR	1/2
EBLV G FOR LD	M	4562–4584	4574–4596	CATCCCTTGGCTGACCCWTCCAC	LD PCR	1/2
EBLV1 L REV LD	G	11221–11197		CTGGAAACCCTCTTACAAGGTCTTG	LD PCR	1
Gen end N E2 REV	G		255–232	GCATTTATTCCGGACAGAATGGAC	5′ end RLM-RACE	2
Gen end N E1 REV	G	254–232		CATTCATCCCTGACAAGATAGAC	5′ end RLM-RACE	1
EBLV-1 L FOR 16	M	10880–10901		CTCTGTTAATGTCAGACTTCTC	3′ end RACE RT	1
EBLV-2 L gen end For 4	M		11417–11438	GACACTTCAATATCTGCTGCAG	3′ end RACE RT	2
EBLV-1 Genomic end L For	M	11555–11576		CCTATTACTTCAAGAAGCAGAC	3′ end RACE PCR	1
EBLV-2 L gen end For 3	M		11532–11552	GAAGCAAGTCATACGAGGAAG	3′ end RACE PCR	2

- Seal plate, vortex briefly and leave for 15 mins at room temperature.
- Centrifuge the plate at 3,000 rpm for 45 mins.
- Invert plate over sink and shake 3 times to remove supernatant.
- Rinse each well with 70 µL 70% ethanol.
- Centrifuge the plate at 3,000 rpm for 15 mins.
- Invert plate over sink and shake 3 times to remove supernatant.
- Keeping the plate inverted, place it on a piece of paper towel and centrifuge upside down for 15–30 seconds at 300 rpm.
- Air dry for 15–10 mins.
- Seal and label the 96-well plate.

16.2.5 Preparing 96-Well Sample Plate(s) for the Genetic Analyzers

Spin samples down if necessary, remove sealing film.

16.2.6 Loading 96-Well Sample Plate(s)

The Genetic Analyzers have a safety feature, which means that no movement will occur inside the machine when the door is open. The machines will not work, and the buttons on the front of the machines will not function, unless the door is properly closed.

For the 3130xl: Press the button on the front of the machine labeled "TRAY." Once the tray has stopped moving, the door can be opened. Inside, the machine has two positions for holding plates. Position A is on the left, and Position B is on the right. The plate assembly will only fit in one way with Column 1 furthest away from the door due to a notch in the base plate of the plate assembly. Place the plate assembly into position. Close the door and wait to see the green light appear on the front of the machine, which indicates it is properly loaded.

For the 3730: Open the stacker drawer of the Genetic Analyzer and then open the In Stack door. Place the plate assemblies into the stacker with the plate(s) orientated so that the notched corner of the plate assembly is at the rear right corner of the stacker. Up to 16 plates can be placed in the "In Stack." Close the "In Stack" door and then the stacker draw.

Note: The In Stack is at the front of the stacker drawer with the "Out Stack" at the rear. Plates to be run are taken from the bottom of the stack and will then be at the top of the "Out Stack" once they have been run.

16.2.7 Checking Run Reagents

Check the polymer and buffers are within expiry dates.

Check there is sufficient volume of polymer in the bottle for the run. Check for bubbles in the tubes (and small tank above) leading from the POP-7. If bubbles are present, an error could occur and your run will stop. To check for bubbles, switch on the light and look inside. Remember to turn off light once checked.

16.2.8 Starting the Run on the 3130xl

In the Data Collection program on the PC, select from the left-hand list 3130xl/Run Scheduler in Plate View. Files which show "Processed" next to them are completed runs. Files which show "Pending" next to them are ready to be run. When you locate your file, it should be "pending."

You will see on the right of the screen two rectangles representing the two plate positions inside the machine. The color of the rectangles means:

- Grey – no plate(s) in holder.
- Yellow – unlinked plate(s) in holder.
- Green – linked plate(s) in holder.

Click once on the position that relates to your plate, Position A (left) or Position B (right). It should turn green. The Plate Record will now have A or B to the left of the name, indicating linkage to the plate position.

- To start your run, click on the green arrow at the top left hand corner of the window.
- A dialog box will appear telling you that you are about to start processing your plate. Click OK.
- Once the run is finished, check that the run has proceeded correctly and the analyzed sequence data is available.

16.2.9 Starting the Run on the 3730

In the Data collection program on the PC, select from the left-hand list 3730/Run Scheduler in Plate View. Find the Plate Record file for the plate to be run first and highlight the file.

Click on Add – This will add it to the list in the Input Stack dialog box with Status as Pending. Keep adding plate records in the same order as the order they will be run, until all those to be run are in the In Stack dialog box.

Plates can be added or removed during instrument operation.

Click on Done to close the Add Plates to In Stack dialog box. Note: The Plate Record at the bottom of the list in the In Stack dialog box is marked "1" and will be run first, the next one second, etc.

If the sequencer is running, nothing further needs to be done. If the sequencer is not running, follow these directions: To start your run, click on the green arrow at the top left hand corner of the window. A dialog box will appear, telling you that you are about to start processing your plate. Click OK.

Once the run is finished, check that the run has proceeded correctly and the analysed sequence data are available.

16.2.10 Analyzing Output Data

The raw data will be generated in different formats depending on the sequencer used, but in general you would expect to obtain a chromatogram or trace from which you can derive the sequence (.seq or .fas file). From an ABI sequencer you would expect ".ab1" or ".abi" file formats (Figure 16.1), while from a Beckman sequencer you may expect a ".scf" file format.

A number of software packages exist for analyzing the data and generating consensus sequences from the replicate forward and reverse outputs (e.g. Seqman in Lasergene). If the sequences are good quality, they will form a contiguous sequence (called a contig). The forward and reverse contiguous sequences are aligned to generate consensus sequences (Figure 16.1).

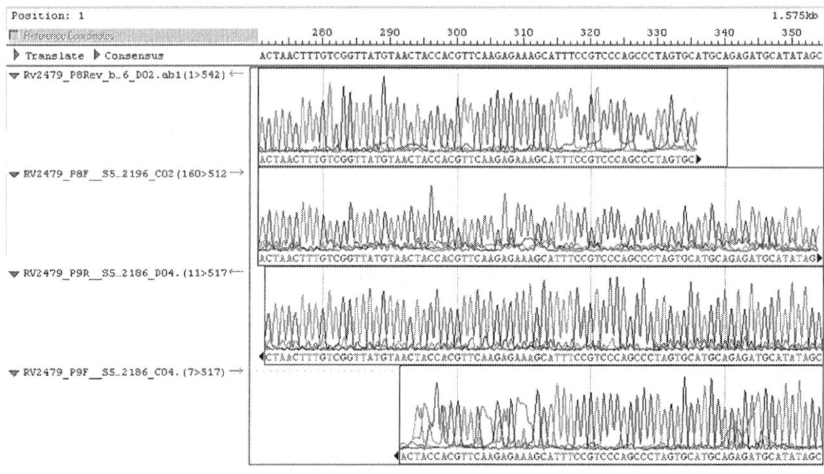

Figure 16.1 Sanger sequencing of overlapping long distance PCR amplicons (forward and reverse) facilitate genome sequencing via "Walking the Genome." The four ABI traces of four overlapping PCR products (RV2479, 1993 EBLV-2 *M. daubentonii* from Switzerland) yield contiguous sequences that are superimposed (aligned) to generate a consensus sequence (top frame) determined using the Seqman Programme (DNASTAR Lasergene 10).

If the consensus sequence saved is more than the required length, it can be trimmed to the correct length by opening in a sequence editing program (e.g. Seqman or Editseq in Lasergene 10 or MEGA[13]).

16.3 DISCUSSION

Sanger sequencing continues to be used widely in lyssavirus diagnostic laboratories for the confirmation of PCR positive material, determining the source of an outbreak, or understanding the molecular evolution of emerging viruses. Sequencing PCR products can now be achieved relatively inexpensively, with the average sequencing run costing less than £5 (~$8).

However, sequencing from PCR products may be error prone, in particular when PCR products are cloned to improve yield or study viral heterogeneity. Hence, the material from which the genome sequence is derived must be reported when publishing viral genomes. In addition to single nucleotide polymorphisms (SNPs), which may occur naturally following passage of the virus *in vitro* or *in vivo*, errors or biases can be introduced into the generated sequences during the PCR or cloning processes.[14] Multiple sequence anomalies have been demonstrated between institutes due to differences in sequencing approaches or sample handling.[15] Next-generation sequencing will highlight similar sequencing anomalies in published data, and has the potential to indicate the presence and prevalence of viral heterogeneity (quasi-species) in the original infected material. Viral sequences which represent only a minor proportion of the viral pool may be amplified during PCR, cloning, or passage, and therefore be misrepresented as the consensus sequence by Sanger sequencing. PCR sequence biases may be reduced by performing the replicate sequencing reactions from replicate PCR reactions rather than relying on a single PCR test. Where possible, viral sequences should be derived from the original host rather than using *in vitro* or *in vivo* passaged material.

All viral sequences should be published, particularly if included in a peer-reviewed journal (e.g. NCBI Nucleotide[16]). For future submissions, we strongly recommend using the most recent version of the standardized sequence submission software, Sequin, obtained from the NCBI website,[17] which will harmonize sequence submissions and encourage submitting laboratories to include all essential data, including collection date, virus species, host species, and country of isolation.[18]

ACKNOWLEDGMENTS

This work was funded by the Department for Environment, Food and Rural Affairs (Defra), UK (grant ROAME SE0427), and by the EU FP7-funded Research Infrastructure Grant European Virus Archive (No.19 228292).

REFERENCES

1. Sanger F, Nicklen S, Coulson AR. DNA sequencing with chain-terminating inhibitors. *Proc Natl Acad Sci U S A*. 1977;**74**(12):5463–7.
2. Sacramento D, Bourhy H, Tordo N. PCR technique as an alternative method for diagnosis and molecular epidemiology of rabies virus. *Mol Cell Probes* 1991;**5**(3):229–40.
3. Fooks AR, Johnson N, Freuling CM, Wakeley PR, Banyard AC, McElhinney LM, et al. Emerging technologies for the detection of rabies virus: challenges and hopes in the 21st century. *PLoS Negl Trop Dis* 2009;**3**(9):e530.
4. Johnson N, Letshwenyo M, Baipoledi EK, Thobokwe G, Fooks AR. Molecular epidemiology of rabies in Botswana: a comparison between antibody typing and nucleotide sequence phylogeny. *Vet Microbiol* 2004;**101**(1):31–8.
5. Smith JS, Orciari LA, Yager PA, Seidel HD, Warner CK. Epidemiologic and historical relationships among 87 rabies virus isolates as determined by limited sequence analysis. *J Infect Dis.* 1992;**166**(2):296–307.
6. Bourhy H, Kissi B, Tordo N. Molecular diversity of the Lyssavirus genus.. *Virology* 1993;**194**(1):70–81.
7. Johnson N, McElhinney LM, Smith J, Lowings P, Fooks AR. Phylogenetic comparison of the Lyssavirus genus glycoprotein and nucleoprotein coding sequences. *Arch Virol* 2002;**147**:2111–23.
8. McElhinney LM, Marston DA, Freuling CM, Cragg W, Stankov S, Lalosevic D, et al. Molecular diversity and evolutionary history of rabies virus strains circulating in the Balkans. *J Gen Virol.* 2011;**92**(Pt 9):2171–80.
9. Tordo N, Poch O, Ermine A, Keith G, Rougeon F. Walking along the rabies genome: is the large G-L intergenic region a remnant gene?. *Proc Natl Acad Sci U S A.* 1986;**83**(11):3914–8.
10. Tordo N, Poch O, Ermine A, Keith G, Rougeon F. Completion of the rabies virus genome sequence determination: highly conserved domains among the L (polymerase) proteins of unsegmented negative-strand RNA viruses. *Virology* 1988;**165**(2):565–76.
11. Marston DA, McElhinney LM, Johnson N, Müller T, Conzelmann KK, Tordo N, et al. Comparative analysis of the full genome sequence of European bat lyssavirus type 1 and type 2 with other lyssaviruses and evidence for a conserved transcription termination and polyadenylation motif in the G-L 3′ non-translated region. *J Gen Virol.* 2007;**88**(Pt 4):1302–14.
12. Delmas O, Holmes EC, Talbi C, Larrous F, Dacheux L, Bouchier C, et al. Genomic diversity and evolution of the lyssaviruses. *PLoS One* 2008;**3**(4):e2057.
13. MEGA. <http://www.megasoftware.net> [accessed 20.2.14].
14. Marston DA, McElhinney LM, Ellis RJ, Horton DL, Wise EL, Leech SL, et al. Next generation sequencing of viral RNA genomes. *BMC Genomics* 2013;**14**(1):444.
15. McElhinney LM, Marston DA, Leech S, Freuling CM, van der Poel WH, Echevarria J, et al. Molecular epidemiology of bat lyssaviruses in Europe. *Zoonoses Public Health* 2013;**60**(1):35–45.
16. NCBI, Nucleotide. <http://www.ncbi.nlm.nih.gov/nucleotide/> [accessed 20.2.14].
17. NCBI, Sequin. <http://www.ncbi.nlm.nih.gov/projects/Sequin/> [accessed 20.2.14].
18. McElhinney LM, Setzkorn C, Marston DA, Freuling CM, Müller T, Fooks AR. Lyssavirus sequence database. *WHO Rabies Bull Eur* 2012;**36**(2):9–11.

CHAPTER SEVENTEEN

Next Generation Sequencing of Lyssaviruses

Denise A. Marston[1], Lorraine M. McElhinney[1,2], Emma Wise[1], Richard J. Ellis[3], Conrad M. Freuling[4], Thomas F. Müller[4], and Anthony R. Fooks[1,2]

[1]Wildlife Zoonosis and Vector-Borne Disease Research Group, Animal Health and Veterinary Laboratories Agency, Addlestone, Surrey, United Kingdom
[2]Institute of Infection and Global Health, University of Liverpool, Liverpool, United Kingdom
[3]Central Sequencing Unit, Animal Health and Veterinary Laboratories Agency, Addlestone, Surrey, United Kingdom
[4]Friedrich-Loeffler-Institut, Federal Research Institute for Animal Health, Institute of Molecular Biology, Greifswald-Insel Riems, Germany

Chapter Contents

17.1 Background 171
17.2 Methodology 172
17.2.1 Viral RNA Extraction 172
17.2.1.1 Tissue Samples 173
17.2.1.2 Virus Culture Samples 173
17.2.2 Preparation of cDNA for Sequencing on the Roche 454 FLX(+) Titanium 174
17.2.2.1 Library Preparation 174
17.2.2.2 Emulsion Polymerase Chain Reaction 176
17.2.2.3 Sequencing on the Roche 454 FLX+ 177
17.2.3 Analyzing and Annotating 454 Genome Sequences 177
17.2.3.1 Obtaining a Consensus Sequence from a De Novo Project 177
17.2.3.2 Obtaining a Consensus Sequence from a Mapper Project 178
17.2.3.3 Checking Gene and Intergenic Lengths 179
17.2.3.4 Obtaining Genomic Termini 181
17.2.4 Depth of Coverage Analysis 181
17.3 Discussion 181
Acknowledgments 183
References 183

17.1 BACKGROUND

Lyssaviruses have small, simple genomes of approximately 12 Kb. Traditionally, this small size has facilitated the acquisition of a complete genome sequence via conventional Sanger sequencing of overlapping

C. Rupprecht & T. Nagarajan (Eds): Current Laboratory Techniques in Rabies Diagnosis, Research, and Prevention, Volume 1.
Doi: http://dx.doi.org/10.1016/B978-0-12-800014-4.00017-2

polymerase chain reaction (PCR) amplicons ("walking the genome"). However, this methodology is extremely time-consuming and expensive. Variation within and between the lyssaviruses species often demands considerable re-optimization of PCR primer sets, particularly for highly divergent or novel lyssaviruses. Moreover, PCR errors can be introduced during Sanger sequencing, often when amplicon cloning is undertaken or low-fidelity polymerases are employed.

Next Generation Sequencing (NGS) is a high-throughput sequencing methodology which generates millions of sequences simultaneously from one sample.[1] Multiple NGS platforms are available; two commonly used platforms are Roche 454 pyrosequencing (454 Life Science) and Illumina (Solexa).[2] These platforms were initially optimized and applied to large mammalian and bacterial genomes, where Sanger methods are prohibitively expensive. Obtaining the hundreds of nanograms of target nucleic acid required for NGS is relatively simple, as nucleic acid is in abundance. Due to the low percentage of viral RNA in relation to host RNA in any given sample, obtaining the quantity of viral RNA required by historical protocols has been problematic.

The advantages of NGS technologies in virology are numerous, and its use is becoming more commonplace,[3,4] particularly to detect and characterize pathogens without prior knowledge of their existence, and in the absence of specific primers.[5] Unfortunately, the virus is often at such low abundance in clinical samples that only one or two reads can be obtained, or at best, a number of unlinked contiguous sequences (contigs), for which traditional Sanger protocols are employed to complete the genome.[6]

In this chapter, we describe a broadly applicable approach to obtaining full genome sequences from clinical or cultured samples, which we have successfully applied to lyssavirus-infected samples.

17.2 METHODOLOGY

The following describes the protocols followed when using the Roche 454 FLX(+) Titanium. However, other platforms are also available.

17.2.1 Viral RNA Extraction

The proper preparation of the infected samples is the key to ensuring the success of obtaining viral genomes via NGS. Sample preparation protocols differ for the variety of tissue samples and viral culture samples employed due to their different properties, but the overall aim is to enrich for viral

RNA, either by depleting host genomic and ribosomal RNA, or by concentrating the culture samples prior to RNA extraction.

17.2.1.1 Tissue Samples

Extract RNA from the sample using TRIzol® (tissue) or TRIzol®LS (liquids) following manufacturer's instructions, and re-suspend in 10 μL molecular-grade water. Removal of cell debris in the sample using a stepwise filtration process involving 0.45 μm polyether sulfone membrane filters (diameter 13 mm) (Millipore) may be performed prior to RNA extraction if preferred, but is not essential if TRIzol or similar phase-separation methodologies are employed.

Make a 1 in 10 dilution of RNA with molecular grade water, and quantify by UV spectrophotometry using the Nanodrop ND-8000 or equivalent.

Perform depletion of genomic RNA and DNA using the RNeasy® plus mini kit (Qiagen), following manufacturer's instructions, eluting in 30 μL RNase-free water. DNA depletion is achieved by following the optional on-column DNase treatment step within the kit protocol.

Perform depletion of ribosomal RNA using Terminator™ 5′-Phosphate-Dependent Exonuclease (Epicentre Biotechnologies) according to the manufacturer's instructions, mixing the total volume (30 μL) of RNA with 3 μL of Buffer A, 0.5 μL RNAsin® Ribonuclease inhibitor (20–40 U/μL, Promega) and 1 μL Terminator (1 U/μL), then incubate at 30°C for 60 mins.

Cool samples on ice for 1 min, then perform one subsequent round of purification using RNeasy plus mini kit (Qiagen), without DNase digestion, following the manufacturer's instructions, and elute RNA in 30 μL RNase free water.

Quantify depleted RNA by UV spectrophotometry using the Nanodrop ND-8000 or equivalent.

RNA samples should be stored at −80°C until use.

Any depleted RNA left over after NGS should be retained and stored at −80°C for future use.

17.2.1.2 Virus Culture Samples

If the virus is at a low concentration, or if downstream procedures require it, an optional step of Polyeythylene glycol (PEG) precipitation may be performed to concentrate the virus prior to RNA extraction. Following manufacturer's instructions, use 1 volume of PEG-it (System Biosciences) to 4 volumes of sample. Re-suspend pellets in tissue culture medium and combine to achieve a final volume of 1 mL.

Extract RNA from the virus culture sample using TRIzol LS (following manufacturer's instructions). Make a 1 in 10 dilution of sample in molecular grade water, and quantify RNA as above.

17.2.2 Preparation of cDNA for Sequencing on the Roche 454 FLX(+) Titanium

The cDNA is generated from the enriched RNA using random or specific primers, and then sequencing primers are attached to each fragment. Individual fragments are then amplified clonally on beads prior to pyrosequencing.

17.2.2.1 Library Preparation

1. RNA sample quantitation is done by fluorometry using the Quant-iT™ RiboGreen® RNA Assay Kit (Life Technologies #R11490) according to the manufacturers instructions (1 μL sample diluted in 50 μL TE, add 50 μL 1:200 (v/v) RiboGreen reagent). Measure fluorescence using a fluorometer with a 480 nm excitation filter and 520 nm emission filter. Compare readings to standard curve.
2. Add up to 200 ng RNA to random primers as follows:
   a. 8 μL RNA sample ≤200 ng.
   b. 2 μL 400 μM Roche Primer "random" (Roche # 11034731001; alternatively, at this stage a cocktail of more specific primers could also be used to increase the relative abundance of the sequences of interest).
   c. Mix and heat to 70°C for 10 mins.
   d. Cool to 4°C.
3. Proceed to first strand synthesis using the Roche cDNA synthesis system (Roche # 11117831001). Add the following:
   a. 4 μL 5× RT-buffer AMV.
   b. 2 μL 0.1 M DTT.
   c. 2 μL 10 mM deoxynucleoside triphosphates (dNTPs).
   d. 0.5 μL Protector RNase Inhibitor (25 U/μL).
   e. 1 μL AMV reverse transcriptase (25 U/μL).
   f. Mix and incubate at 25°C for 10 mins and then at 42°C for 60 mins.
   g. Cool to 4°C.
4. Proceed immediately to second strand synthesis using the Roche cDNA synthesis system (Roche # 11117831001). Add the following:
   a. 15 μL 5× 2nd strand synthesis buffer.
   b. 36 μL Molecular Biology Grade Water.

   c. 1 μL 10 mM dNTPs.
   d. 3 μL 2nd strand enzyme.
   e. Mix and incubate at 16°C for 2 hrs.
   f. Add 10 μL T4 DNA polymerase.
   g. Mix and incubate at 16°C for a further 5 mins.
   h. Add 8.5 μL 0.2 M EDTA (pH 8).
5. Purify double-stranded cDNA:
   a. Add cDNA synthesis reaction to 150 μL Ampure XP beads (Beckman Coulter # A63880).
   b. Mix and incubate at room temperature for 5 mins.
   c. Place on a magnetic stand until solution clears, leaving beads against the side or bottom of the tube.
   d. Remove and discard supernatant without disturbing beads.
   e. Keeping the tube on the magnet, wash the beads 3 times with 500 μL 70% ethanol, removing all ethanol each time.
   f. Allow the beads to air dry for 5 mins.
   g. Remove the tube from the magnet and add 16 μL of 10 mM Tris-HCl, pH 7.5 and mix.
   h. Incubate at room temperature for 5 mins.
   i. Place on a magnetic stand until the solution clears, leaving beads against the side or bottom of the tube.
   j. Transfer supernatant containing cDNA to a new tube.
6. End-repair cDNA using a Roche Rapid Library Preparation Kit (Roche # 05608228001). Add the following to the cDNA:
   a. 2.5 μL RL 10× PNK Buffer.
   b. 2.5 μL RL ATP.
   c. 1 μL RL dNTP.
   d. 1 μL RL T4 Polymerase.
   e. 1 μL RL PNK.
   f. 1 μL RL Taq Polymerase.
   g. Mix and incubate at 25°C for 20 min, 72°C for 20 min followed by cooling to 4°C.
7. Ligate 454 sequencing primers using a Roche Rapid Library Preparation Kit (Roche # 05608228001), or using optional Multiplex identifiers (Roche # 05619211001). Add the following to the completed end-repair reaction:
   a. 1 μL of RL Adaptor or RL MID Adaptor.
   b. 1 μL of RL Ligase.
   c. Mix and incubate at 25°C for 10 mins.

8. Purify library fragments for sequencing as follows:
   a. Aliquot 125 μL Ampure XP beads into a 1.5 mL tube.
   b. Place on a magnetic stand until the solution clears, leaving the beads against the side or bottom of the tube.
   c. Remove and discard the supernatant without disturbing the beads.
   d. Add 26 μL TE and 140 μL sizing solution and mix.
   e. Add adapted library (~27 μL) from Step 7c.
   f. Mix and incubate at room temperature for 5 mins.
   g. Place on a magnetic stand until the solution clears, leaving the beads against the side or bottom of the tube.
   h. Remove and discard the supernatant without disturbing the beads.
   i. Add 53 μL TE and 140 μL sizing solution.
   j. Mix and incubate at room temperature for 5 mins.
   k. Place on a magnetic stand until the solution clears, leaving the beads against the side or bottom of the tube.
   l. Remove and discard the supernatant without disturbing the beads.
   m. Repeat Steps 8i to 8l.
   n. Keeping the tube on the magnet, wash the beads 3 times with 500 μL 70% ethanol, removing all ethanol each time.
   o. Allow the beads to air dry for 5 mins.
   p. Remove the tube from the magnet, add 22 μL of TE, and mix.
   q. Incubate at room temperature for 5 mins.
   r. Place on a magnetic stand until the solution clears, leaving the beads against the side or bottom of the tube.
   s. Transfer the supernatant containing the purified library to a new tube.
9. Assess library quality and concentration as follows:
   a. Run 1 μL aliquot on an Agilent Bioanalyzer High Sensitivity DNA chip.
   b. Based on the average fragment size and concentration determine the number of DNA molecules per μL, using the following formula: (Concentration [pg/μL] / Average size [bp]) $\times 9.17 \times 10^8$.
   c. Dilute libraries to 1×10^7 molecules per μL.
10. Pool up to 12 libraries (each with a different MID adapter added at Stage 7a) for efficient multiplexed sequencing.

17.2.2.2 Emulsion Polymerase Chain Reaction

DNA library fragments are amplified clonally on beads using Lib-L emPCR kits (Roche #05619149001). The DNA is mixed with the beads, which are then suspended in an oil emulsion, prior to amplification.

Typically, mixing three molecules of DNA per bead provides a suitable level of enrichment (maximizing the number of beads carrying amplified DNA, while reducing the possibility of having mixed molecules of DNA attached to the same bead), and leads to optimal sequence output. Manufacturers' instructions are followed for emPCR and bead recovery.

17.2.2.3 Sequencing on the Roche 454 FLX+

Recovered beads are deposited in PicoTiter plates and sequenced on the Roche 454 FLX+ instrument according to the manufacturers' instructions. Reads of up to 900 bp can be achieved using this approach. The instrument runs for approximately 24 hours, including the initial data processing. Reads are assembled using the GS Assembler (Newbler) software, using standard settings. This performs *de novo* assembly, and often this approach leads to a single viral consensus contig. Using the methods given here, viral sequencing reads account for between 0.5 and 10% of the output. Where a very closely related virus is available, it can be used as a reference to which sequence reads can be mapped with GS Mapper.

When the consensus contig sequence is used in GS Reference Mapper (Roche), further sequence reads may be obtained from the original raw data. If a suitable reference sequence is available, this is used in the GS Reference Mapper program with the 454 reads to obtain a consensus sequence without the initial use of the GS *de novo* assembly software. If no genome reference sequence is available, a BLAST search, with the individual contig sequences from the *de novo* assembly will indicate which contigs are viral sequences (use the megablast algorithm on the nucleotide database[7]). Due to the increased difficulty of mapping reads at the ends of contigs, it may be necessary occasionally to splice the 3′ and/or 5′ UTR sequences from a reference genome sequence onto the termini of the contig sequence of interest. Using the spliced consensus sequence in the GS Reference Mapper program may result in extra reads being mapped to the contig, resulting in a complete, or almost complete, genome sequence.

17.2.3 Analyzing and Annotating 454 Genome Sequences

The following protocols are undertaken using the software provided by Roche with the 454 sequencer and sequence editing programs provided in DNAStar package, Lasergene 10. Other editing programs are available.

17.2.3.1 Obtaining a Consensus Sequence from a De Novo Project

Open the *de novo* assembled project, click on "project" then "GS reads." This gives you the sff file ID number, the multiplex ID number, and an indication of how successful the sequencing was accomplished (percentage of reads assembled should be 70% or higher).

Click on "alignment results" and look at the contig sequences listed on the left. If the top contig is approximately 11–12kb in length (if analyzing a lyssavirus genome), this will be a single contig containing the virus genome. If the contigs are smaller, the viral reads may still be present, but in a number of smaller contigs.

Rename "454AllContigs.fna" file with the virus ID at the beginning.

Open SeqMan (DNAStar, Lasergene 10) and drag the .fna file into the "unassembled sequences" box to open the contig files. In "options," uncheck the "don't add single sequences" option, then click "assemble."

If the viral sequences are all in a single contig (always named contig00001), then simply rename contig0001 from the *de novo* folder and save the consensus sequence. If the viral sequences are in a number of contigs, add an appropriate reference genome sequence to bring them together into a single contig.

At this point you can enter the sequence into BLAST, if desired, to identify similar sequences.

The majority of sequences do not have the genomic ends covered by reads, and occasionally there are gaps in between contigs. To complete the genome, either re-sequence the sample on the NGS platform, to cover the regions without reads, or use a PCR-based strategy to amplify across the gaps, and use the Sanger sequencing technique to obtain the sequence.

17.2.3.2 Obtaining a Consensus Sequence from a Mapper Project

If a closely related sequence is available, this can be used instead of *de novo* analysis. Open the Mapper program, create a new mapper project, and call it "ID…mapped".

In "project," click on "reference" then on the "+" icon. Navigate to a relevant consensus sequence saved as a fasta file. Click on "GS reads," "+," and locate the sff folder relevant for your virus. Between one and four sff files will be available; use shift to highlight these and click "select." Check the "use multiplex filtering" box, click on the arrow next to "scheme, no multiplexing" and choose "RLMIDs." Uncheck all boxes except the RL used for the virus of interest. Click "OK."

Click the "Start" button, located far right. Analysis should take less than 1 min. When complete, the "alignment results" tab will be bold.

Click on "alignment results," then click on "contig0001." The top sequence is the imported reference sequence, and below will be the aligned NGS reads.

Scroll through the alignment to get an idea of the depth of coverage and if there are any gaps in the sequence (this will be obvious as there will be more than just the one contig00001 on the left). The project overview indicates the sff file ID, the multiplex ID, the percentage of viral specific reads, and the total number of viral specific reads.

Open the .fna file in EditSeq.

The above steps ensure optimal mapping has been carried out, and should be repeated any time the consensus sequence is modified (e.g. when genomic ends have been obtained by conventional PCR and sequencing and are incorporated into the genome), as reads may exist in the raw reads, but are not mapped. Be aware that the resulting contig will be the consensus sequence of the reads, therefore any corrections made may be lost on a subsequent mapped contig file and will need to be redone.

Save this reference sequence as a .fas file.

17.2.3.3 Checking Gene and Intergenic Lengths

Print out:

1. The consensus sequence in EditSeq to annotate.
2. If the lyssavirus is a known species, perform an alignment of the consensus sequence with other genomes from the same species. Annotate the start and stop regions of both using annotation from the Genbank file of the known lyssavirus genome.

Using the print outs from above, annotate the Editseq file with start and stop codons. Highlight each region in Editseq and note the lengths. Gene lengths MUST be divisible by 3. If the gene length differs from other viruses of same species, then:

1. Look at the genome alignment (or individual gene alignment) and identify indels (indicated by dashes).
2. In EditSeq, copy the coding sequence and paste it into a new DNA file. Select all and translate, and find where stop codons are introduced to determine where the problem is.

If intergenic regions differ, first check that the termination signal is GAAAAAAA. Often this is only 5 or 6 As because NGS platforms cannot read polymers easily. Look at raw data reads and identify number of 7 As, and look at flowgrams. Otherwise, look at the genome alignment to determine indels. It is common to have different length intergenic regions (Table 17.1).

Table 17.1 Lengths of Coding and Non-Coding Regions of Lyssavirus Genomes

	RABV	LBV	MOKV	DUVV	EBLV-1	EBLV-2	ABLV	ARAV	KHUV	IRKV	WCBV	SHIBV	BBLV	IKOV
3′ UTR*	70	70	70	70	70	70	70	70	70	70	70	70	70	70
N protein	**1353**	**1353**	**1353**	**1356**	**1356**	**1356**	**1353**	**1356**	**1356**	**1356**	**1353**	**1353**	**1356**	**1353**
N–P	90-1	101	100-102	90	90	101	94	85	95	93	64	98	91	66
P protein	**894**	**918**	**912**	**897**	**897**	**894**	**894**	**894**	**894**	**897**	**894**	**918**	**894**	**870**
P–M	88	75	80	83	83	88	89	85	72	82	133	76	86	74
M protein	**609**	**609**	**609**	**609**	**609**	**609**	**609**	**609**	**609**	**609**	**609**	**609**	**609**	**609**
M–G	211-5	204	203-204	191	211	210 (205)	207-209	210	208	214	206	205	210	209
G protein	**1575**	**1569**	**1569**	**1602**	**1575**	**1575**	**1578-81**	**1581**	**1581**	**1575**	**1578**	**1569**	**1575**	**1575**
G–L	522	578-588	546-563	562-3	560	512	508-509	514	504	569	862	613	496	569
L protein	**6384**	**6384**	**6384**	**6384**	**6384**	**6384**	**6384**	**6384**	**6384**	**6384**	**6384**	**6384**	**6384**	**6381**
5′ UTR	131	145	112–114	130–131	131	131	131	130	130	131	125	150	129	126
Genome	**11 923-8**	**12006-16**	**11 940-57**	**11 975-6**	**11 966**	**11 930**	**11 918**	**11918**	**11903**	**11980**	**12278**	**12045**	**11900**	**11902**

Full genomes of BBLV (Bokeloh Bat Lyssavirus, JF311903), RABV (Rabies Virus, M31046; EU293111; EU293115; EU293113; EU293116), DUVV (Duvenhage Virus, EU293120; EU293119), EBLV-1 (European Bat Lyssavirus type-1, EU293109, EU293112, EF157976), IKOV (Ikoma Lyssavirus, JX193798), ABLV (Australian Bat lyssavirus, NC_003243; AF081020), KHUV (Khujand Virus, EF614261), IRKV (Irkut Virus, FJ905105, EF614260), EBLV-2 (European Bat Lyssavirus type-2, EF157977, EU293114), ARAV (Aravan Virus, EF614259), LBV (Lagos Bat Virus, EU293108; EU293110), MOKV (Mokola Virus, EU293118, EU293117), SHIBV (Shimoni Bat Lyssavirus, GU170201) and WCBV (West Caucasian Bat Virus, EF614258) were derived from NCBI Genbank.

*6387 in SHBRV.

Check any positions queried by lower case letters in the GS program. Edit the EditSeq file according to your findings and save.

17.2.3.4 Obtaining Genomic Termini

The lyssavirus genomic termini are extremely conserved; in particular the first and last 9 nucleotides have complete complementarity across all lyssaviruses,[8] furthermore the conservation remains stringent until residue 25.[9] Pan-lyssavirus primers situated at the leader and trailer extremities, leader: LYS001F (5′ACGCTTAACGAMAAA3′) and trailer: LYSEND (5′ACGCTTAACAAAWAAA3′)[9] can be used with panlyssavirus primers JW6UNI (5′CARTTVGCRCACATYTTRT3′) and LgeneFor (5′CTCACTGGATMAGRTTRATITACAA3′), respectively, to obtain PCR products which can be sequenced directly using Sanger sequencing.[10] Use the remaining depleted RNA as the template.

17.2.4 Depth of Coverage Analysis

The depth of viral read coverage from which a single genome contig is obtained can vary depending on the starting material and sample preparation employed. Tissue culture supernatant prepared samples achieved the highest read coverage and depth of coverage, as the sample is amplified virus. However, culturing virus is not always possible. Furthermore, culturing virus in mammalian cells will result in adaptation of the virus resulting in sequence modification.

Sequences generated on the 454 platform, due to the way the incorporation of bases are detected by the intensity of light emission, inherently have issues with long homopolymeric repeats. On the whole, a consensus sequence with 10 or more reads, when mapped using an available genome from the same species, can be confirmed confidently by checking the flowgram data. However, novel viruses, or viruses with a limited number of reads, may require additional confirmation by Sanger sequencing PCR products, which span the regions in question.

17.3 DISCUSSION

The methodology described above enables the sequencing from brain tissue, and cell cultured samples without the use of specific viral enhancement, ultracentrifugation or filtration to identify novel viruses, or obtain genome sequences for known virus species. This methodology does not require prior knowledge of the virus being extracted, and uses a methodology that is widely used to extract viral RNA (so archival RNA samples can be investigated retrospectively). Instead of enriching for virus-specific RNA, host genomic DNA and rRNA is depleted to increase the percentage of viral specific RNA in the sample.

There are a large number of RNA extraction methodologies and kits available that can be divided broadly into two categories: (1) Phenol/chloroform phase separation-based extraction, and (2) Column or bead based extraction. We have compared TRIzol as a phenol-based system and RNeasy plus mini kit (Qiagen) as a column-based system. In our hands, the use of spin columns to extract the RNA from tissue removed the majority of RNA in process, resulting in less than the 200 ng minimum quantity of total RNA recommended for fragmentation and, more importantly, too few viral reads were obtained to provide useful contigs. The RNeasy plus mini kit (Qiagen) was used successfully as part of the depletion protocol after initial extraction with TRIzol with good results.[4] Complete genome sequence can be obtained by NGS even when less than the recommended 200 ng of RNA is available for library preparation in TRIzol-extracted tissue samples. These observations are important to illustrate that the methods for library preparation can be refined further for virus sequencing.

The requirement to deplete tissue culture supernatant-derived RNA was investigated using two 100 mL supernatant samples.[4] Half the RNA was subjected to gDNA and rRNA depletion, whereas the other was not. The additional preparation time required to deplete the sample was not rewarded with a significant improvement of either proportion of viral-specific reads or read depth. Therefore, 454 sequencing directly from a TRIzol-extracted RNA is the simplest and most effective way to obtain a viral genome sequence from PEG-precipitated tissue culture supernatant samples.

In addition to genome sequencing, NGS technologies can be applied to "Deep Sequence" clinical material for a plethora of uses, including determining viral heterogeneity, or the effect of the immune system or pharmaceutical drugs on viruses.[11] These deep sequencing methodologies employ the use of viral PCR amplicons to obtain a depth of coverage sufficient to detect variants which occur down to 0.1% frequency, and may be applied to lyssavirus-infected material to gain a greater understanding of the role that viral heterogeneity ("quasispecies") plays in virus adaptation and evolution.

The platform described in this chapter (Roche 454) has been shown to be reliable and consistent. However, the RNA prepared using the methodologies described here can be used for any NGS platform and have been used successfully on the Illumina platform. The application of NGS sequencing of viruses has begun a new era of virus discovery and characterization of novel viruses, and will revolutionize phylogeographic studies in all fields of virology in the coming years.

All viral sequences should be published, particularly if included in a peer-reviewed journal (e.g. NCBI Nucleotide[12]). For future submissions, we strongly recommend using the most recent version of the standardized sequence submission software, Sequin, obtained from the NCBI website,[13] which will harmonize sequence submissions and encourage submitting laboratories to include all essential data, including collection date, virus species, host species, and country of isolation.

ACKNOWLEDGMENTS

This work was funded by the Department for Environment, Food and Rural Affairs (Defra), UK (grant ROAME SE0427), and by the EU FP7-funded Research Infrastructure Grant European Virus Archive (No. 19 228292).

REFERENCES

1. Schuster SC. Next-generation sequencing transforms today's biology. *Nat Methods* 2008;**5**(1):16–18.
2. Shendure J, Ji H. Next-generation DNA sequencing. *Nat Biotechnol* 2008;**26**(10):1135–45.
3. Radford AD, Chapman D, Dixon L, Chantrey J, Darby AC, Hall N. Application of next-generation sequencing technologies in virology. *J Gen Virol* 2012;**93**(Pt 9):1853–68.
4. Marston DA, McElhinney LM, Ellis RJ, Horton DL, Wise EL, Leech SL, et al. Next Generation Sequencing of viral RNA genomes. *BMC Genomics* 2013;**14**(1):444.
5. Tse H, Tsang AK, Tsoi HW, Leung AS, Ho CC, Lau SK, et al. Identification of a novel bat papillomavirus by metagenomics. *PLoS One* 2012;**7**(8):e43986.
6. Ge X, Li Y, Yang X, Zhang H, Zhou P, Zhang Y, Shi Z. Metagenomic analysis of viruses from bat fecal samples reveals many novel viruses in insectivorous bats in China. *J Virol* 2012;**86**(8):4620–30.
7. Basic Local Assignment Search Tool (BLAST): <http://www.ncbi.nlm.nih.gov/blast/Blast.cgi?CMD=Web&PAGE_TYPE=BlastHome> [accessed 23.02.14].
8. Marston DA, Ellis RJ, Horton DL. Kuzmin IV, Wise EL, McElhinney LM, et al. Complete genome sequence of Ikoma lyssavirus. *J Virol* 2012;**86**(18):10242–3.
9. Kuzmin IV Wu X, Tordo N, Rupprecht CE. Complete genomes of Aravan, Khujand, Irkut and West Caucasian bat viruses, with special attention to the polymerase gene and non-coding regions. *Virus Res* 2008;**136**(1–2):81–90.
10. Marston DA, McElhinney LM, Ali YH, Intisar KS, Ho SM, Freuling C, et al. Phylogenetic analysis of rabies viruses from Sudan provides evidence of a viral clade with a unique molecular signature. *Virus Res* 2009;**145**(2):244–50.
11. Henn MR, Boutwell CL, Charlebois P, Lennon NJ, Power KA, Macalalad AR, et al. Whole genome deep sequencing of HIV-1 reveals the impact of early minor variants upon immune recognition during acute infection. *PLoS Pathog* 2012;**8**(3):e1002529.
12. NCBI, Nucleotide: <http://www.ncbi.nlm.nih.gov/nucleotide/> [accessed 23.02.14].
13. NCBI, Sequin: <www.ncbi.nlm.nih.gov/projects/Sequin/> [accessed 23.02.14].

CHAPTER EIGHTEEN

Genetic Characterization of Rabies Viruses by *In Situ* Hybridization

Susan A. Nadin-Davis
Centre of Expertise for Rabies, Canadian Food Inspection Agency, Ottawa Laboratory Fallowfield, Ottawa, Canada

Chapter Contents

18.1 Introduction 185
18.2 Materials 186
18.3 Methods 187
18.3.1 Generation of Control Tissues 187
18.3.2 Section Preparation 188
18.3.3 Production and Evaluation of Probes 188
18.3.4 *In Situ* Hybridization 188
18.3.5 Interpretation of Results 189
18.4 Discussion 189
18.4.1 Experimental Tips 189
18.4.2 Alternative Materials and Methods 189
18.4.3 Future Considerations 190
Acknowledgments 191
References 191

18.1 INTRODUCTION

Identification of the rabies virus (RABV) variant responsible for a diagnosed case can be achieved by a variety of methods when applied to infected fresh brain tissue as reviewed elsewhere.[1] Traditional antigenic methods of variant identification, in which panels of monoclonal antibodies (MAbs) that bind differentially to distinct epitopes on viral proteins, continue to be useful in routine diagnostic settings, while methods identifying differences in the nucleotide sequences of viral genes have become a powerful tool, particularly in the research setting. The latter most frequently employs products of a reverse transcription-polymerase

C. Rupprecht & T. Nagarajan (Eds): Current Laboratory Techniques in Rabies Diagnosis, Research, and Prevention, Volume 1.
Doi: http://dx.doi.org/10.1016/B978-0-12-800014-4.00018-4

chain reaction (RT-PCR) for direct sequence determination, followed by comparison of this sequence to a series of reference sequences.[1] These methods are, however, often unsuccessful when applied to formalin-fixed tissues. The fixation process alters the conformational epitopes required for binding by many antibodies, while the damage caused to RNA molecules during fixation and macromolecular cross-linking reduces amplification efficiency significantly. If RT-PCR products can be produced at all, they are usually small in size (150–300 bp), and can therefore yield only limited sequence information. While amplification of multiple overlapping gene regions followed by sequencing could be entertained, such a viral typing process is extremely laborious.[2]

An alternative approach to variant typing of RABVs present in formalin-fixed tissues that overcomes some of these limitations employs multiple probes, each specific for a particular viral variant, for *in situ* hybridization (ISH). This method successfully discriminates RABV variants that circulate across Canada.[3] The strategy requires probes of sufficient length and sequence variability such that they bind preferentially to RNA corresponding to particular viral variants, but are sufficiently small so that target sequence is accessible. Based on studies illustrating the limited sequence conservation of certain regions of the lyssavirus P gene,[4] probes targeting these highly variable coding regions were shown to be suitable for variant discrimination. In the protocol described here, digoxigenin (DIG)-labeled single-stranded RNA probes are generated by transcription from a collection of plasmids, each containing an insert of viral variant-specific P gene sequence flanked by SP6 and T7 RNA polymerase promoter sequences. Each probe is tested separately for binding to a section of a brain tissue sample. The probe type that results in pronounced colorimetric signal after detection using anti-DIG conjugate identifies the variant. An internal control for this assay employs a probe targeting conserved N gene sequence that detects a wide range of RABVs as described in Chapter 9. Generation of signal using this probe ensures that conditions appropriate for viral RNA detection are employed.

18.2 MATERIALS

1. Reagents
   a. Refer to Section 9.2.1 of Chapter 9.
   b. For digestion of plasmid DNAs various restriction endonucleases are used, as detailed in Table 18.1.

Table 18.1 Description of Selected Clones used to Produce Strain-Specific Probes for ISH

Clone Designation[a]	Restriction Site used for Pre-Transcription Cleavage	Length of RABV Sequence in Probe[b]	Probe Coordinates[c]
pGEM3-RVP-AFX	Bsu36I	244	1661–1904
pGEM3-RVP-RAC	EcoRI (MCS)[d]	248	1902–2149
pGEM3-RVP-WSK	Bsu36I	557	1904–2460
pGEM3-RVP-ERA	HindIII (MCS)	375	1593–1967
pGEM3-RVP-BBCAN1	MunI	277	1682–1958
pGEM3-RVP-BBCAN2	EaeI	344	1719–2062
pGEM3-RVP-BBCAN3	HaeII	357	1706–2062
pGEM3-RVP-LACAN (H)	EaeI	344	1719–2062
pGEM3-RVP-LACAN (R)	EaeI	256	1649–1904
pGEM3-RVP-SHCAN	MunI	381	1682–2062
pGEM3-RVP-MYCAN (E)	EaeI	345	1718–2062
pGEM3-RVP- MYCAN (W)	EaeI	186	1719–1904

[a]All clones were generated in the vector pGEM3zf(+) and contained a restricted fragment of the P gene. The designation indicates the viral variant from which the construct was generated. For full details on clone generation and information on additional constructs refer to Nadin-Davis et al. (2003).[3]
[b]Actual length of probe may be slightly larger due to inclusion of short segments of vector sequence.
[c]For comparative purposes, coordinates refer to the corresponding base positions of the reference PV strain, GenBank accession # M13215.
[d]MCS, site in the plasmid multiple cloning region. All other sites occur within the RABV P gene sequence.

2. Equipment: Refer to Section 9.2.2 of Chapter 9.
3. Biological materials: Plasmid DNA of RABV constructs as detailed in Table 18.1.
4. Laboratory animals: Mice (for generation of positive and negative control brain tissue sections).

18.3 METHODS

18.3.1 Generation of Control Tissues

To generate positive control material, mice are inoculated intra-cerebrally with 10% suspensions of RABV-infected brain tissue. Each RABV to be included in the probe panel should be propagated in this manner. Once an animal displays clinical signs it is euthanized, the brain is removed, and divided into two halves. One half is reserved for confirmation of RABV infection using a standard method (e.g. direct fluorescent antibody test, DFAT) and the other half is fixed in 3% paraformaldehyde at room temperature for up to 48 hrs. The tissue is then processed and paraffin-embedded at a temperature not exceeding 60°C. Brains from un-inoculated mice are similarly treated to generate material for use as negative controls.

18.3.2 Section Preparation

Refer to Section 9.3.2 of Chapter 9.

18.3.3 Production and Evaluation of Probes

Each RNA probe is generated by transcription from one of the plasmids described in Table 18.1 using SP6 RNA polymerase, except for the ERA strain clone for which T7 RNA polymerase is used because the insert of that construct is in the opposite orientation.

1. Digest 10 μg plasmid DNA with the appropriate restriction enzyme. An aliquot can be electrophoresed through an agarose gel to confirm complete digestion.
2. Add 1 μL proteinase K (20 μg/mL) and continue incubation for 15 min to remove any nuclease activity. Add 1 volume phenol–chloroform, mix periodically over 2–3 minutes, and spin at 12,000 *g* for 3 minutes. Recover the aqueous phase.
3. Precipitate the restricted DNA by addition of 0.1 volume 3 M sodium acetate and 2 volumes ethanol. Incubate at −70°C for 30 minutes or −20°C overnight. Spin at 12,000 *g* for 15 minutes, wash the pellet with 70% ethanol, and dry. Dissolve in sterile water to a final concentration of 1 μg/μL and store at −20°C.
4. Set up a transcription reaction using the DIG Northern starter kit as described in Step 4 in Section 9.3.3 of Chapter 9.
5. Evaluate labeling efficiency for each probe as described in Step 5 in Section 9.3.3 of Chapter 9.

18.3.4 *In Situ* Hybridization

The ISH procedure proceeds in three main steps: slide preparation, probe hybridization, and probe detection, as described in Section 9.3.4 of Chapter 9. For each tissue to be employed for viral typing, multiple sections are required to allow separate hybridization of each strain-specific probe with one section. An additional section is hybridized with the N gene probe to confirm a positive result. Positive and negative control tissue sections for each probe should also be performed, and consequently the total number of sections required for a single run can become considerable depending on how many variants are being investigated.

18.3.5 Interpretation of Results

Each run of the assay must be evaluated independently. Negative controls must present minimal staining of all tissue types. Positive controls (N and P gene probes) must exhibit purple stained intracytoplasmic inclusions, varying in size from fine particles to large inclusion bodies, in the perikarya of a proportion of all neurons, while staining of all other tissue elements should be minimal. For each RABV-positive test tissue, the N gene probe must generate positive staining while only one P gene probe should produce significant staining and thereby identify the viral variant responsible for infection. There can be significant variations in the extent of staining depending on the viral load within the section. Positive sections may be scored as 1+ to 3+ depending on staining intensity, similar to the scoring of results that are obtained using DFAT.

18.4 DISCUSSION

18.4.1 Experimental Tips

The use of the N gene probe in this procedure ensures that the conditions employed for ISH are appropriate for RABV RNA detection in the test sample. As indicated in Chapter 9, the biggest challenge to successful performance of ISH is ensuring probe access to its target sequence by optimization of the proteinase K treatment of the tissue. Although the conditions described here can be used as a preliminary guide, some modification of these parameters, including altering enzyme concentration and incubation time, may be needed to optimize results, and this has to be investigated empirically with individual samples. Most of the probes utilized in this protocol vary in length between 200–400 bases, a length that has been found to yield optimal results in this laboratory.

Refer to Chapter 9 for additional information on limitations, time lines, and further precautions to be taken when applying this method.

18.4.2 Alternative Materials and Methods

In this procedure, variant-specific probes were used to identify and discriminate between RABVs that circulate in Canada and the laboratory-adapted ERA strain used for oral vaccination. Application of this approach to samples collected in other parts of the world is feasible, but requires generation of probes suitable for identifying RABVs or other lyssaviruses that circulate within the targeted geographical zone. In Europe, multiple lyssaviruses have been documented, but the most frequently encountered include RABV and two European bat lyssaviruses, EBLV-1 and EBLV-2. Riboprobes generated from N gene clones of these viruses have been reported to discriminate between these three viral species when employed for ISH, using methods similar to those described here.[5] Further application of this approach to typing of viruses that circulate in other regions could be entertained. However, generating clone collections for each set of RABVs of interest could be time consuming. An alternative means of producing probes employs RT-PCR products generated with primers that incorporate the sequences for SP6 and/or T7 RNA polymerases.[6] Purified amplicons can then be used directly for transcription and have the advantage of allowing very specific regions of the genome to be targeted without the need to identify suitable restriction sites for cloning and transcription termination.

Critical to the success of this methodology is the identification of genome regions that exhibit the appropriate genetic diversity required for differentiating the viruses of interest. While N gene probes allow distinction between lyssavirus species,[5] more divergent areas of the genome must be targeted for finer resolution of lyssaviruses at the intra-species level. An in-depth comparison of the lyssavirus P gene sequence identified a highly modular structure[4] that reflects the P protein's multi-functional properties. In particular, conserved amino acid (aa) sequences, identified as conserved domains CD1 (aa 1–150) and CD2 (aa 201–245) were separated by a more divergent central region with especially high variation noted in two variable domains, VD1 (aa 61–80) and VD2 (aa 134–180). These latter two domains, which represent some of the least conserved coding regions of the entire viral genome, are encoded by residues spanning 1696–2053 of the PV reference strain. As illustrated in Table 18.1, the variant-specific probes employed generally target this divergent region of the genome.

18.4.3 Future Considerations

The application of *in situ* PCR followed by detection of amplicon using a DIG-labeled probe for detection of RABV in formalin-fixed tissues has been reported.[7] If reproducible, this approach could be of value for viral typing applications in the future by use of variant-specific primers and/or probes.

ACKNOWLEDGMENTS

I thank M. Sheen and Drs. G. Mitchell and C. Fehlner-Gardiner for critical review of this work.

Additional Information

Refer to the Additional Information section of Chapter 9 for details on the composition of buffers and solutions and examples of major supplies and equipment required.

Disclaimer: While sources of some materials are suggested, this does not constitute official endorsement of specific products.

REFERENCES

1. Nadin-Davis SA. Molecular Epidemiology. In: Jackson AC, editor. *Rabies* 3rd ed.. Oxford: Academic Press; 2013. p. 123–77.
2. Wacharapluesadee S, Ruangvejvorachai P, Hemachudha T. A simple method for detection of rabies viral sequences in 16-year old archival brain specimens with one-week fixation in formalin. *J Virol Methods* 2006;**134**:267–71.
3. Nadin-Davis SA, Sheen M, Wandeler AI. Use of discriminatory probes for strain typing of formalin-fixed, rabies virus-infected tissues by in situ hybridization. *J Clin Microbiol* 2003;**41**:4343–52.
4. Nadin-Davis SA, Abdel-Malik M, Armstrong J, Wandeler AI. Lyssavirus P gene characterisation provides insights into the phylogeny of the genus and identifies structural similarities and diversity within the encoded phosphoprotein. *Virology* 2002;**298**:286–305.
5. Finnegan CJ, Brookes SM, Johnson L, Fooks AR. Detection and strain differentiation of European bat lyssaviruses using in situ hybridisation. *J Virol Methods* 2004;**121**:223–9.
6. Warner CK, Whitfield SG, Fekadu M, Ho H. Procedures for the reproducible detection of rabies virus antigen mRNA and genome in situ in formalin-fixed tissues. *J Virol Methods* 1997;**67**:5–12.
7. Praveena PE, Jayakumar R, Balachandran C, Thirumurugan G, Raj GD, Manohar BM. Detection of rabies virus genes by in-situ polymerase chain reaction. *Vet Res Communs* 2007;**31**:775–81.

CHAPTER NINETEEN

Oligonucleotide Microarray: Applications for Lyssavirus Speciation

Akbar Dastjerdi[1], Anthony R. Fooks[2], and Nicholas Johnson[1]
[1]Animal Health and Veterinary Laboratories Agency, Addlestone, Surrey, United Kingdom
[2]University of Liverpool, Department of Clinical Infection, Microbiology and Immunology, Liverpool, United Kingdom

Chapter Contents

19.1 Introduction 193
19.2 Methodology 195
19.2.1 Oligonucleotide Design 195
19.2.2 Microarray Fabrication: In-House Printed Microarray 196
19.2.3 Microarray Fabrication: *In Situ* Synthesis (Inkjet Printing) Technology 196
19.2.4 Nucleic Acid Extraction and DNase Digestion 197
19.2.5 Reverse Transcription-Polymerase Chain Reaction 198
19.2.6 Labeling Amplified Target 198
19.2.7 Microarray Hybridization 199
19.2.8 Post Hybridization Washes and Image Capture 200
19.2.9 Data Analysis 200
19.3 Discussion 201
Acknowledgments 202
References 202

19.1 INTRODUCTION

Members of the genus *Lyssavirus* cause fatal encephalitis. Infection occurs following the bite of a rabid animal. The vast majority of human rabies cases are caused by the type species, rabies virus (RABV). This virus species is maintained in non-flying mammals belonging to the order *Carnivora*, most importantly the domestic dog (*Canis lupus familiaris*). The virus is also present in most species of New World bats, particularly the

C. Rupprecht & T. Nagarajan (Eds): Current Laboratory Techniques in Rabies Diagnosis, Research, and Prevention, Volume 1.
Doi: http://dx.doi.org/10.1016/B978-0-12-800014-4.00019-6

common vampire bat (*Desmodus rotundus*) that transmits RABV to humans and other mammals, such as cattle.[1] RABV is present worldwide, with the exception of parts of Western Europe, Australia, New Zealand and numerous islands. Until the late 20th century, the genus contained six further viruses that had caused occasional cases of rabies in humans, domestic animals, and wildlife. This has now expanded with the addition of a further seven virus species mainly isolated from insectivorous bat species and African wildlife. Another lyssavirus sequence has been detected in a bent-winged bat (*Miniopterus schreibersii*) in Spain.[2] Many of these viruses have caused spill-over infection from the presumed reservoir species to domestic animals, wildlife, and humans.[3] When confronted with a case of rabies, it is now desirable to discriminate the causative agent to the species level. Oligonucleotide microarrays can be used to achieve this objective.

Oligonucleotide microarrays have evolved from Southern blotting, where target DNA is attached to a solid support and used as "bait" for binding a known labeled DNA fragment (probe) whose presence in the target DNA is to be determined. The first microarrays, where multiple oligonucleotides were attached to a solid support such as a microscope slide, were reported in the late 1980s. These microarrays were used to measure variation in gene expression in response to cellular changes in cancerous cells or modulation in response to cell events such as proliferation.[4] A simplified schematic of microarray detection of target nucleic acid is shown in Figure 19.1.

The development of microarrays for the detection of viruses (pathogen microarray) was a response to the inability to culture some virus species and the limitations of polymerase chain reaction (PCR) to detect multiple viruses. However, to increase the sensitivity and make microarrays practical for unknown disease investigation, random sequence amplification of target nucleic acid was needed to increase the abundance of target sequences and simultaneously label DNA to detect hybridization to the probes. During the 2000s, a consortium developed the Defra BioChip for the purpose of detecting viruses causing infection in animals and plants. This array was used to confirm RABV as the causative agent in a case of human rabies encephalitis.[5] A spinoff from this pan-viral array was a smaller array (GEO accession GSE14705) encompassing oligonucleotides specific for seven lyssavirus species designed from the sequence of the nucleoprotein gene.[6] Alternative approaches have used re-sequencing to detect lyssaviruses[7] within a microarray specific for rhabdoviruses.

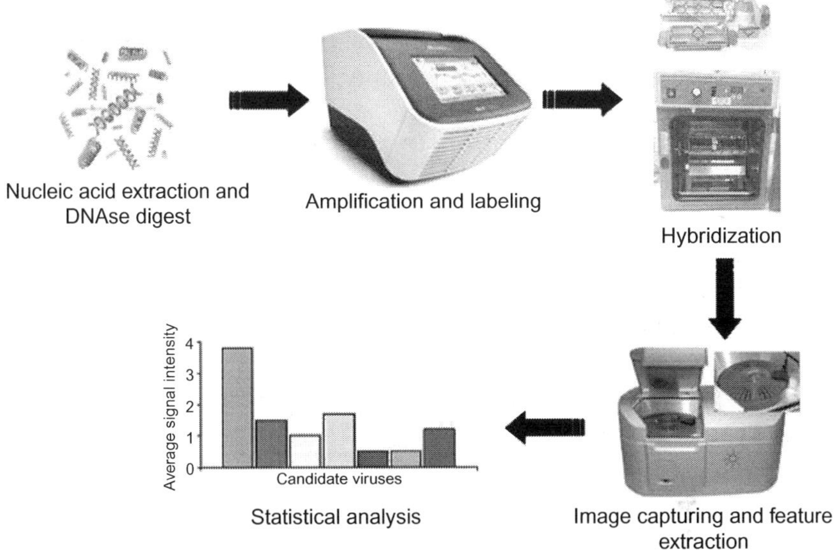

Figure 19.1 Schematic of a pathogen microarray process and analysis. Extracted nucleic acid from clinical samples are randomly amplified, labeled (fluorescent dyes), and hybridized to oligonucleotide probes on a microarray. The microarray is then scanned and analyzed using a variety of computational software. *(Scanner picture courtesy of Agilent Technologies.)*

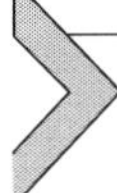

19.2 METHODOLOGY

19.2.1 Oligonucleotide Design

The approach taken to design oligonucleotide probes for a pathogen microarray is dependent on the intended use of the array, for virus discovery or targeted detection and speciation, and the capacity of the array. Wang and colleagues[8] have used complete viral genomes available in GenBank[9] to develop a pan-viral microarray. Their approach resembles design of oligonucleotide probes on conserved genomic regions. This minimizes the number of probes required to capture virus species in a virus genus or family. The methodology also provides the added advantage of detecting novel viruses. The AlleleID software developed by Premier Biosoft International,[10] used routinely in our laboratories for the purpose of probe design, implements a function in which viral sequences are aligned and a minimal set of probes designed on the sequence alignment.

As an alternative, and in the absence of such software, probes can be designed on the consensus nucleotide sequence of a DNA sequence alignment, an approach used to develop a genotyping array for lyssaviruses.[6]

19.2.2 Microarray Fabrication: In-House Printed Microarray

The in-house microarray fabrication consists of printing pre-synthesised oligonucleotides on a surface area such as a microscope slide. Limitations such as initial capital investment for the printing equipment, batch-to-batch variation, and low capacity due to size of spots, render the in-house printing technology obsolete.

19.2.3 Microarray Fabrication: *In Situ* Synthesis (Inkjet Printing) Technology

The *in situ* synthesis technology offered by Agilent Technologies[11] synthesizes 60-mer length oligonucleotides, base-by-base, from digital sequence files. The precise inkjet process enables the delivery of extremely small, accurate volumes (picoliters) of the chemicals to be spotted. Standard phosphoramidite chemistry used in the reactions allows for very high coupling efficiencies to be maintained at each step in the synthesis of the full-length oligonucleotide. Multiple microarrays can be printed simultaneously in a single manufacturing run. The technology allows more than a million oligonucleotides to be synthesized onto the surface of a single slide (Figure 19.2).

The procedures described below to process lyssavirus-suspect samples for microarray analysis should be considered a guideline, and the latest

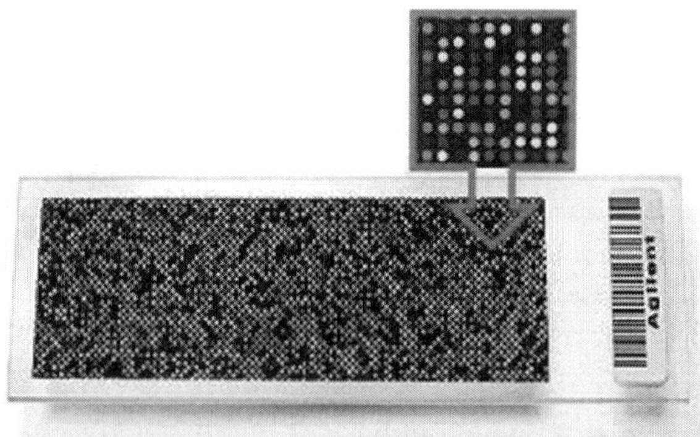

Figure 19.2 A depiction of a high-density microarray with a capacity to accommodate more than one million *in situ* synthesized oligonucleotides. *(Image courtesy of Agilent Technologies.)*

manufacturers' instructions should be consulted regarding products and procedures described. Furthermore, for an alternative protocol on random amplification, labeling, and hybridization, refer to publications by the Center for Infection and Immunity, Columbia University.[12]

19.2.4 Nucleic Acid Extraction and DNase Digestion

TRIzol® Reagent (Life Technologies[13]) is used routinely in our laboratories to extract nucleic acid from various tissues. To do this, approximately 25 mg of tissue, cut to small pieces, is added to 1 mL of TRIzol Reagent. The mix is then vortexed vigorously for about 1 min, centrifuged briefly, and incubated at room temperature for 5 min. Alternatively, the tissue can be homogenized in TRIzol Reagent inside an M tube (Miltenyi Biotec[14]) using gentleMACS™ Dissociator as per manufacturer instructions. Chloroform (0.2 mL) is then added to the homogenate, the tube is capped securely, vortexed, or shaken vigorously by hand for 15 seconds and incubated for 3 minutes at room temperature. The tube is then transferred to a refrigerated microcentrifuge and spun at 12,000 × *g* for 15 min at 4°C. The aqueous phase of the sample is removed by angling the tube at 45° and pipetting the solution out and placing into a new tube for RNA isolation. Avoid drawing any of the interphase or organic layer into the pipette when removing the aqueous phase. To isolate RNA, 1 µL (the equivalent of 20 µg) of UltraPure Glycogen (Life Technologies) and 0.5 mL of 100% isopropanol are added to the aqueous phase, incubated at room temperature for 10 min, and centrifuged at 12,000 × *g* for 10 minutes at 4°C. The supernatant is carefully removed from the tube, leaving only the RNA pellet, and the pellet is washed with 1 mL of 75% ethanol and vortexed briefly. The tube is centrifuged at 7,500 × *g* for 5 minutes at 4°C, the wash discarded, and the RNA pellet air dried for 5–10 minutes at room temperature. As alternatives, commercially available manual or automated nucleic acid isolation kits, such as the manual QIAamp Viral RNA Mini Kit (Qiagen[15]) or EZ1 RNA Universal Tissue Kit (Qiagen) on the EZ1 Advanced XL robotic workstation, can be used to extract nucleic acid from the aqueous phase of the TRIzol extracted sample. The extracted nucleic acid is re-suspended in nuclease-free water and quantified using a Nanodrop (Agilent Technologies). The sample is then digested with Amplification Grade DNase I (Life Technologies) to remove DNA from the sample according to the manufacturers' protocol. In brief, 1–2 µL of DNase I (1 U/µL) and 1 µL of 10 × DNase I Reaction Buffer are added to 200–300 ng of nucleic acid and the final volume adjusted to 10 µL with water. The mix is then incubated for 30 minutes at 37°C before DNase I is inactivated by the addition of 1 μL of 25 mM EDTA (pH 8.0) and heating at 65°C for 10 minutes.

19.2.5 Reverse Transcription-Polymerase Chain Reaction

To generate complementary DNA (cDNA), 1 μL of 25 mM deoxynucleoside triphosphate (dNTP) and 1 μL of 40 μM primer A, 5′ GTT TCC CAG TCA CGA TAN NNN NNN NN 3′ are added to the DNase I-digested nucleic acid. The mixture is then heated at 65°C for 5 min and chilled on ice immediately. Then 4 μL of 5 × First Strand buffer (Life Technologies), 1 μL 0.1 M DTT, 1 μL RNasin Ribonuclease inhibitor (20–40 U/μL, Promega[16]), and 1 μL Superscript III RT (200 U/μL) are added to the mix, incubated at 50°C for 60 min, and then heated to 70°C for 15 minutes. Second-strand DNA synthesis and nucleic acid amplification are conducted following the methods of Wang *et al.* (2002).[8] Briefly, cDNA is heated to 94°C for 2 minutes and cooled rapidly to 10°C. A 10 μL mix of 2 μL 5 × Sequenase buffer (Affymetrix[17]), 0.3 μL Sequenase DNA polymerase, and 7.7 μL water is added and heated slowly from 10°C to 37°C over an 8-minute period and held at 37°C for another 8 minutes followed by 94°C for 8 minutes. This double-stranded DNA template (5 μL) is amplified in a PCR that is composed of 5 μL AccuTaq LA DNA polymerase 10 × buffer (Sigma-Aldrich®[18]), 0.5 μL 25 mM dNTP mix, 1 μL primer B, 5′ GTT TCC CAG TCA CGA TA 3′, and 0.5 μL AccuTaq LA DNA polymerase (5 U/μL) to a final volume of 50 μL with water. If probes are designed on a conserved genomic region of ≤1 Kb, virus specific primers, with or without primer B, can also be used in the PCR. The PCR mix is initially heated to 98°C for 30 sec, 68°C for 5 minutes, followed by 35 cycles of 94°C for 30 seconds, 50°C for 1 minute, 68°C for 1 minute, a final extension of 2 minutes at 68°C, and then held at 10°C.

19.2.6 Labeling Amplified Target

Aminoallyl dUTP is incorporated into the amplified target nucleic acid in 20 cycles of PCR. The PCR consists of 5 μL AccuTaq LA DNA polymerase 10 × buffer, 0.5 μL AccuTaq LA DNA polymerase (5 units/μL), 1 μL amino allyl dNTP mix [10 mM amino allyl dUTP (Fermentas[19]), 2.5 mM dTTP, and 12.5 mM each of dGTP, dCTP and dATP], 1 μL 100 pmol/μL amino-primer B, amino-C6 5′ GTT TCC CAG TCA CGA

TA 3′, 5 μL PCR amplicon, and water to a volume of 50 μL. The PCR conditions are as outlined above. The PCR product is purified using MinElute PCR Purification kit (Qiagen) according to the manufacturer instructions, but replacing wash buffer with 70% ethanol and DNA eluted in 10 μL of water. The eluted DNA is coupled to Alexa Fluor 647 reactive dye (Life Technologies) and dissolved in 18 μL of DMSO. The coupling is performed for one hour at room temperature in the dark, and consists of 4 μL dye, 6 μL sodium bicarbonate (25 mg dissolved in 1 mL of water), and 10 μL purified aminoallyl-labeled DNA. The final volume of coupling reaction is then adjusted to 50 μL with water, and the unincorporated dye removed using illustra AutoSeq™ G-50 columns (GE Healthcare[20]) according to the manufacturer's protocol before hybridization.

19.2.7 Microarray Hybridization

The hybridization protocol described here is for use with microarrays produced using *in situ* synthesis technology (Agilent Technologies) in a format of 8 × 15K or 8 × 60K. For in-house printed microarray slides, refer to Gurrala *et al.*[6] for a detailed protocol on hybridization and subsequent washes. For each fluorescently labeled PCR amplicon, 2 μL Human Cot-1 DNA (1.0 mg/mL), 4.5 μL 10 × aCGH Blocking Agent (Agilent Technologies), and 22.5 μL 2 × HI-RPM Hybridization Buffer (Agilent Technologies) are mixed inside a microfuge tube containing 16 μL labeled DNA. The mix is transferred to a thermal cycler and incubated at 95°C for 3 minutes, 37°C for 30 minutes, and then spun briefly. A clean gasket slide is loaded into the Agilent SureHyb chamber base, with the gasket label facing up and aligned with the rectangular section of the chamber base (Figure 19.3). The hybridization sample mixture is slowly dispensed

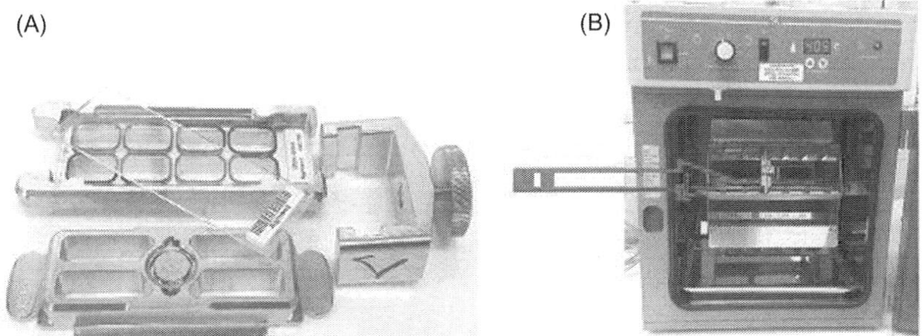

Figure 19.3 Agilent SureHyb chamber, gasket slide, microarray slide, chamber cover, and the clamp (A); and a hybridization oven (B).

onto the gasket well in a "drag and dispense" manner. Once all gasket wells are loaded, a microarray slide "active side" is laid down onto the gasket slide and the SureHyb chamber cover placed onto the sandwiched slide and the clamp hand-tightened firmly onto the chamber. The assembled slide chamber is lodged in the rotator rack inside a hybridization oven set to 65°C and 10 rpm and hybridized for 16 hours.

19.2.8 Post Hybridization Washes and Image Capture

One slide-staining dish is completely filled with Oligo aCGH/ChIP-on-Chip Wash Buffer 1 (Agilent Technologies) at room temperature. A slide rack and a magnetic stir bar are then put into a 2nd slide-staining dish, filled with enough room temperature Oligo aCGH/ChIP-on-Chip Wash Buffer 1 to cover the slide rack, and placed on a magnetic stir plate. A 3rd slide-staining dish is filled approximately three-quarters full with Oligo aCGH/ChIP-on-Chip Wash Buffer 2 (warmed to 37°C) on a magnetic stir plate with heating element and a magnetic stir bar. The temperature of Agilent Oligo aCGH/ChIP-on-Chip Wash Buffer 2 is maintained at 37°C and monitored using a thermometer.

The hybridization chamber is removed from the hybridization oven and disassembled by putting the hybridization chamber assembly on a flat surface and loosening the thumbscrew. With gloved fingers, the microarray-gasket sandwich is removed from the chamber base and submerged into the 1st slide-staining dish. The sandwich is pried open by slipping one of the blunt ends of the forceps between the slides and then gently twisting the forceps to separate the slides. The microarray slide is removed and quickly put into the slide rack in the 2nd slide-staining dish and stirred to get good but not vigorous mixing for 5 minutes. The slide rack is then transferred to the 3rd slide-staining dish containing Agilent Oligo aCGH/ChIP-on-Chip Wash Buffer 2 at 37°C, and stirred for 1 minute. The slide rack is removed slowly (5–10 seconds), trying to minimize droplets on the slide. The slide is then placed carefully onto the slide ledge, transferred to a microarray scanner (e.g. Agilent C Scanner), and scanned at 2 μm resolution and red PMT gain set at 100%. A "Feature Extraction" software is then used to extract digital values from signal intensities.

19.2.9 Data Analysis

Several statistical packages (e.g. DetectiV,[21] E-predict[22]) have been used to analyze output data from high-density microarrays. We use DetectiV

Table 19.1 Detection of RABV CVS Strain from Mouse Brain Using an Updated Defra BioChip Microarray

Virus Name	No. of Probes	T Value	P Value	Average
Rabies virus	760	30.43741	6.38E − 134	3.881148
Norovirus Hu/Ulm70/2003/DE	16	7.917662	4.89E − 07	1.713929
Influenza A virus (A/ostrich/South Africa/KK98/98(H6N8))	12	12.9697	2.61E − 08	1.521023
Hantavirus SN7	16	6.924717	2.43E − 06	1.221591
Human echovirus 18	70	5.561723	2.36E − 07	1.088964
Hepatitis B virus	188	4.776702	1.80E − 06	0.573116
Porcine lymphotropic herpesvirus 1	102	4.965092	1.40E − 06	0.566951

The RABV probes had the highest log^2 average signal intensity, 3.88, compared to those of probes for other viruses on the microarray.

software for this purpose. To normalize data, output data from a control array (e.g. RNA from a known uninfected animal simultaneously processed for microarray) is subtracted from those of the test array using an in-house script. The average value for each set of probes is then calculated. The log^2 is taken, sorted on magnitudes, and presented (Table 19.1). Virus or viruses with the highest signal averages are considered as potential candidates and subjected to further confirmatory tests for verification.

19.3 DISCUSSION

The main advantage of an oligonucleotide microarray is in its ability to detect multiple and novel pathogens within a single test. Technical improvements in non-specific amplification and removal of host RNA improve sensitivity, a major limiting factor when direct comparison is made with technologies such as virus-specific reverse transcription-polymerase chain reaction (RT-PCR). Furthermore, improved probe design software has also contributed to a higher sensitivity, and has simplified the design of oligonucleotides in response to the emergence of new pathogens. Addition of new probes to a high-density microarray is now rapid and relatively inexpensive, allowing timely response to changes in disease epidemiology.

For lyssaviruses, current gold standard tests do not discriminate between different species, even when virus is recoverable from clinical samples. Automated sequencing of PCR amplicons, or next generation sequencing, offer potential alternatives to speciation, but these can be costly, and demand high levels of expertise and considerable capital investment. Oligonucleotide microarrays offer an alternative that demands relatively lower expertise and is affordable by smaller laboratories. This is particularly important where more than one lyssavirus is known to be present and a patient has been diagnosed with rabies. In recent years, human deaths due to infection with European bat lyssavirus type 2 (EBLV-2), Duvenhage virus (DUVV), and Irkut virus have all required intensive investigation to confirm the causative agent to provide reassurance to public health authorities that RABV was not responsible.

The main disadvantage of oligonucleotide microarray, particularly in its dissemination to parts of the world where other lyssaviruses and RABV cause disease, is clearly cost. Initial investment is required to purchase specific equipment for nucleic acid amplification, hybridization, and scanners to measure signal intensity. Consumable and reagents costs, including the cost of microarray slides, are also a barrier to introduction, in addition to the cost of training and maintenance of skills for staff to perform the assay. Microarray detection of viruses is considerably slower when compared to other technologies, such as real-time RT-PCR, and it is this technology that is coming to dominate innovations in rabies diagnosis with gradual spread into the developing world.

In conclusion, oligonucleotide microarrays have been applied to the detection of RABV and other members of the genus lyssavirus,[6,23] and have the ability to identify the lyssavirus species in a particular sample. However, the barriers to its widespread application, other than within a pan-viral or pan-pathogen platform, are considerable, and it is unlikely that oligonucleotide microarray will find widespread application in the detection of RABV as a diagnostic test.

ACKNOWLEDGMENTS

This study was funded through Defra, UK, grants SV3500 and SE0537.

REFERENCES

1. Banyard AC, Hayman D, Johnson N, McElhinney LM, Fooks AR. Bats and lyssaviruses. *Adv Virus Res* 2011;**79**:239–89.
2. Aréchiga Ceballos N, Vázquez Morón S, Berciano JM, Nicolás O, Aznar López C, Juste J, et al. Novel lyssavirus in bat, Spain. *Emerg Infect Dis* 2013;**19**:793–5.

3. Johnson N, Vos A, Freuling C, Tordo N, Fooks AR, Müller T. Human rabies due to lyssavirus infection of bat origin. *Vet Microbiol* 2010;**142**:151–9.
4. Schulze A. Downward J. Navigating gene expression using microarrays – a technology review. *Nature Cell Biol* 2001;**3**:E190–5.
5. Fooks AR, Johnson N, Freuling CM, Wakeley PR, Banyard AC, McElhinney LM, et al. Emerging technologies for the detection of rabies virus: challenges and hopes in the 21st century. *PLoS Negl Trop Dis* 2009;**3**:e530.
6. Gurrala R, Dastjerdi A, Johnson N, Nunez-Garcia J, Grierson S, Steinbach F, et al. Development of a DNA microarray for simultaneous detection and genotyping of lyssaviruses. *Virus Res* 2009;**144**:202–8.
7. Dacheux L, Berthet N, Dissard G, Holmes EC, Delmas O, Larrous F, et al. Application of broad-spectrum resequencing microarray for genotyping rhabdoviruses. *J Virol* 2010;**84**:9557–74.
8. Wang D, Coscoy L, Zylberberg M, Avila PC, Boushey HA, Ganem D, et al. Microarray-based detection and genotyping of viral pathogens. *Proc Natl Acad Sci USA* 2002;**99**:15687–92.
9. GenBank: <https://www.ncbi.nlm.nih.gov/genbank> [accessed 19.02.14].
10. Premier BioSoft: <http://www.premierbiosoft.com> [accessed 19.02.14].
11. Agilent Technologies: <http://www.agilent.com> [accessed 19.02.14].
12. Columbia University, The Center for Infection and Immunity: <http://cii.columbia.edu> [accessed 19.02.14].
13. Life Technologies: <http://www.lifetechnologies.com> [accessed 19.02.14].
14. Miltenyi Biotec: <http://www.miltenyibiotec.com> [accessed 19.02.14].
15. Qiagen: <http://www.qiagen.com> [accessed 19.02.14].
16. Promega: <http://www.promega.com> [accessed 19.02.14].
17. Affymetrix: <http://www.affymetrix.com> [accessed 19.02.14].
18. Sigma-Aldrich: <http://www.sigmaaldrich.com> [accessed 19.02.14].
19. Fermentas Molecular Biology Products: <http://www.thermoscientificbio.com/fermentas> [accessed 19.02.14].
20. GE Healthcare Life Sciences: <http://www.gelifesciences.com> [accessed 19.02.14].
21. Watson M, Dukes J, Abu-Median AB, King DP, Britton P. DetectiV: visualization, normalization and significance testing for pathogen-detection microarray data. *Genome Biol* 2007;**8**:R190.
22. Urisman A, Fischer KF, Chiu CY, Kistler AL, Beck S, Wang D, et al. E-Predict: a computational strategy for species identification based on observed DNA microarray hybridisation patterns. *Genome Biol* 2005;**6**:R78.
23. Xi J, Guo H, Feng Y, Xu Y, Shao M, Su N, et al. Differentiation of the seven major lyssavirus species by oligonucleotide microarray. *J Clin Microbiol* 2012;**50**:619–25.

CHAPTER TWENTY

Demonstration of Lyssavirus Nucleic Acids by Pyrosequencing

Paola De Benedictis and Cristian De Battisti
FAO Reference Centre for Rabies, Istituto Zooprofilattico Sperimentale delle Venezie, Legnaro (Padova), Italy

Chapter Contents

20.1 Introduction 205
20.2 Materials 206
 20.2.1 Reagents and Equipment 206
 20.2.2 Diagnostic Samples 206
20.3 Methods 206
 20.3.1 Preliminary Phase – Population of the Local Database (IdentiFire Library) 206
 20.3.2 Phase 1 – RNA Extraction, OneStep Reverse Transcription-Polymerase Chain Reaction, Gel Electrophoresis Analysis 208
 20.3.3 Phase 2 – Pyrosequencing Reaction and Result Analysis 209
20.4 Discussion 210
 20.4.1 Experimental Tips 210
 20.4.2 Critical Parameters and Troubleshooting 210
 20.4.3 Precautions 213
 20.4.4 Alternative Methods 213
 20.4.5 Time Consideration 213
 20.4.6 Limitations 213
 20.4.7 Future Considerations 214
Acknowledgments 214
References 216

20.1 INTRODUCTION

Pyrosequencing is a novel DNA sequencing technology, and is the first alternative to the conventional Sanger method. This technique is based on the sequencing-by-synthesis principle[1] and on the real-time luminometric detection of pyrophosphate (PPi), which is released during

C. Rupprecht & T. Nagarajan (Eds): Current Laboratory Techniques in Rabies Diagnosis, Research, and Prevention, Volume 1.
Doi: http://dx.doi.org/10.1016/B978-0-12-800014-4.00020-2

nucleotide incorporation.[1,2] A specific enzymatic cascade is a consequence of the release of PPi (as described in Figure 20.1), and a light signal is generated, eventually visualized as a peak signal in the pyrogram (corresponding to electropherogram in dideoxy sequencing). Its magnitude is proportional to the number of nucleotides incorporated. Pyrosequencing uses two analytical protocols: Single Nucleotide Polymorphism (SNP) genotyping and Sequence Analysis (SQA). These protocols have been used mainly to classify a given number of pathogens, to identify secondary structures, and to detect mutations, as well as in DNA methylation analysis, in multiple sequencing, in species identification, and in whole genome sequencing.[1,3]

Pyrosequencing has been applied recently to lyssavirus detection and typing[4] by targeting on the 3′ terminus of the N gene. The protocol developed for this purpose, One-Step reverse transcription polymerase chain reaction (RT-PCR) followed by pyrosequencing, has been validated largely on field samples collected from rabies-suspect humans and other animals.

20.2 MATERIALS

20.2.1 Reagents and Equipment

Information on the reagents and the equipment necessary to perform the protocol are reported in the Additional Information section.

20.2.2 Diagnostic Samples

Diagnostic samples include brain tissues, salivary glands, oral swabs or saliva, skin biopsies, and cell culture supernatants.

20.3 METHODS

20.3.1 Preliminary Phase – Population of the Local Database (IdentiFire Library)

The final identification of the sequences obtained is based on their alignment on a local database specifically populated with the sequences of interest (the IdentiFire library).

- Download reference sequences from the public databases (i.e. GenBank) as txt file format. Select further reference sequences available in the laboratory.

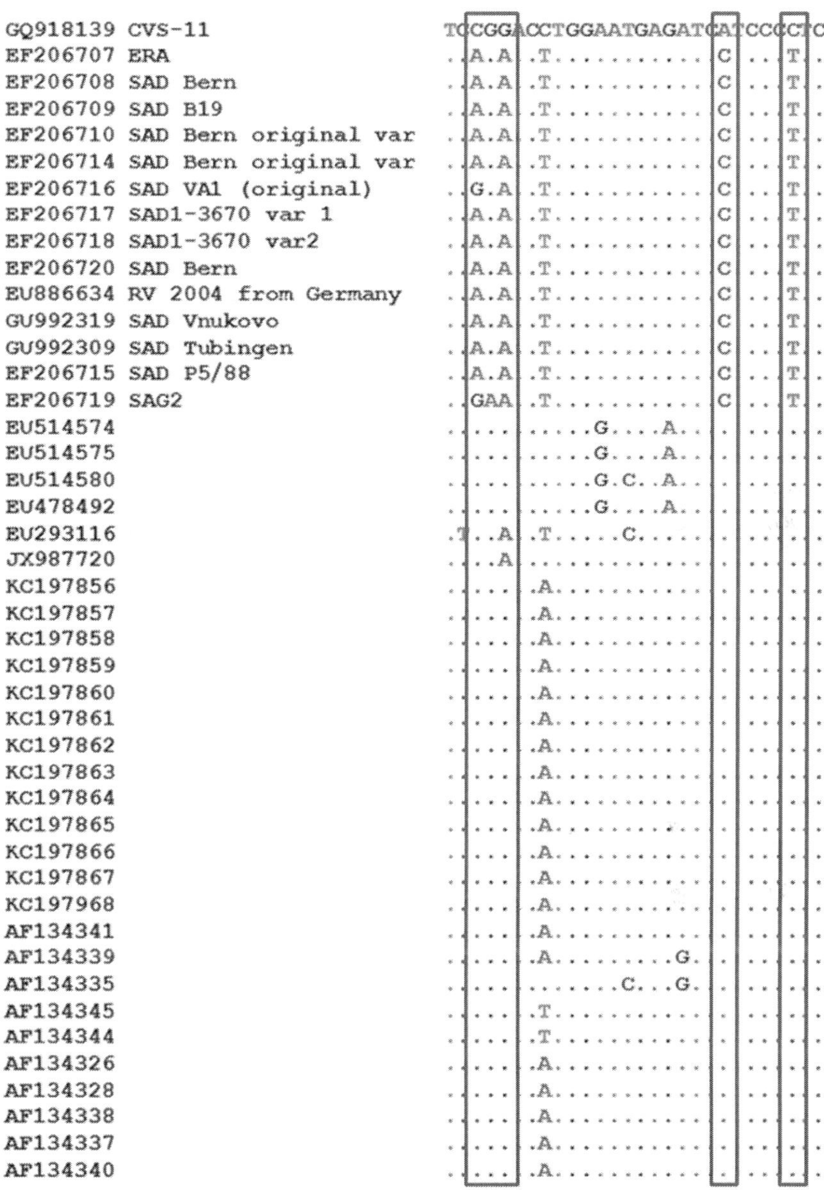

Figure 20.1 A primed DNA template and four enzymes involved (namely DNA-polymerase, ATP-sulfurylase, Luciferase, Apyrase) are placed in a well of a microtiter plate. DNA-polymerase catalyzes stepwise the incorporation of the nucleotide(s) into the newly synthesized nucleic acid strand. As a result of the incorporation, a pyrophosphate (PPi) molecule is released and subsequently converted to ATP by ATP-sulfurylase. A light signal is produced in the Luciferase reaction during which a luciferin molecule is oxidized. At the end of each reaction cycle, the exceeding nucleotides are continuously degraded by Apyrase.

- Align the sequences using Mega4 software.[5]
- Cut the sequences to obtain a short fragment including the 30–40 nucleotides target of the protocol (corresponding to position 76 to 116 of the GenBank accession number GQ918139).
- Create the local database by saving all sequences available in a single file FASTA format.

The lyssavirus-specific primer set targets a relatively conserved region of the N gene.[4] The primer set designed, tested *in silico* or *in vitro*, was able to anneal to all sequences analyzed belonging to all 12 species and to the new/putative lyssaviruses, namely Bokeloh bat lyssavirus, Ikoma lyssavirus and Lleida Bat Lyssavirus.

20.3.2 Phase 1 – RNA Extraction, OneStep Reverse Transcription-Polymerase Chain Reaction, Gel Electrophoresis Analysis

Samples are prepared for RNA extraction as follows:

- Dilute and homogenize 1 gram of tissue 1:3 (w/v) in sterile Phosphate Buffered Saline (PBS); saliva and cell culture supernatants are used undiluted.
- Centrifuge the tube for a few seconds.
- Add 100 μL of supernatant to 300 μL of lysis buffer as provided by the Extraction kit (NucleoSpin RNA kit, Macherey-Nagel, Duren, Germany).

Organs are homogenized manually using a mortar and pestle, with the possible addition of sterile quartz powder for samples containing a high amount of connective tissues (i.e. skin biopsies and salivary glands). Alternatively, an automated homogenizer (TissueLyser II, Qiagen, Hilden, Germany) can be used, so that a thorough high-speed sample shaking in plastic tubes with beads can be obtained. The advantage of using automated systems resides not only in reducing the risk of poor-quality homogenized sample occurrence (related to human error), but also in the potential simultaneous treatment of multiple biological samples.

In this case, samples are prepared according to the following procedure:

- Dilute 1 gram of tissue 1:3 (w/v) in lysis buffer as provided by the Extraction kit (NucleoSpin RNA kit, Macherey-Nagel, Duren, Germany) in a plastic tube (1.5 mL) containing 1 stainless steel bead of 5 mm diameter. This passage will limit the potential contamination of workrooms with infectious virus particles.

- Set TissueLyser II (Qiagen, Hilden, Germany) at 3 minutes for 30 Hz to crumble the samples.

Viral RNA is extracted using the Nucleospin RNA kit according to the manufacturer's protocol (with slight modifications; see Additional Information) (Macherey-Nagel, Duren, Germany). Briefly, 100 microliters of sample suspension have to be used for the extraction (see Section 20.2.2). The RNA is eluted in a final volume of 60 μL and stored at −80°C.

For One-Step RT-PCR, use the One-Step RT-PCR kit, according to the manufacturer's instructions (Qiagen, Hilden, Germany), with the following final concentration:

- 400 nM of each primer (2 μL), 400 nM of deoxynucleoside triphosphates (dNTPs) mix (2 μL), 1 × of PCR Buffer 5 × (10 μL), 1 μL of OneStep RT-PCR Enzyme Mix.
- 20U of RNAse inhibitor 40 U/μl (0.5 μL).
- 5 μL of isolated RNA.
- Water (27.5 μL) to reach a final volume of 50 μL.

The following protocol was used: 30 min at 50°C, 15 min at 95°C followed by 45 cycles at 94°C for 30 sec, 52°C for 30 sec, and 72°C for 40 sec. Biotinylated amplicons are detected by polyacrylamide gel electrophoresis (PAGE) followed by silver staining.

20.3.3 Phase 2 – Pyrosequencing Reaction and Result Analysis

Pyrosequencing reaction is performed using the Pyromark ID platform (Biotage, Uppsala, Sweden).

Biotinylated amplicons are bound to Streptavidin-coated Sepharose beads, and the beads are captured with the Vacuum Tool on the Vacuum Workstation, according to the manufacturer's instructions, thus generating single-stranded DNA suitable for pyrosequencing. Single-stranded biotinylated DNAs are then transferred to 40 μL annealing buffer containing pyrosequencing primers at a final concentration of 0.5 μM, and then placed into the PyroMark ID instrument. Volumes of the PyroMark Gold reagents are provided by the software (based on the number of samples to be analyzed), and are pipetted into the dispensing cartridge that is further placed in the PyroMark platform.[6]

A preliminary quality control of the pyrogram is available immediately (as Good, Uncertain, Failed, Not Analyzed), and is based on colorimetric indications on the IdentiFire report (blue, yellow, red and white, respectively).

Based on local alignment, each query sequence is matched to one or more reference sequences belonging to the local database created previously. A score value is assigned to each query sequence, taking into account the percentage of homology between the query and the reference sequence(s), the quality of the pyrogram, and the intensity of the peaks obtained for each query sequence. Results are analyzed in a few seconds by the software and reported as an IdentiFire report (IdentiFire Software, Biotage, Uppsala, Sweden; see Figure 20.2). According to the results obtained on field samples using the protocol described, score values greater or at least equal to 80 are considered as satisfactory for a correct classification.

20.4 DISCUSSION

20.4.1 Experimental Tips

Particular attention must be taken when considering the storage conditions of the pyrosequencing reaction. Enzymes and substrate mixtures have to be stored correctly to avoid any spurious signal and/or anomalous pyrogram. The correct storage condition for enzymes and substrate mixtures is −20°C.

20.4.2 Critical Parameters and Troubleshooting

Several challenges are known to occur when developing a pyrosequencing protocol.[1,3] One of the most common critical parameters for viral typing protocols is due to primer design for both PCR and sequencing. Sequencing primers must be checked for self-looping, primer-dimer, and cross-hybridization. The protocol developed and applied so far has never encountered such gaps.

The correct choice and identification of the sequences that will populate the IdentiFire library is a crucial and preliminary activity that will affect the success of the pyrosequencing protocol. The library must contain at least one reference sequence (a comprehensive panel of all species of interest) to identify correctly and univocally the unknown generated sequences (also referred to as query sequences). A continuous update of the IdentiFire library is strongly recommended, so that reference sequences of newly discovered viral species may be included as soon as they are available in the public domain. In fact, the absence of the correspondent reference sequence in the local database can generate a mistaken identification due to a forced alignment with the most similar sequence,

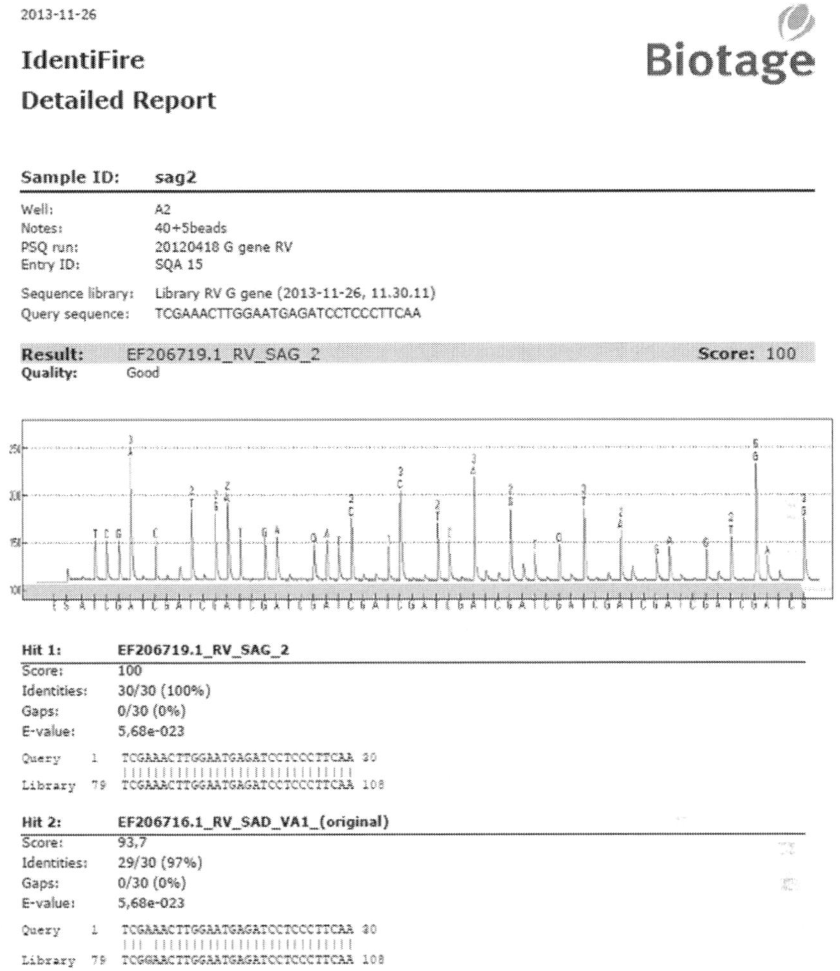

2013-11-26

IdentiFire
Detailed Report

Sample ID: **sag2**

Well:	A2
Notes:	40+5beads
PSQ run:	20120418 G gene RV
Entry ID:	SQA 15
Sequence library:	Library RV G gene (2013-11-26, 11.30.11)
Query sequence:	TCGAAACTTGGAATGAGATCCTCCCTTCAA

Result: EF206719.1_RV_SAG_2 **Score:** 100
Quality: Good

Hit 1: **EF206719.1_RV_SAG_2**

Score:	100
Identities:	30/30 (100%)
Gaps:	0/30 (0%)
E-value:	5,68e-023

```
Query    1  TCGAAACTTGGAATGAGATCCTCCCTTCAA 30
            ||||||||||||||||||||||||||||||
Library 79  TCGAAACTTGGAATGAGATCCTCCCTTCAA 108
```

Hit 2: **EF206716.1_RV_SAD_VA1_(original)**

Score:	93,7
Identities:	29/30 (97%)
Gaps:	0/30 (0%)
E-value:	5,68e-023

```
Query    1  TCGAAACTTGGAATGAGATCCTCCCTTCAA 30
            ||| ||||||||||||||||||||||||||
Library 79  TCGGAACTTGGAATGAGATCCTCCCTTCAA 108
```

Figure 20.2 IdentiFire detailed report as generated by the IdentiFire Software (Biotage, Uppsala, Sweden). Example of a correct classification of a sample as "rabies virus" (score 100). The quality of the pyrogram has been recognized as "Good" and is clearly mentioned in the report. Information available on the report are the sample ID and corresponding query sequence, protocol adopted (*Notes*), run ID (*PSQ run*), number and type of enzymatic reactions setup (*Entry ID*), and the IdentiFire library used for the analysis.

often resulting in a poor score value on the IdentiFire report. Similarly, this finding (score values of less than 80) could be suggestive of an incorrect performance of the technical procedure as well as of a poor quality of the sample of origin (see Figure 20.3). In case of doubtful IdentiFire

2013-11-26

IdentiFire
Detailed Report

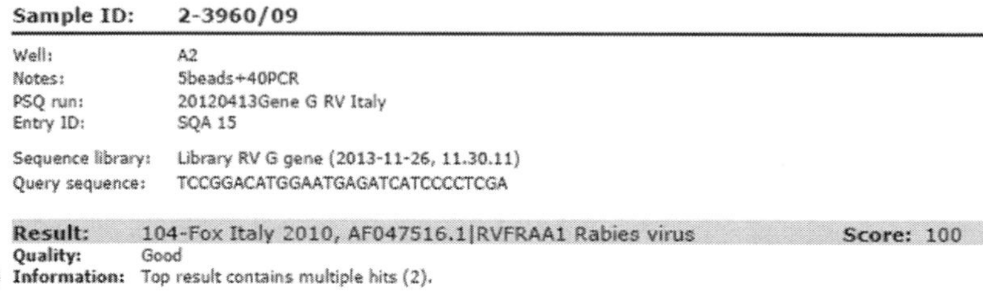

Sample ID: **2-3960/09**

Well:	A2
Notes:	5beads+40PCR
PSQ run:	20120413Gene G RV Italy
Entry ID:	SQA 15
Sequence library:	Library RV G gene (2013-11-26, 11.30.11)
Query sequence:	TCCGGACATGGAATGAGATCATCCCCTCGA

Result: 104-Fox Italy 2010, AF047516.1|RVFRAA1 Rabies virus **Score: 100**
Quality: Good
Information: Top result contains multiple hits (2).

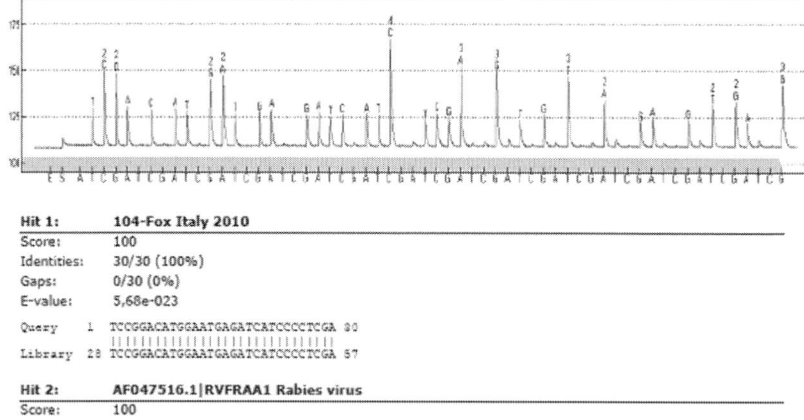

Hit 1:	**104-Fox Italy 2010**
Score:	100
Identities:	30/30 (100%)
Gaps:	0/30 (0%)
E-value:	5,68e-023

```
Query    1  TCCGGACATGGAATGAGATCATCCCCTCGA 30
            ||||||||||||||||||||||||||||||
Library 28  TCCGGACATGGAATGAGATCATCCCCTCGA 57
```

Hit 2:	**AF047516.1|RVFRAA1 Rabies virus**
Score:	100
Identities:	20/20 (100%)
Gaps:	0/20 (0%)
E-value:	3,75e-012

```
Query     1  TCCGGACATGGAATGAGATC 20 (30)
             ||||||||||||||||||||
Library 391  TCCGGACATGGAATGAGATC 410
```

Figure 20.3 IdentiFire detailed report as generated by the IdentiFire Software (Biotage, Uppsala, Sweden). Example of a doubtful classification of a sample as Mokola virus (score value 68.1 and uncertain quality of the pyrogram). It was a *Rhabdoviridae* not belonging to the *Lyssavirus* genus (*LeDantec* virus), which was amplified by the One-Step RT-PCR. The absence of reference sequences (in the public database GenBank and, consequently, in the local database) has led to a forced alignment with the most similar sequence in the IdentiFire library. A warning signal also appears in case the pyrogram proves to be of low quality, thus suggesting the occurrence of possible drawbacks.

reports, a conventional Sanger sequencing of the entire 603 bp targeted by the One-Step RT-PCR protocol is recommended.

Some other troubleshooting is closely related to the hardware of the instrument or of the pyrogram plots (wide peaks, nonspecific peaks, low signal, etc.). Further details on possible technical issues are reported in the trouble shooting section on the Qiagen website.[7]

20.4.3 Precautions

The pyrosequencing procedure does not include steps that are potentially dangerous for the operator. Good laboratory practices (GLPs) normally adopted in laboratories performing molecular biology tests are recommended.

20.4.4 Alternative Methods

An improved pyrosequencing technology is based on the use of the upgraded platform (PyroMark Q24, Qiagen, Hilden, Germany) with advanced reagents. This improved technology can lead to longer sequence fragments (140 and more nucleotides) from the obtained amplicons of 603bp, thus enhancing the accuracy of some unclear results and expanding the field applications of this technique.

As mentioned earlier, an alternative method for viral RNA confirmation is the Sanger sequencing method. Amplicons obtained using biotinylated as well as non-biotinylated primers could be sequenced correctly according to dideoxy sequencing, thus amplicons generated for pyrosequencing purposes could also be analyzed conventionally.

20.4.5 Time Consideration

Using the method described, a maximum time span of about 5–6 hours (depending on the number of samples to run) from the submission to the final typing of a positive field sample is needed to analyze a suspect sample in laboratory conditions. The amplified product is obtained in about 3 hours, thus analyzed by gel electrophoresis in one additional hour (Phase 1). Starting from the amplified products, a pyrosequencing analysis requires less than 2 hours to obtain the final result (Phase 2).

20.4.6 Limitations

Despite the several advances of the pyrosequencing technique (flexibility, automation, accuracy, etc.), some limitations have reduced its application to specific fields.[3] The pyrosequencing protocol described herein has proved to have only one limitation: short sequence fragments obtained in Phase 2, which could not be further used for phylogenetic or phylogeographic analysis. On the other hand, amplicons produced during Phase 1 could be sequenced immediately by using the classical procedure (Sanger method), thus obtaining a 603 bp fragment that will be useful for further analyses if needed.

20.4.7 Future Considerations

Although the limited read length obtainable by applying this technique has often been considered as a limiting factor, the performances of the protocol described has never been affected. However, the currently available improved technology (described in Section 20.4.4) may enhance the general accuracy of some unclear results.

ACKNOWLEDGMENTS

Several partners have contributed to the development and field validation of this protocol and are gratefully acknowledged. Among them, partners from: Pasteur Institute (Paris, France) – Laurent Dacheux, Hervé Bourhy, Jean-Claude Manuguerra and Rachel Lavenir; from ISC-III (Madrid-Spain) – Juan Echevarrìa Majo – and more. All the staff at the FAO Reference Centre for Rabies, IZSVe-Italy, are acknowledged for technical help and advice, with a special mention going to Francesca Ellero, who revised the manuscript.

Additional Information

Reagents

- Sample preparation: PBS (Phosphate Buffered Saline). Per 1 L mix the following: 0.2 g KH_2PO_4, 2.9 g Na_2HPO_4 +$2H_2O$, 0.2 g KCl, 8 g NaCl. Adjust pH to 7.2–7.4 with HCl. Add distilled water to 1,000 mL. Sterilize in an autoclave at 121°C for 20 min. Store at +4°C for a maximum of 1 year.
- Isolation of RNA: NucleoSpin RNA kit (Macherey-Nagel, Duren, Germany). One modification was applied to the manufacturers' instructions indicated in the user manual. β-Mercaptoethanol reagent was not used for operator safety reasons.
- One-Step RT-PCR: One-Step RT-PCR kit (Qiagen, Hilden, Germany), RNasin® Plus RNase Inhibitor (Promega, Madison, USA), primers (Eurofins MWG Operon, Ebersberg, Germany). Please note that Position* referred to the challenge virus strain (CVS) of rabies virus (RABV) (GenBank accession number GQ918139).
  - RabForPyro AACACYYCTACAATGGA* Position 59–75
  - RabRevPyro-biot 1 TCCAATTNGCACACATTTTGTG Position* 662–641
  - RabRevPyro-biot 2 TCCARTTAGCGCACATYTTATG Position* 662–641
  - RabRevPyro-biot 3 TCCAGTTGGCRCACATCTTRTG Position* 662–641

- Electrophoresis analysis
  - Fixing solution for acrylamide gel. Per 1 L solution: 100 mL Methanol (absolute) and 15 mL Nitric Acid (70%). Add 885 mL of double distilled water (ddH_2O). Store at room temperature for a maximum of 6 months.
  - Silver Nitrate (0.4%) for acrylamide gel. 2 g Silver nitrate in 500 mL ddH_2O. Filter the solution. Store for a maximum of 6 months.
  - Developing solution for acrylamide gel. Per 1 L solution: 30 g. Dissolve 30 g Sodium carbonate in 1 L ddH_2O. At the first use add 700 μL Formaldehyde (36%). Store at room temperature for a maximum of 6 months.
  - Acetic Acid (5%). Per 1 L solution: 50 mL pure glacial Acetic Acid (100%) in 950 mL ddH_2O. Store for a maximum of 6 months.
  - Acrylamide solution (7%). Per 500 mL solution: 50 mL TBE 10 ×, 87.5 mL Acrylamide/bis acrylamide 29:1 (40%) and 50 mL Glycerol (100%) in 313 mL ddH_2O. Store at room temperature for a maximum of 1 year.
  - APS (ammonium persulfate) 10%. For 10 mL solution: 1 g Ammonium persulfate in 10 mL ddH_2O. Store at room temperature for a maximum of 1 year.
  - TBE 10 ×. Per 1 L solution: 108 g Trizma base, 55 g Boric Acid, 40 mL EDTA 0.5 M pH 8. Add ddH_2O to 1 L volume. Autoclave at 121°C for 20 minutes. Store at +4°C for a maximum of 1 year.
  - Gel loading buffer 10 ×. Per 10 mL solution: dissolve 0.2 g Ficoll and 1 mL TBE 10 × in 2 mL ddH_2O, using a thermal bath at 37°C. Add 6 mL Glycerol (100%), 1 mL bromophenol blue and Xylene cyanol (0.5%). Store at +4°C for a maximum of 1 year.
  - Marker. For 800-μL solution: 100 μL Marker VI (Roche), 300 μL gel loading buffer 10 × and 400 μL TBE 1 ×. Store at +4°C for a maximum of 1 year.
- Pyrosequencing reaction: PyroMark Gold Q96 SQA Reagents, PyroMark Annealing Buffer, PyroMark Binding Buffer, PyroMark Denaturation Solution, PyroMark Wash Buffer. Qiagen (Hilden, Germany) currently provides all reagents.

Equipment

Equipment is listed in Table 20.1.

Table 20.1 Equipment Listing

Phase	Equipment	Source
Sample preparation	TissueLyser II	Qiagen, Hilden, Germany
	Stainless steel beads 5 mm	Sigma Aldrich, St Louis, Mi, USA
Phase 1	GeneAmp PCR System 9700	Applied Biosystems, Foster City, CA, USA
	Mini PROTEAN Tetra Cell	Biorad, Hercules, CA, USA
Phase 2	Tmix Thermal mixer	Analytikjena Bio solutions, Jena, Germany
	PyroMark ID	Biotage, Uppsala, Sweden; acquired by Qiagen, Hilden, Germany
	PyroMark Vacuum Workstation	Biotage, Uppsala, Sweden; acquired by Qiagen, Hilden, Germany
	Vacuum Prep Worktable	Biotage, Uppsala, Sweden; acquired by Qiagen, Hilden, Germany
	Vacuum source Laboport	KNF Neuberger, Freiburg, Germany
	EXPY Shaking Incubator	Explera, Jesi, Italy
	PyroMark Q96 Cartridge (3)	Biotage, Uppsala, Sweden; acquired by Qiagen, Hilden, Germany
	PyroMark Q96 Plate Low (100)	Biotage, Uppsala, Sweden; acquired by Qiagen, Hilden, Germany
	PyroMark Vacuum Prep Filter Probe (100)	Biotage, Uppsala, Sweden; acquired by Qiagen, Hilden, Germany
	PyroMark Q96 Vacuum Prep Trough	Biotage, Uppsala, Sweden; acquired by Qiagen, Hilden, Germany

REFERENCES

1. Ronaghi M. Pyrosequencing sheds light on DNA sequencing. *Genome Res* 2001; **11**(1):3–11.
2. Nyrén P. The history of pyrosequencing. *Methods Mol Biol* 2007;**373**:1–14.
3. Frakuddin MD, Chowdhury A, Nur Hossain MD, Mannan KSB, Mazumdar RM. Pyrosequencing-principles and applications. *Int Journ Life Science & Pharma Res* 2012;**2**(2):65–76.

4. De Benedictis P, De Battisti C, Dacheux L, Marciano S, Ormelli S, Salomoni A, et al. Lyssavirus detectionand typing using pyrosequencing. *J Clin Microbiol* 2011;**49**(5):1932–8.
5. Tamura K, Dudley J, Nei M, Kumar S. MEGA4: Molecular Evolutionary Genetics Analysis (MEGA) software version 4.0. *Mol Biol Evol* 2007;**24**(8):1596–9.
6. Cummings PJ, Ahmed R, Durocher JA, Jessen A, Vardi T, Obom KM. Pyrosequencing for microbial identification and characterization. *J Vis Exp* 2013;**2**(78):e50405.
7. Qiagen: <http://www.qiagen.com > [accessed 20.02.14].

Demonstration of Viral-Induced Changes in Tissues/Organs

CHAPTER TWENTY ONE

Rabies Diagnosis: MR Imaging

Jiraporn Laothamatas[1], Witaya Sungkarat[1], and Thiravat Hemachudha[2]

[1]Advanced Diagnostic Imaging Center, and Department of Radiology, Faculty of Medicine Ramathibodi Hospital, Mahidol University, Bangkok, Thailand
[2]WHO Collaborating Center for Research and Training on Viral Zoonoses, and Department of Medicine, Faculty of Medicine, Chulalongkorn University, Bangkok, Thailand

Chapter Contents

21.1 Introduction 221
21.2 Materials 222
  21.2.1 Magnetic Resonance Imaging Examination 222
  21.2.2 Human and Animal Subjects 222
21.3 Methods 223
  21.3.1 Magnetic Resonance Imaging Techniques 223
    21.3.1.1 Structural Data 223
    21.3.1.2 Susceptibility Imaging Techniques 223
    21.3.1.3 Metabolite Imaging Techniques 223
    21.3.1.4 Molecular Imaging Techniques 225
    21.3.1.5 Magnetic Resonance Perfusion Techniques 225
    21.3.1.6 Voxel-Based Group Analysis 226
    21.3.1.7 Application of MRI Techniques in Rabies: Utility and Limitations During the Early and Late Stages 226
21.4 Discussion 230
Acknowledgments 230
References 231

21.1 INTRODUCTION

Rabies is unique among viral encephalitides in terms of minimal neuropathological changes, reflecting different underlying pathogenetic mechanisms.[1] The neuroimaging of choice in studying rabies, therefore, is magnetic resonance imaging (MRI). Conventional MRI techniques for detailing abnormal structures and signal intensity, MR angiography and venography for vascular patency, and MR perfusion for tissue perfusion, susceptibility

C. Rupprecht & T. Nagarajan (Eds): Current Laboratory Techniques in Rabies Diagnosis, Research, and Prevention, Volume 1.
Doi: http://dx.doi.org/10.1016/B978-0-12-800014-4.00021-4

imaging for bleeding and calcification, have been of limited value in rabies. MR diffusion and diffusion tensor imaging (DTI) for cytotoxic edema, integrity of the white matter tracts, and the status of the interstitial spaces including the blood brain barrier (BBB),[2] meanwhile, can give more useful data. Voxel-based group analysis (against normal controls) is a promising new and useful quantitative technique for evaluation of the cyto-architecture of the brain using DTI techniques. Determination of the presence of iron as an inflammatory marker using susceptibility imaging techniques can also be conducted simultaneously.[3] Estimated MR T2* voxel-based group analysis may contribute as a sensitive and specific tool for quantitative iron deposits.[4] MR spectroscopy for chemical metabolite evaluation[5] can be added as an adjunct.

Clinical features of furious and paralytic forms, usually associated with genotype 1, are classic manifestations in rabid humans and dogs. Differences in the amount of viral load in the brain (more abundantly in furious cases) and marked intensity of inflammation especially at the brainstem in the case of paralytic rabies, may contribute to the diversity of clinical outcome.[1,6,7] DTI studies in non-comatose rabid dogs showed compromised neural tract integrity only in the brainstem of paralytic dogs.[1,8] The BBB remains intact in both forms. Cytotoxic brain edema was noted exclusively in the brains of paralytic dogs. Advanced imaging technology with quantitative assessment should be helpful for early diagnosis and management of rabies.

21.2 MATERIALS

21.2.1 Magnetic Resonance Imaging Examination

Studies should be performed with high-field MR imaging. Animal studies and advanced MRI techniques in this report were imaged by a 3.0-Tesla Achieva MRI scanner (Philips Medical System, Best, The Netherlands). MR imaging of the brain and upper spinal cord was performed with Philips SENSE Flex-M coils in axial diffusion weighted imaging (DWI) and DTI, sagittal and coronal T2-weighted fluid-attenuated inversion recovery (FLAIR), and T1-weighted (T1W) before and after gadolinium administration, 0.1 mmol/kg.

21.2.2 Human and Animal Subjects

During transfer and examination, the subjects can be sedated with ketamine and pentobarbitone sodium. Vital signs must be monitored carefully. Ventilator support and inotropic agents should be provided whenever required during the whole process, in particular in the case of paralytic rabies, where respiration is usually compromised.

21.3 METHODS

21.3.1 Magnetic Resonance Imaging Techniques

MR imaging is the technique using radiofrequency pulses in a magnetic field and the specific T1 and T2 relaxation times of hydrogen atoms in different molecules such as in water, fat, and proteins, and the presence of susceptibility materials in different organs.[9] There is no radiation exposure with this technique, but there will be heat created inside the body by the specific absorption rate (SAR) related to magnet field strength and study protocols.[10] There are conventional MRI techniques to demonstrate structural data of both normal and abnormal findings, and advanced imaging techniques allowing the study of the brain and neural axis beyond basic structure, such as molecular imaging studies, tissue perfusion, chemical metabolite, and iron accumulation in the brain tissue. The detailed descriptions of these techniques are covered in the following sections.

21.3.1.1 Structural Data

Structural data are related to the underlying basic anatomy, and any changes due to conditions such as infection, edema, masses, hemorrhage, and infarction, by utilizing conventional techniques, such as T1W, T2-weighted (T2W) by using spin echo, gradient echo and inversion recovery techniques, and their modifications for fat suppression and fast imaging acquisition. The T2W image is good for abnormal signal intensity detection. The FLAIR image is the T2W image with subtraction of high signal intensity of the CSF in the ventricle and cortical sulci and cisterns. It allows better detection of an abnormality adjacent to the ventricle and cortical sulci and in the subarachnoid space (Figure 21.1).[11]

21.3.1.2 Susceptibility Imaging Techniques

Susceptibility imaging techniques, using gradient pulse sequence methods and susceptibility-weighted imaging (SWI) techniques, are used to detect the presence of susceptibility materials, such as blood products and heavy metals. Gradient echo techniques compose only the magnitude aspect of the susceptibility, while SWI composes both magnitude and phase shift due to susceptibility effects in the tissue. It can be used to calculate the presence of iron in organs such as the liver, myocardium, and the brain, to study hemorrhagic components of the tissue (Figure 21.2).[12] It can also be applied in the assessment of inflammatory changes in the brain.[13]

21.3.1.3 Metabolite Imaging Techniques

Metabolite imaging techniques or MR spectroscopy is a chemical shift imaging of hydrogen and other atoms, such as phosphorus, sodium, and potassium,

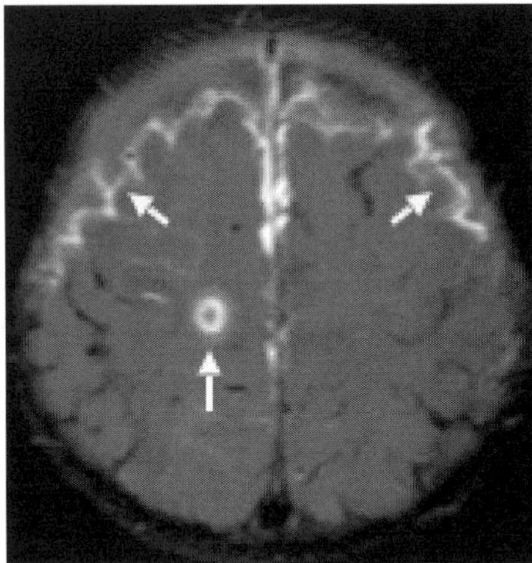

Figure 21.1 Post-gadolinium axial T2 FLAIR with fat suppression demonstrates gyral hypersignal of leptomeningeal inflammation along bilateral anterior frontal cortical sulci and a small abscess at the right precentral gyrus (short white arrows).

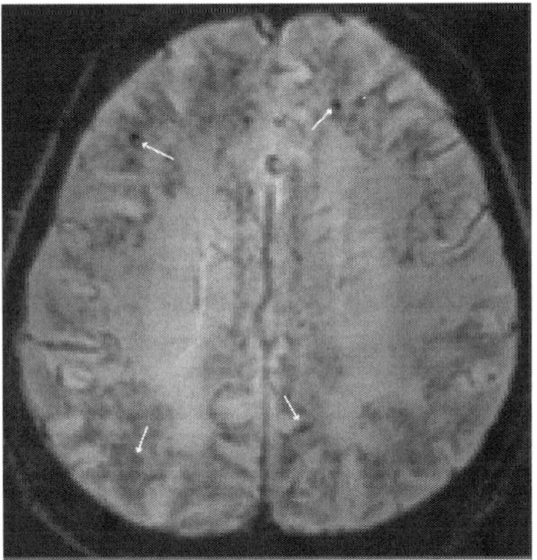

Figure 21.2 MRI of the brain of a 33-year-old male case of viral encephalitis demonstrates a non-significant abnormal signal intensity T2 FLAIR image but extensive scattered minute foci of markedly hyposignal SWI due to the susceptibility effect of blood products seen in the axial SWI image (small white arrows).

in different molecules, shown as a spectral pattern. The most commonly utilized method is proton MR spectroscopy.[5] In the brain, each metabolite spectral peak represents a specific brain marker. As such, each occurs in a specific location of the parts per million (ppm) spectral graph. For example, in a normal brain, the N-acetyl aspartase (NAA) peak represents a neuronal marker seen at 2 ppm. The choline peak represents cell membrane destruction and synthesis observed at 3.2 ppm. The creatine peak represents energy metabolism detected at 3 ppm and the myo-inositol peak represents an astrocyte marker seen at 3.5 ppm. The glutamine and glutamate peaks represent neurotransmitter markers seen at 2.2–2.4 ppm. The lactate peak, seen at 1.3 ppm, will be detected with anaerobic metabolism in anoxic, hypoxic and ischemic conditions such as infarction or tumor necrosis. It is commonly used in tumor imaging and can be used as an adjunct in encephalitis imaging.

21.3.1.4 Molecular Imaging Techniques

Molecular imaging techniques comprise DWI and DTI.[2,14] They are used for mapping the mobility of water molecules in biological tissue *in vivo*, noninvasively. It is used mainly in the brain. DWI measures the mobility of water molecules in the tissue qualitatively as an increase, decrease, or the same, compared to normal tissue. The study can measure quantitative changes by using apparent diffusion coefficients (ADC). It can differentiate cytotoxic edema from vasogenic edema, and acute demyelinating processes from gliosis in the brain.

DTI is an MR diffusion technique enabling the measurement of the restricted diffusion of water in tissues in several directions to produce neural tract images. At least six different orientations of the diffusion-sensitizing gradients are used. The greater the orientations of the diffusion-sensitizing gradients that the study assigns, the better the estimation of the white matter tract directions. Its measurement unit is the fractional anisotropy (FA), ranging from 0–1. The higher the FA, the more compact and better oriented the white matter tracts are, compared to decreased FA in the less organized and disrupted white matter tracts. Moreover, the DTI mean diffusivity (MD), similar to ADC in pathological detection, can be derived as a byproduct.

21.3.1.5 Magnetic Resonance Perfusion Techniques

At the core of MR perfusion techniques,[15] perfusion is the process of delivering oxygen and nutrients to the tissue by the blood circulation. The technique is used to study the blood flow, blood volume, and mean transit time of tissue both qualitatively and quantitatively. It also allows evaluation of the leakiness of the tissue due to angiogenesis of the lesion, mainly in tumor imaging. It can be performed by using intravenous contrast agents and the non-contrast arterial spin-labeling technique. Perfusion techniques are used mainly in cerebrovascular disease and tumor imaging.

21.3.1.6 Voxel-Based Group Analysis

Voxel-based group analysis (against normal controls) based on whole-brain probabilistic tractography map morphometry is a new method to perform quantitative group comparison analysis of imaging parameters, such as the DTI data shown as FA and MD maps[16] and susceptibility images, shown as an accumulation of blood products or iron in the brain. To perform this technique, a substantial collection of normal brain data of different age groups are crucial because they will serve as the normal controls for comparison with the diseased brain. The most utilized analyses are those for DTI FA and MD maps. T2 and FLAIR signal maps can also be performed. The susceptibility imaging map for iron accumulation detection and calculations can demonstrate that inflammation has been introduced recently.[13] To build a robust brain template, the whole-brain probabilistic DTI tractography maps of normal controls must be calculated to create brain maps representing relative axonal density, then co-registered and averaged. Based on a whole-brain probabilistic tractography map against the template, all warping and registration parameters can be calculated and then applied to individual FA and MD maps. Voxel-based group analyses of encephalitis against normal brains were performed on FA and MD maps as shown in Figures 21.3 and 21.4. Figure 21.5 demonstrates voxel-based group analysis of susceptibility (estimated T2*) imaging for blood product accumulation in the brain of young boxers in Thailand.

21.3.1.7 Application of MRI Techniques in Rabies: Utility and Limitations During the Early and Late Stages

Conventional MR techniques (as in Sections 21.3.1.1, 21.3.1.2, and 21.3.1.5) are targeted as diagnostic imaging tools to exclude other non-rabies encephalitides, such as those associated with herpes simplex, Japanese encephalitis viruses, and immune/paraneoplastic origin. To evaluate the bleeding or hemorrhagic complication, susceptibility imaging is used (Section 21.3.1.2). To confirm or exclude metabolic or toxic encephalopathy, vascular, or tumor causes from rabies, metabolite imaging is used (Section 21.3.1.3).

Structural data studies (Section 21.3.1.1) have limited value in the diagnosis of rabies, but can provide evidence of immune responses at the peripheral nerve, brachial plexus, and spinal nerve roots at the bitten segment during the prodromal phase (Figure 21.6A). Nonspecific subtle hypersignal T2 signal in the brain and spinal cord can also be detected (Figure 21.6B and C).

During the early or conscious phase of the disease, abnormal hypersignal T2 change along the upper spinal cord and brainstem, temporal

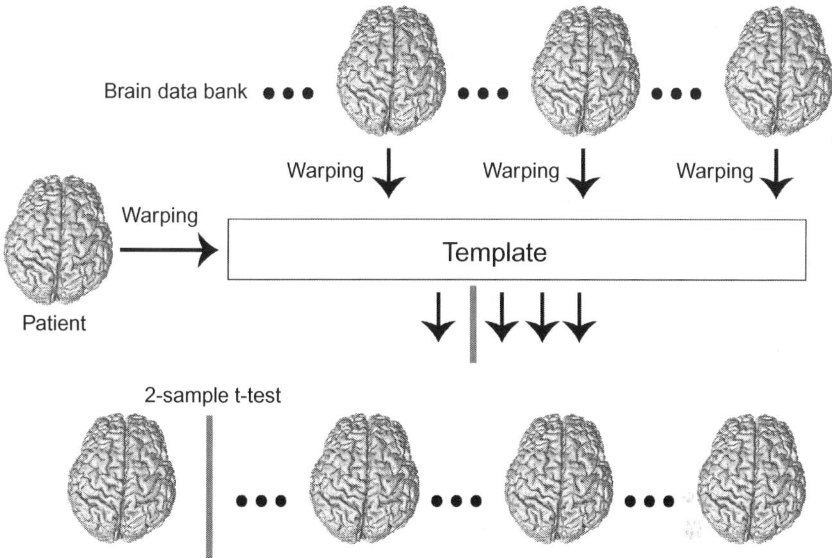

Figure 21.3 Demonstration of a method to obtain a voxel-based morphometric group analysis against an age-matched normal brain data bank. A quantitative comparison between these two data can be obtained by using a 2-sample t-test to detect statistically significant differences between these sets (using SPM). *(Source: http://www.fil.ion.ucl.ac.uk/spm).*

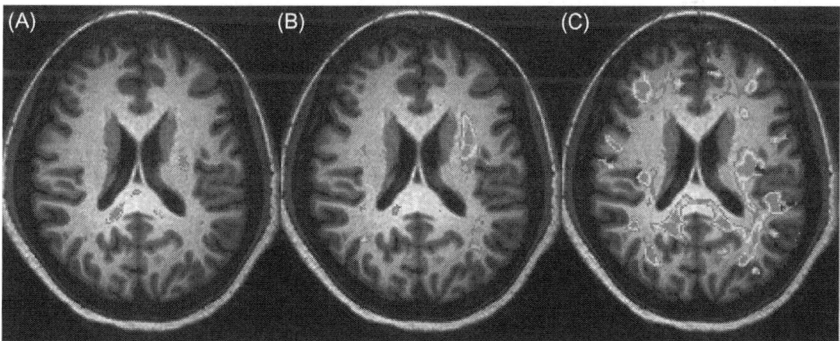

Figure 21.4 Voxel-based morphometry group analysis of DTI: increased MD probabilistic brain map in viral encephalitis patients compared with normal age-matched control ($p < 0.05$) with different clinical outcome as complete recovery in A, partial recovery in B, and disable in C demonstrates direct correlation of the degree of abnormality and clinical outcome seen as more progressive severe brain damage in the case of partial recovery and disable respectively.

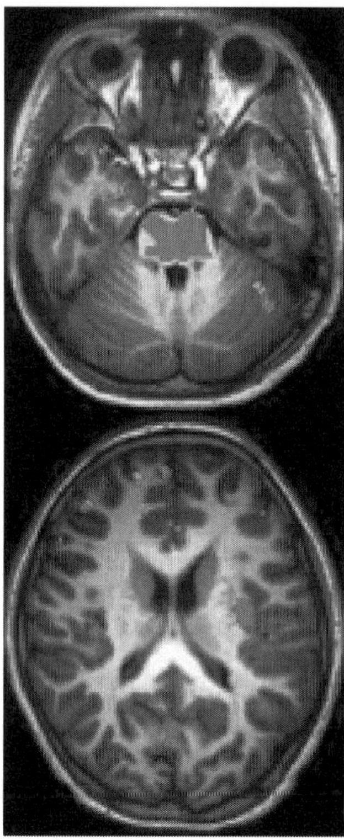

Figure 21.5 Voxel-based group analysis morphometry probabilistic map of the brain of young boxers compared to the normal age-matched control using susceptibility imaging technique ($p < 0.001$, $N = 11$ vs. 33) demonstrating increased iron deposition at the brainstem, temporal, occipital, and frontal and parietal cortical gray-white junction due to intracranial hemorrhage caused by chronic repeated head injury. *(Data from normal Thai brain mapping research, Advanced Diagnostic Imaging Center, Faculty of Medicine Ramathibodi Hospital, Mahidol University: AIMC.mahidol.ac.th & Med.mahidol.ac.th).*

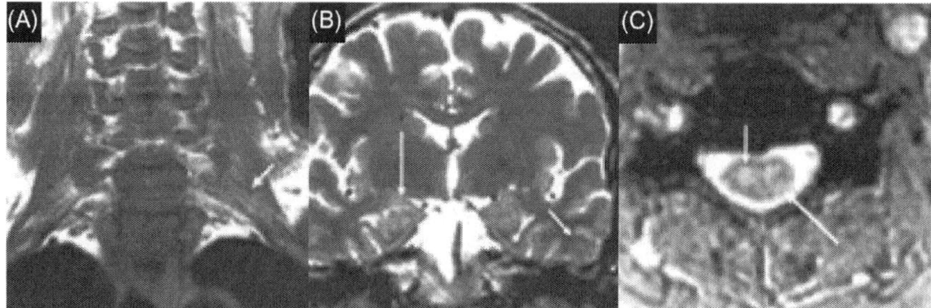

Figure 21.6 A 50-year-old male with furious rabies encephalitis during the prodromal phase. (A) Coronal T1W post-gadolinium MR image of the brachial plexus demonstrating enhancing left brachial plexus (white arrows in A). (B and C) Coronal and axial T2-weighted image of the brain and cervical cord demonstrating ill-defined nonenhancing mild-to-moderate hypersignal T2 change at the amygdala and hippocampi (long white arrow in B) and adjacent temporal cortical gray matter (short white arrow in B) and central gray and left posterolateral white matter column (short and long white arrows in C) *(A is reproduced with permission from Laothamatas et al. (2003)[17]).*

lobes, and cerebral hemispheres is demonstrated, with a preserved BBB.[7] Slight-to-moderate degree of differences in signal intensity and area of distribution can be demonstrated in furious and paralytic dog brains by conventional MR imaging (more disturbances in paralytic). Advanced imaging techniques (Sections 21.3.1.4 and 21.3.1.6) clearly demonstrated micro- and macro-cellular damages and integrity of white matter tracts and the BBB[6–8] (see details in Figure 21.7).

During the comatose phase, MR abnormalities are widespread and intense, involving the entire cerebral hemispheres, brainstem, cervical cord, and cranial and spinal nerves, with breakdown of the BBB and

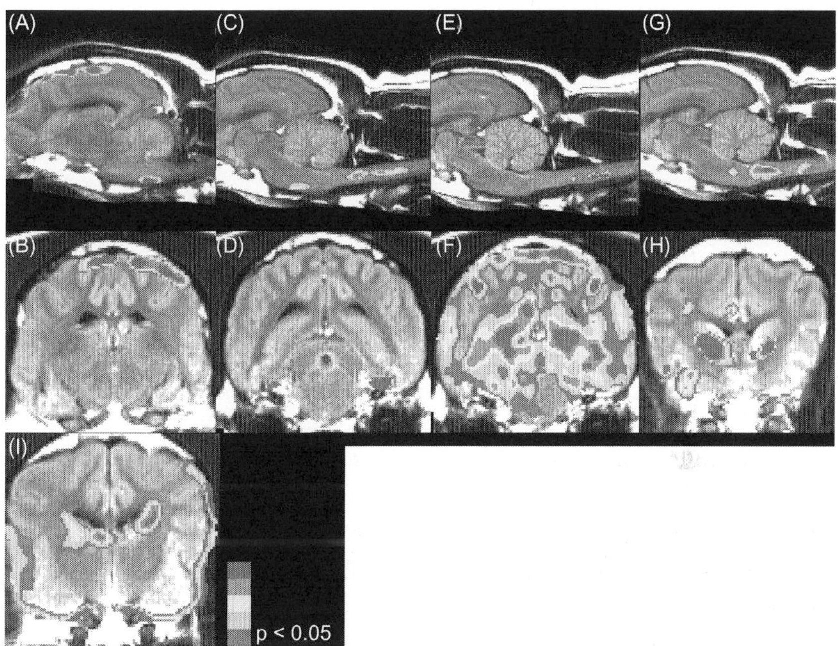

Figure 21.7 MR images of rabid dogs in an early phase (A-to-I from voxel-wise group analyses, based on DTI probabilistic tractography density map normalization): (A and B) Mild increased FLAIR signals at the brainstem and frontoparietal cortex in furious rabid dogs ($p < 0.05$, N=2, normal control = 7); (C and D) More intense increased FLAIR signals in the cervical cord and temporal lobes in paralytic rabid dogs ($p < 0.05$, N = 4, normal control = 7); (E) Decreased FA; while (F) Increased FA values in furious rabid dogs ($p < 0.05$, N = 2, normal control = 8); (G) Decreased FA in the cervical cord; while (H) Increased FA values in the right temporal lobe, basal ganglia, and cingulated gyrus in paralytic rabid dogs ($p < 0.05$, N = 4, normal control = 8) compatible with condition compressing the white matter tracts; and (I) Decreased MD values at the basal ganglia in paralytic dogs suggesting cytotoxic edema ($p < 0.05$, N = 4, normal control = 8). There were no significant differences in increased MD values in both rabid dog groups, as well as decreased MD values in furious rabid dogs.

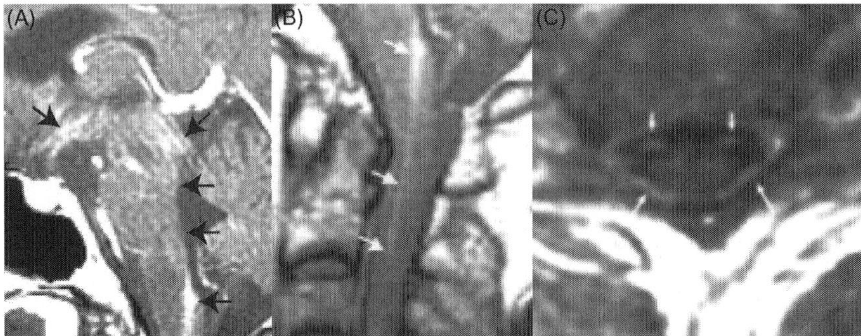

Figure 21.8 Conventional MR images of the brain and cervical spines of comatose patients with paralytic rabies with post-gadolinium sagittal and axial T1-weighted images at the brainstem (A) and upper cervical cord (B and C); black arrows in A demonstrate enhancement along the floor of the aqueduct of Sylvius and fourth ventricle and hypothalamus, midbrain, and tectal plate and periaqueductal gray matter; white arrows in B demonstrate enhancement along the upper cervical cord and enhancing dorsal and ventral intrathecal cervical nerve roots in C.

blood nerve barriers in both forms (human cases shown in Figure 21.8).[17] Superimposed insults from hypoxia/anoxia, toxic, and metabolic complications, and from the disease process itself, are likely explanations.

21.4 DISCUSSION

Previously considered as a technique of limited value, MRI has become valuable not only in excluding other forms of encephalitis (of sporadic and endemic origins), but also in elucidating mechanisms that underlie different manifestations in rabies. When combined with in-depth investigations of post-mortem tissue, such *in vivo* MR techniques would further detail the ongoing process during life. Susceptibility image techniques (voxel-based group analysis), as shown in Figure 21.5, should be extremely useful and applicable for grading inflammatory reactions and extent in rabies, once hemorrhage has been excluded by structural studies.

ACKNOWLEDGMENTS

Research by the authors described in this chapter was supported by grants from the Advanced Diagnostic Imaging Center, Ramathibodi Foundation and Mahidol University Faculty of Medicine Ramathibodi Hospital, and a Research Chair Grant, National Science and Technology Development Agency (Thailand).

REFERENCES

1. Hemachudha T, Ugolini G, Wacharapluesadee S, Sungkarat W, Shuangshoti S, Laothamatas J. Human rabies: neuropathogenesis, diagnosis, and management. *Lancet Neurol.* 2013;**12**(5):498–513.
2. Nucifora PGP, Verma R, Lee S-K, Melhem ER. Diffusion-tensor MR imaging and tractography: Exploring brain microstructure and connectivity. *Radiology* 2007;**245**(2):367–84.
3. Haacke EM, Mittal S, Wu Z, Neelavalli J, Cheng Y-CN. Susceptibility-weighted imaging: Technical aspects and clinical applications, Part 1. *Am J Neuroradiol* 2009;**30**(1):19–30.
4. Haacke EM, Cheng NY, House MJ, Liu Q, Neelavalli J, Ogg RJ, et al. Imaging iron stores in the brain using magnetic resonance imaging. *Magn Reson Imaging* 2005;**23**(1):1–25.
5. Peter BB, Doris DML. In vivo proton MR spectroscopy of the human brain. *Prog Nucl Magn Reson Spectrosc* 2006;**49**(2):99–128.
6. Shuangshoti S, Thepa N, Phukpattaranont P, Jittmittraphap A, Intarut N, Tepsumethanon V, et al. Reduced viral burden in paralytic compared to furious canine rabies is associated with prominent inflammation at the brainstem level. *Vet Res* 2013;**9**(31):1746–8.
7. Laothamatas J, Wacharapluesadee S, Lumlertdacha B, Ampawong S, Tepsumethanon V, Shuangshoti S, et al. Furious and paralytic rabies of canine origin: Neuroimaging with virological and cytokine studies. *J Neurovirol* 2008;**14**(2):119–29.
8. Laothamatas J, Sungkarat W, Hemachudha T. Neuroimaging in rabies. *Adv Virus Res* 2011;**79**:309–27.
9. Jacobs MA, Ibrahim TS, Ouwerkerk R. MR Imaging: Brief overview and emerging applications. *Radiographics* 2007;**27**(4):1213–29.
10. Bottomley PA. Turning up the heat on MRI. *J Am Coll Radiol* 2008;**5**(7):853–5.
11. Rosas H, Wippold II FJ. West Nile Virus: Case report with MR imaging findings. *Am J Neuroradiol* 2003;**24**(7):1376–8.
12. Akhlaghpoor S, Ghahari A, Morteza A, Khalilzadeh O, Shakourirad A, Alinaghizadeh MR. Quantitative T2* magnetic resonance imaging for evaluation of iron deposition in the brain of beta-thalassemia patients. *Clin Neuroradiol* 2012;**22**(3):211–7.
13. Urrutia P, Aguirre P, Esparza A, Tapia V, Mena NP, Arredondo M, et al. Inflammation alters the expression of DMT1, FPN1 and hepcidin, and it causes iron accumulation in central nervous system cells. *J Neurochem* 2013;**126**(4):541–9.
14. De Backer ME, Nabuurs RJA, Van Buchem MA, Van der Weerd L. MR-based molecular imaging of the brain: The next frontier. *Am J Neuroradiol* 2010;**31**(9):1577–83.
15. AG S. Cerebral MR Perfusion Imaging. Principles and Current Applications. New York: Georg Thieme Verlag 2001; 2003.
16. Liu Z, Zhu H, Marks BL, Katz LM, Goodlett CB, Gerig G, et al. Voxel-wise group analysis of DTI. *Proc IEEE Int Symp Biomed Imaging* 2009:807–810.
17. Laothamatas J, Hemachudha T, Mitrabhakdi E, Wannakrairot P, Tulayadaechanont S. MR imaging in human rabies. *Am J Neuroradiol* 2003;**24**(6):1102–9.

PART THREE

Rabies Biologics

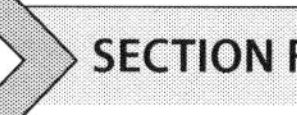

SECTION F

Rabies Vaccines for Humans or Other Animals

CHAPTER TWENTY TWO

Attenuated Vaccines for Veterinary Use

Ad Vos, Grit Neumann, Boris Hundt, and Andreas Neubert
IDT Biologika GmbH, Dessau-Rosslau, Germany

Chapter Contents

22.1 Introduction 237
22.2 Materials 238
  22.2.1 Reagents 238
  22.2.2 Equipment and Facilities 238
  22.2.3 Biological Materials 238
    22.2.3.1 Selection of Seed Virus and Seed Cells *238*
22.3 Methods 240
  22.3.1 Establishment of Master Seed Lot 240
  22.3.2 Upstream Process 240
    22.3.2.1 Seed Train *240*
    22.3.2.2 Production Batch *241*
  22.3.3 Downstream Process 242
    22.3.3.1 Stability/Stabilizers *242*
    22.3.3.2 Concentration *242*
  22.3.4 Potency 243
22.4 Discussion 243
References 244

22.1 INTRODUCTION

An attenuated vaccine is a vaccine developed by reducing the virulence of the pathogen; it remains replication-competent after administration. Basically, the attenuated virus administered is unable to replicate sufficiently to cause disease in the recipient, but will still induce an immune response that can protect against infection before the virus is cleared by the immune system. Because the vaccine virus is replication-competent, there is a risk of reversion to virulence after administration, albeit small. Therefore, the use of live attenuated rabies virus (RABV) vaccines in domestic animals is not recommended due to the potential residual pathogenicity of the strains used.

C. Rupprecht & T. Nagarajan (Eds): Current Laboratory Techniques in Rabies Diagnosis, Research, and Prevention, Volume 1.
Doi: http://dx.doi.org/10.1016/B978-0-12-800014-4.00022-6

Cases of vaccine-induced rabies have been reported in both domestic and wild animals using such rabies vaccines.[1–5] In spite of these potential risks, live attenuated virus vaccines offer considerable advantages over inactivated (killed) vaccines. Attenuated vaccines: (i) activate a wide range of immune responses, (ii) induce long-lasting and rapid onset of immunity, (iii) often reduce the need for booster vaccinations, (iv) do not need adjuvants, (v) can be produced at relatively low cost, and (vi) can be administered by the oral route. Hence, all currently available vaccines used for oral vaccination of free-roaming animals against rabies are live modified virus vaccines. The RABV can be attenuated via conventional methods like passaging the virus through tissue cultures (cells), embryonated eggs, or animals. However, nowadays, new methods are available that enable the selection of more genetically homogeneous vaccine strains. Attenuated mutants can be selected by monoclonal antibodies (MAbs) or developed by site-directed mutagenesis in genes involved with virulence using reverse genetics (see Chapter 24). The vaccine production process itself is comprised of a seed train, virus propagation (upstream processing), clarification and formulation (downstream processing), and final product release. The whole process is accompanied by intensive testing of all raw materials, as well as in-process and final product.

22.2 MATERIALS

22.2.1 Reagents

Reagents that are required for the production of a live attenuated RABV vaccine are shown in Table 22.1.

22.2.2 Equipment and Facilities

Relevant equipment and facilities required for the production of a live attenuated RABV vaccine as described in this chapter are shown in Tables 22.2 and 22.3.

22.2.3 Biological Materials

22.2.3.1 Selection of Seed Virus and Seed Cells

The fact that virus strains can vary in their virulence and immunogenic properties in the different animal species targeted underscores the need for using well-established and characterized strains for routine production. Most animal rabies vaccines are prepared from derivatives of Pasteur's original 1882 strain (Challenge Virus Standard (CVS), Pitman-Moore) and

Table 22.1 Reagents for the Production of Live Attenuated Rabies Vaccine

Reagent	Purpose
Growth media containing 5–10% serum and 2% L-Glutamine	Cell propagation
Propagation media with 2–10% serum	Virus propagation
Glutamine (200 mM stock solution)	Media additive
Newborn calf serum or fetal bovine serum	Media additive
Buffer solution (PBS)	Washing/rinsing of cell cultures
Trypsin	Cell detachment (for cell-culture upscaling/passaging)
Stabilizer with gelatin, saccharose, and balanced salt solution	Vaccine stabilization
Sodium chloride solution	Dilution buffer

Table 22.2 Equipment for the Production of Live Attenuated Rabies Vaccine

Equipment	Purpose
Incubators (35–37°C)	Cell incubation
Microscope	Cell counting, cell evaluation, CPE evaluation
Cell culture flasks	Cell cultivation/passaging after thawing
Roller bottle (RB)	Cell passaging for seed train
Cell factories (CF)	Cell cultivation for production batch
Harvest containers	Storage of harvest/intermediate products
Mixing vessels	Vaccine formulation
Refrigerators	Storage of harvest/intermediate products
Freezers	Storage of seed lots/intermediate product/final product

Table 22.3 Facilities Needed for the Production of Live Attenuated Rabies Vaccine

Facilities
Quality control Area I for raw materials
Storage areas (−20°C, 2–8°C, room temperature) for raw materials
Medium preparation area
Cell preparation area
Virus propagation area (BSL2 advisable)
Quality Control Area II for in-process control and final product
Storage area for virus bulk and final product (≤ −20°C)

strains isolated in the 20th century (e.g. Flury High Egg Passage (HEP); Flury Low Egg Passage (LEP); Street alabama Dufferin (SAD); and Evelyn Rokitniki Abelseth (ERA)). Different vaccine substrates for RABV vaccines are available. Initially, RABV vaccines were propagated in nervous tissue, followed by primary cell cultures.[6] In the early 1960s, continuous cell lines became available for the production of vaccines. Although some animal rabies vaccines are still produced in nerve tissue or primary cells, these products will not be considered here. The production and use of these vaccines should be discouraged because these products are not only associated with certain safety issues, but also raise concerns from an animal welfare point of view. The most widely used continuous cell lines for animal rabies vaccine are baby hamster kidney-21 (BHK-21) and Vero cells. Both cell lines can be grown in stationary and suspension culture (using microcarriers or in free-suspension).

22.3 METHODS

This chapter provides information on the manufacturing process until final formulation and filling. These last steps are influenced by the final use of the product (parenteral or oral vaccination). Oral vaccines are almost exclusively liquid and are filled in blister or sachet and subsequently incorporated in a bait matrix. Vaccines for parenteral use are filled in single- or multiple-dose vials in liquid or freeze-dried form.

22.3.1 Establishment of Master Seed Lot

For the preparation, establishment, and characterization of a seed lot system (Master and Working Seed Virus [MSV/WSV] and Master and Working Cell Seed [MCS/WCS]), international guidelines/recommendations should be followed (e.g. 9 CFR §113.52 and EP §5.2.4). In this chapter, the production of a SAD-derived vaccine strain is described. The MSV and WSV (up to four passages from MSV) are stored at $\leq -70°C$ in suitable containers, like cryotubes. Hence, the final product is up to five passages from the MSV. BHK-21 continuous cell line is used as the cell substrate. Cell seed (MCS/WCS) is also stored in cryotubes in liquid nitrogen. The virus for the final product is usually propagated on cells up to 20 passages from MCS. Seed lots are extensively tested and stored at two different locations to avert risks.

22.3.2 Upstream Process

22.3.2.1 Seed Train

The seed train describes the process of propagating WCS cell material up to starting the vaccine production batch using cell culture flasks and roller bottles. A WCS cryovial (about 4 × 10^6 cells/mL in 1.8 mL) is seeded into a cell culture flask (175 cm^2) using 100 mL of the following growth medium: MEM supplemented with 5–10 % fetal bovine serum (FBS) and 2 % L-glutamine. The cells are incubated at 37°C without the use of CO_2, and a half or full medium exchange is done after 24 h. Typically, a confluent monolayer for the first passage is reached after 3–5 days. Afterwards, the cells are passaged into the next scale applying the following scheme, which is also used for further expansion steps:

- Decant the growth medium.
- Rinse the cell monolayer with buffer solution at room temperature (175 cm^2 flask with 20 mL, roller bottle with 100 mL).
- Decant the rinsing buffer.
- Addition of Trypsin or recombinant protease (175 cm^2 flask with 2 mL, roller bottle with 5 mL) pre-warmed to 37°C.
- Incubation at 37°C until cells are disaggregated from the surface (under microscopic control); typically this takes 15–30 min.
- Pooling of the cells and viable cell counting.
- Cell seeding in the next production system using a starting cell number of 2–5 × 10^4 cells/cm^2 (split rates of 1:3 to 1:15 possible).

For the next expansion step, the cells from the 175 cm^2 flask are typically seeded into an 850 cm^2 roller bottle with 300 mL growth medium. The incubation (revolution for roller bottles 15–25 rounds per h) and passaging is done as described above, typically 5–6 passages and 15–20 days are needed to get cell material for starting a production batch of 64 CF10 (CF, cell factories).

22.3.2.2 Production Batch

The cell material needed for starting the production batch of 64–128 CF10 (alternative: 210–240 roller bottles) is harvested from 100–200 roller bottles, and the cell number of the pool is determined. Subsequently, the cell suspension is pre-diluted to 1–3 × 10^5 cells/mL and is transferred to the virus production labs. The virus from WSV is thawed at room temperature and added at an MOI of 10^{-3} to 10^{-4} to the cell suspension (time of infection 0 h). The infected cell suspension is then seeded in CF10 (2,200 mL growth medium) with a cell number of 3.5–10 × 10^4 cells/cm^2 (alternative: roller bottles with 300 mL growth medium). The seeded culture vessels are incubated at 35°C for 4 days. At this time (92–96 h p.i.) the virus is harvested by simply exchanging the cell culture supernatant against fresh medium. Up to two more harvests are possible with harvest intervals of 24–72 h.

22.3.3 Downstream Process

For parenteral vaccines, the viral harvest collected should be clarified immediately to eliminate cell debris; this is typically done using 1 or 3 µm-depth filters. The (clarified) virus harvests are pooled and collected in sterile 35–36 liter containers that can be stored at ≤−20°C for a maximum of two years. Samples from each bag are collected before freezing and tested for sterility (50 mL) and virus titer (>10 mL). For final vaccine formulation, the selected virus pools are mixed under sterile conditions with buffer solution, stabilizer, and antibiotics in a 500 L mixing vessel. The amount of each of these components in the final formulation should be watched carefully: virus material between 26–36%, stabilizer between 32–36%, buffer solution between 30–39%, and antibiotics approximately 1% of total volume. The titer of the pooled virus material prior to mixing should be calculated from the individual virus pools used to make sure that the final virus titer of the vaccine is not falling below the releasing titer of the final product. After final formulation, the vaccine can be stored <+20°C under light protection for a maximum period of 24 hours until the vaccine formulation is filled in the appropriate containers (vials, blisters, etc.) and further processed. Instead of a liquid vaccine, the vaccine can also be freeze-dried. This step would add to the thermal stability of the vaccine. However, this production step will increase production costs considerably, and the freeze-drying process will also result in an initial loss of titer.

22.3.3.1 Stability/Stabilizers

In case of attenuated live vaccines, the great importance of obtaining the necessary infective virus titers in the final product, and preserving these titers until administration is obvious, especially for oral rabies vaccines that are distributed in the environment and exposed to the prevailing climatic conditions. Hence, a stabilizer is needed. Presently, gelatin-based stabilizers provide the best results at both sub-zero and elevated temperatures.

22.3.3.2 Concentration

In case the required titer to induce protective immunity in the target species exceeds the titer that can be reached during upstream processing, a concentration step may be required. The antigen is concentrated and purified using ultrafiltration. A method of choice for concentration of live RABV is tangential flow filtration using either hollow fibers or flat sheet cassettes. For flat sheet cassettes, 300 or 500 kDa should be used, and for hollow fibers 500 kDa MWCO is the best choice. It has to be noted that antigen prior to concentration needs to be clarified to remove cell debris, and to some extent, host cell proteins and DNA, to protect the ultrafiltration membranes from clogging.

22.3.4 Potency

The immunogenicity of live modified RABV vaccines can easily be determined through standard viral titration methods. The titer of the final product should be above the minimum effective dose for the respective target species. Furthermore, a safety margin of at least 0.5 log for the pooled harvest virus titer should be considered, to overcome possible losses of infectious titers during downstream processing.

22.4 DISCUSSION

There are many points to consider for optimizing or adapting the described manufacturing process to produce a vaccine meeting particular requirements. For example, for some vaccine strains it is better to delay the time of infection from direct infection to 48 hours. Another point to consider is that newborn calf serum (NCS) may be the better choice for virus yield, but with FBS, higher cell yields can be obtained. Hence, during the establishment of the manufacturing process, critical process steps should be identified and, if necessary, adjusted accordingly. The following timelines need to be considered for the establishment of the manufacturing process: seed lot establishment (1 to 1.5 years, including testing), seed train (15 to 20 days), production batch-upstream (10 to 14 days), production batch-downstream (10 to 14 days), and final product testing and release (6 to 8 weeks).

As mentioned in the introduction, rabies vaccine can also be manufactured in suspension culture whereby cells are often grown on microcarriers in bioreactors. Both cell lines, BHK-21 and Vero, can be adapted for growth as suspension culture. Suspension culture has certain advantages, such as larger production volumes, continuous medium perfusion, and certain parameters, including pH, oxygen, and carbon dioxide levels, which can be monitored and controlled continuously. Cells are separated from the microcarriers by trypsination. Possible production systems here include disposable wave motion-based reactors or a stirred-tank reactor. Another alternative to cell factories is a more consistent fixed bed reactor.

Also, the use of serum-free growth and maintenance culture medium has certain advantages: reduced transmissible spongiform encephalopathy (TSE)-risk and the potential contamination risk of other extraneous agents of bovine origin. However, it is often observed that under these conditions (suspension and serum-free), virus yield is lower. Finally, the production facility should operate under the appropriate biosafety procedures and practices. For example, staff must receive rabies pre-exposure prophylaxis, and their antibody titers should be checked at least annually. While it is a live replication-competent virus vaccine, not only should production staff be vaccinated, but also all persons who have direct contact with the final product (labeling, packaging, etc.).

REFERENCES

1. Bellinger DA, Chang J, Bunn TO, Pick JR, Murphy M, Rahija R. Rabies induced in a cat by high-egg-passage Flury strain vaccine. . *J Am Vet Med Assoc* 1983;**183**:997–8.
2. Esh JB, Cunningham JG, Wiktor TJ. Vaccine-induced rabies in four cats. *J Am Vet Med Assoc* 1982;**180**:1336–9.
3. Barnard BJ, Geyer HJ, De Koker WC. Neurological symptoms in a cat following vaccination with high egg passage Flury rabies vaccine of chicken embryo origin. *Onderstepoort J Vet Res* 1977;**44**:195–6.
4. Fehlner-Gardiner C, Nadin-Davis S, Armstrong J, Muldoon F, Bachmann P, Wandeler A. Era vaccine-derived cases of rabies in wildlife and domestic animals in Ontario, Canada, 1989–2004. *J Wildl Dis* 2008;**44**:71–85.
5. Müller T, Bätza HJ, Beckert A, Bunzenthal C, Cox JH, Freuling CM, et al. Analysis of vaccine-virus-associated rabies cases in red foxes (Vulpes vulpes) after oral rabies vaccination campaigns in Germany and Austria. *Arch Virol* 2009;**154**:1081–91.
6. Kissling RE. Growth of rabies virus in non-nervous tissue culture. *Proc Soc Exptl Biol Med* 1958;**98**:223–5.

CHAPTER TWENTY THREE

Pox Viral Vectored Vaccines for Rabies

Jacqueline Weyer[1] and Louis H. Nel[1,2]
[1]Centre for Emerging and Zoonotic Diseases, National Institute for Communicable Diseases, Sandringham, Gauteng, South Africa
[2]Department of Microbiology and Plant Pathology, University of Pretoria, Pretoria, Gauteng, South Africa

Chapter Contents

23.1 Introduction 245
23.2 Materials 246
  23.2.1 Reagents 246
  23.2.2 Cells, Viruses, and Other Biological Materials 247
  23.2.3 Equipment 247
23.3 Methods 248
  23.3.1 Biological Safety Considerations 248
  23.3.2 Preparation of Vaccinia Virus Transfer Vector Containing a Rabies Virus Glycoprotein Gene 248
  23.3.3 Preparation of Recombinant Vaccinia Viruses 249
  23.3.4 Isolation of Homogenous Recombinant Vaccinia Virus Clones Expressing Rabies Virus Glycoprotein using Mycophenolic Acid Resistance Screening 250
  23.3.5 Expression Analysis 250
  23.3.6 Culturing of Recombinant Vaccinia Virus for Animal Studies 251
23.4 Discussion 251
References 252

23.1 INTRODUCTION

Recombinant poxviruses have been used extensively with respect to rabies for nearly three decades.[1] In 1984, the Wistar Institute developed a recombinant vaccinia virus (or V-RG) expressing the G gene of the Evelyn–Rokitnicki–Abelseth (or ERA) rabies virus (RABV) vaccine strain, which has been widely used in oral immunization programs for raccoons, coyotes, and gray foxes in North America, and red foxes in several European countries.[2–8] As proof of concept, the efficacious application of V-RG in bats was also demonstrated.[9–10] In addition, a highly host-restricted avian poxvirus (i.e. canarypox) called ALVAC, expressing two copies of the ERA G gene,

C. Rupprecht & T. Nagarajan (Eds): Current Laboratory Techniques in Rabies Diagnosis, Research, and Prevention, Volume 1.
Doi: http://dx.doi.org/10.1016/B978-0-12-800014-4.00023-8

has also been available for the vaccination of cats in North America.[11–15] Apart from these two constructs with commercial availability, a myriad of recombinant poxviruses expressing RABV antigens have been constructed over the years as possible vaccine candidates.[16–26] With the ease of construction of recombinant vaccinia viruses, it has provided a useful tool over time for studying the immunogenicity of viral antigens.[27] This has been demonstrated in studies testing the immune responses of mice against RABV nucleoprotein-expressing vaccinia viruses.[23–25] In a study by Weyer and co-workers,[26] the vaccinia model was used to determine the cross-protective responses, or lack thereof, for different lyssavirus G proteins.

Despite the success of recombinant poxvirus vaccines for rabies and other diseases, concerns remain regarding the use of replication competent vectored vaccines (such as recombinant vaccinia).[28] Nevertheless, poxvectored vaccines, with a high volume of use for vaccination against rabies, have displayed a great level of safety.[29–30] Because V-RG is distributed in bait form usually by aerial dispersal (but also by hand), concerns exist for non-intentional contact with non-target species, but these concerns may have been overstated with only single cases of untoward effects reported.[30–33] Although the use of ALVAC as an example of a host-restricted poxvirus has been successful, the investigation of other replication-deficient poxviruses such as Modified Vaccinia Virus Ankara revealed limited utility, especially as oral vaccines.[19]

This chapter will focus on the description of generic methodology for construction of a recombinant vaccinia virus expressing a RABV G gene and the preparation thereof for animal model studies. This was done using the thymidine kinase (TK) gene as insertion site for RABV G and mycophenolic acid (MPA) resistance as selectable marker.[34,35] Protocols for construction of other recombinant poxviruses will differ with respect to the point of insertion of the foreign DNA in the poxvirus genome and the selectable marker used for the purification of recombinant viruses for which there are a number of options.

23.2 MATERIALS

23.2.1 Reagents

- Minimal Essential Medium (MEM), glutamine, vitamin supplement, antibiotics, and fetal bovine serum and Dulbecco's phosphate buffered saline (PBS) for cell culture.
- Cell culture vessels including flasks (different volumes), 6-well and 24-well plates, and serological pipettes.

- Mycophenolic acid, xanthine, and hypoxanthine for selection of recombinant viruses (cell culture grade).
- Anti-rabies virus serum (polyclonal or monoclonal) for detection of G (primary antibody).
- Flourescein isothiocyanate-labeled conjugate for detection of primary antibody.
- Evan's Blue counterstain, 0.001%.
- Acetone.
- Bacterial growth medium, such as Luria Bertani broth (with required antibiotic for selection).
- Microcentrifuge tubes, culture tubes, and petri dishes.
- Recombinant *Escherichia coli* selection reagents (e.g. IPTG and xanthine for blue white selection).
- Polymerase chain reaction (PCR) reagents and consumables for amplification of RABV G gene.
- Liposomal transfection reagent.
- Poxvirus lysis buffer.[36]

23.2.2 Cells, Viruses, and Other Biological Materials

Monolayer cell cultures of Vero cells (e.g. CCL81 available from the American Type Culture Collection[37]) are grown at 37°C and 5% CO_2 in MEM supplemented with 4mM glutamine, 2× MEM vitamin solution and antibiotics (100μg/mL penicillin, 100μg/mL streptomycin and 250μg/mL amphotericin), and 10% fetal bovine serum using standard cell culture techniques.

A vaccinia virus isolate is required as parental vaccine strain. Various strains of vaccinia virus are available from the American Type Culture Collection,[37] but others may have to be requested from investigators directly.

A vaccinia virus transfer vector containing fragments homologous to the vaccinia virus TK gene flanking a cloning site with a poxvirus promoter sequence for the expression of the foreign gene sequence, and an *Escherichia coli* MPA resistance gene under regulation of a poxvirus promoter sequence, is used as in Weyer et al. (2008).[26]

23.2.3 Equipment

The following equipment is typically required for the protocol described in this chapter:

- Personal protective equipment (PPE), including gloves and laboratory coats (use of PPE will be based on site-specific risk assessment; see Section 23.3.1).

- Biosafety cabinet.
- Incubators, 37°C with 5% CO_2 and 37°C (shaking for bacterial cultures).
- Water baths for incubation at different temperatures.
- Micropipettes (1–1,000 μL), serological pipettors.
- Bench top centrifuges for volumes 1–50 mL, refrigerated.
- Fridge (4–8°C) and freezer (−20°C and −70°C).
- Inverted light microscope for visualization of cell culture and monitoring of viral infection.
- Fluorescence microscope, with 20× and 100× magnification, for visualization of fluorescence antibody tests for expression analysis.

23.3 METHODS

23.3.1 Biological Safety Considerations

Experimentation with recombinant poxviruses using attenuated or naturally host-restricted poxviruses (e.g. fowl poxvirus or Modified Vaccinia Virus Ankara) should be performed observing good laboratory practice and biosafety level 2 (BSL 2) laboratory conditions. No vaccination is required for working with these vectors. Handling of vaccinia virus does, however, pose safety concerns, and cases of laboratory-acquired infection have been documented. Vaccination with smallpox vaccine is recommended for workers handling vaccinia virus. All experiments using vaccinia and other poxviruses should be conducted after a site-specific risk assessment including all procedures.[35]

23.3.2 Preparation of Vaccinia Virus Transfer Vector Containing a Rabies Virus Glycoprotein Gene

- PCR amplification of the gene of interest. Amplify the RABV G gene from source material using standard reverse transcription PCR (reverse transcription (RT) PCR) protocols.[38] PCR primer sequences and conditions for the amplification of the RABV G have been described in previous studies, such as Tuchiya et al. (1992).[39]
- Molecular cloning of the gene of interest. The PCR amplicon must either be sub-cloned in a cloning vector, or appropriate restriction enzyme sites may be incorporated into the PCR amplicon by including these sequences at the 5′ ends of the PCR primers. The PCR

amplicon must finally be cloned into the expression cassette of a vaccinia virus transfer vector using standard molecular cloning protocols not described here.[40–42] If a non-directional cloning approach was used, the orientation of the inserted gene should be verified in the recombinant plasmid. This is done using restriction enzyme digestion and/or molecular sequencing of the region.

- Verification of the recombinant vaccinia virus transfer vector. Recombinant vaccinia virus transfer vector plasmids that contain the RABV G gene in the appropriate orientation to the poxvirus promoter using restriction enzyme digestion (or conventional Sanger sequencing of the expression cassette) are then amplified and purified using standard operating procedures (SOPs) in overnight *E.coli* cultures.[40]

23.3.3 Preparation of Recombinant Vaccinia Viruses

The procedures for the construction of recombinant vaccinia viruses are well described.[34,35,43] In summary:

- Preparation of cell cultures. Confluent, overnight monolayers of Vero cell culture are prepared in 6-well cell culture plates according to standard cell culture procedures.
- Inoculation of cell cultures with vaccinia virus. The monolayers are then inoculated with parental vaccinia virus at a multiplicity of infection of 0.01. Briefly, the cell culture medium is aspirated from the monolayers. The vaccinia virus culture is diluted in cell culture medium with 2% fetal calf serum in a maximum volume of 0.5 mL. The inoculum is pipetted onto the monolayers and the plates incubated at 37°C and 5% CO_2 for 60 minutes to allow for the absorption of the virus. The vaccinia virus will have to be cultured and titrated before use.[44]
- During the incubation period, the purified recombinant plasmids are used to prepare transfection complexes by using a commercial liposomal-based transfection reagent as suggested by the manufacturer (alternatively calcium chloride precipitation of the DNA may also be used, if preferred). Additional optimization may be required to attain suitable levels of transfection as instructed by the manufacturer. Using vaccinia transfer vectors that express *E. coli* green fluorescent protein facilitate the assessment of transfection efficiency and also aids in the selection of recombinants.[17]

23.3.4 Isolation of Homogenous Recombinant Vaccinia Virus Clones Expressing Rabies Virus Glycoprotein using Mycophenolic Acid Resistance Screening

- Following the incubation period, the inoculums are aspirated from the monolayers, the monolayers washed twice with sterile PBS (Dulbecco), and the transfection complexes added to the cell cultures as recommended by the manufacturer of the transfection reagent. Finally, the cultures are fed with fresh cell culture medium containing 2% serum (i.e. maintenance cell culture medium) and returned to incubation as before for 48 hours.
- After a 48-hour incubation period the virus is harvested by three freeze–thaw cycles. The cell lysates are centrifuged briefly to pellet cellular debris.
- The supernatants are then diluted in a ten-fold serial dilution on fresh monolayers of Vero cultures and under MPA selection. The latter is achieved by the addition of 25 μg/mL MPA, 250 μg/mL xanthine, and 15 μg/mL hypoxanthine to the standard maintenance cell culture medium. The cultures are then incubated for 48 hours as before.
- Classical virus plaque purification is performed by picking individual plaques and repeating the passage as described before.
- PCR screening of the recombinant clones may be performed by extracting the vaccinia virus DNA as described before[36] and following standard PCR procedures using oligonucleotides targeting the TK region (i.e the site of insertion of the glycoprotein gene).[28] Agarose gel electrophoresis of the amplification products will reveal the presence of a larger amplification fragment (representing the recombinant expression cassette) and/or a smaller fragment (representing the non-recombinant expression cassette). The aim is to repeat the plaque purification step until homogenous recombinant vaccinia viruses are isolated.

23.3.5 Expression Analysis

A recombinant clone expressing the RABV G should be selected. Expression of the G may be determined via various techniques including Western blotting. Here we describe the use of indirect immunofluorescence as a rapid technique for determining antigen expression.

- Seed Vero cell culture as before, but using 6-well cell culture plates lined with sterilized cover slips (alternatively 4- or 8-well chamber slides may be used).

- Inoculate the confluent overnight monolayers with harvested supernatants (as before) of different clones (i.e. plaques).
- Incubate the inoculated cultures as before for 48 hours.
- Following the incubation period, aspirate the cell culture medium and remove the cover slips from the culture plates. Fix the cells on the cover slips in ice-cold acetone for 20 minutes.
- Following the fixation, air dry the cover slip followed by overlaying the cultures with anti-rabies serum (may be polyclonal or monoclonal for glycoprotein epitopes). The latter is diluted in PBS containing 0.001% Evan's blue counterstain. The slips are then incubated at 37°C in a humid chamber for one hour.
- Following the incubation, an appropriate FITC-labeled conjugate may be used if the primary antibody is not labeled. The cover slips are flooded with conjugate after the primary antibody has been washed off (wash twice with PBS) and incubated as before for 30 minutes.
- The cover slips are washed again as before, but followed by a rinse in distilled water. The washed cover slips are left to air dry, after which they are mounted on microscope slides and visualized under a UV microscope (typically using 200× magnification). Apple green fluorescence, which is cell-associated, would be an indication of expression of the recombinant protein. Controls that constitute cell culture that is not infected, and cell culture infected with parental vaccinia virus, should be included to aid in distinguishing between background of non-specific fluorescence.

23.3.6 Culturing of Recombinant Vaccinia Virus for Animal Studies

The recombinant vaccinia virus expressing the RABV G is then cultured to attain the suitable volume of virus to conduct the planned experimentation. Standard protocols for culturing, harvesting, and titration of recombinant vaccinia virus are described elsewhere.[44] For animal experimentation, the virus preparation should be purified (e.g. semi-purification through sucrose cushion centrifugation).

23.4 DISCUSSION

Poxvirus-based vaccines for rabies have played a major role in the control of wildlife rabies in Europe and North America during the course of the past three decades. Methods for the generation of recombinant poxviruses are well established, and as shown in this chapter, such constructs can be prepared with ease by competent personnel working in an appropriately equipped facility. In addition, vaccines based on attenuated poxviruses have been shown to be safe over the many years of use in humans and animals. Recombinant poxviruses also remain a most useful tool for the study of the properties of specific antigens in a live viral vector system. Nevertheless, as far as the use of recombinant poxvirus vaccines for rabies is concerned, the current climate also favors the exploration of new recombinant and attenuated RABVs, as well as adenovirus-based vaccines.

REFERENCES

1. Weyer J, Rupprecht CE, Nel LH. Poxvirus-vectored vaccines for rabies – a review. *Vaccine* 2009;**27**:7198–201.
2. Kieny MP, Lathe R, Drillien R, Spehner D, Skory S, Schmitt D, et al. Expression of rabies virus glycoprotein from a recombinant vaccinia virus. *Nature* 1984;**312**:163–6.
3. Wiktor TJ, MacFarlan RI, Reagen KJ, Dietschold B, Curtis PJ, Wunner WH, et al. Protection from rabies by a vaccinia virus recombinant containing the rabies virus glycoprotein recombinant virus vaccine. *PNAS* 1984;**81**:7194–8.
4. Blancou J, Kieny M-P, Lathe R, Lecocq JP, Pastoret PP, Soulebot JP, et al. Oral vaccination of fox against rabies using a live recombinant vaccinia virus. *Nature* 1986;**322**(6077):373–5.
5. Rupprecht CE, Wiktor TJ, Johnston DH, Hamir AN, Dietschold B, Wunner WH, et al. Oral immunization and protection of raccoons (Procyon lotor) with a vaccinia-rabies glycoprotein recombinant virus vaccine. *PNAS* 1986;**83**:7947–50.
6. Tolson DN, Charlton KM, Casey A, Knowles MK, Rupprecht CE, Lawson KF, Campbell JB. Immunization of foxes against rabies with a vaccinia recombinant virus expressing the rabies glycoprotein. *Archives of Virology* 1988;**102**:297–301.
7. Cliquet F, Aubert M. Elimination of terrestrial rabies in Western European countries. *Dev Biol (Basel)* 2004;**119**:185–204.
8. Rupprecht CE, Hanlon CA, Slate D. Oral immunization of wildlife against rabies: opportunities and challenges in prevention and control. *Dev Biol (Basel)* 2004;**119**:173–84.
9. Aguilar-Setién A, Brochier B, Tordo N, De Paz O, Desmettre P, Péharpré D, et al. Experimental rabies infection and oral vaccination in vampire bats (Desmodus rotundus). *Vaccine* 1998;**16**(11/12):1123–6.
10. Aguilar-Setién A, Campos YL, Cruz ET, Kretschmer R, Brochier B, Pastoret P-P. Vaccination of vampire bats using recombinant vaccinia-rabies virus. *J Wild Dis* 2002;**38**(3):539–44.
11. Cadoz M, Meignier B, Plotkin S, Strady A, Taylor J, Tartaglia J, et al. Immunization with canarypox virus expressing rabies glycoprotein. *The Lancet* 1992;**339**(8807):1429–32.
12. Poulet H, Minke J, Pardo MC, Juillard V, Nordgren B, Audonnet J-C. Development and registration of recombinant veterinary vaccines, the example of the canarypox vector platform. *Vaccine* 2007;**25**:5606–12.
13. Taylor J, Trimarchi C, Weinberg R, Languet B, Guillermin F, Desmettre P, Paoletti E. Efficacy studies on a canarypox-rabies recombinant virus. *Vaccine* 1991;**9**(3):190–3.

14. Taylor J, Meignier B, Tartaglia B, Languet B, Van der Hoeven J, Franchini G, et al. Biological and immunogenic properties of a canarypox-rabies recombinant, ALVAC-RG (vCP65) in non-avian species. *Vaccine* 1995;**13**(6):539–49.
15. Fries LF, Tartaglia J, Taylor J, Kauffman EK, Meignier B, Paoletti E, Plotkin S. Human safety and immunogenicity of a canarypox-rabies glycoprotein recombinant vaccine: an alternative poxvirus vector system. *Vaccine* 1996;**14**(5):428–34.
16. Aspden K, Passmore J-A, Tiedt F, Williamson A-L. Evaluation of lumpy skin disease virus, a capripoxvirus, as a replication-deficient vaccine vector. *J Gen Virol* 2003;**84**:1985–96.
17. Wallace DB, Weyer J, Nel LH, Viljoen GJ. Improved method for the generation and selection of homogenous lumpy skin disease virus (SA-Neethling) recombinants. *J Virol Method* 2007;**146**(1–2):52–60.
18. Amman R, Rohde J, Wulle U, Conlee D, Raue R, Martinon O, Rziha H-J. A new rabies vaccine based on a recombinant orf virus (parapoxvirus) expressing the rabies virus glycoprotein. *J Virol* 2013;**83**(3):1618–30.
19. Weyer J, Rupprecht CE, Mans J, Viljoen GJ, Nel LH. Generation and evaluation of a recombinant modified vaccinia virus Ankara vaccine for rabies. *Vaccine* 2007;**25**:4213–22.
20. Esposito JJ, Knight JC, Shaddock JH, Novembre FJ, Baer GM. Successful oral rabies vaccination of raccoons with raccoon poxvirus recombinants expressing rabies virus glycoprotein. *Virol* 1988;**165**(1):313–6.
21. Lodmell DL, Sumner JW, Esposito JJ, Bellini WJ, Ewalt LC. Raccoon poxvirus recombinants expressing the rabies virus nucleoprotein against lethal rabies virus infection. *J Virol* 1991;**65**:63400–5.
22. Esposito J, Brechling K, Baer G, Moss B. Vaccinia virus recombinant expressing rabies virus glycoprotein protect against rabies. *Virus Genes* 1987;**1**(1):7–21.
23. Sumner JW, Fekadu M, Shaddock JH, Esposito JJ, Bellini WJ. Protection of mice with vaccinia virus recombinants that express the rabies nucleoprotein. *Virol* 1991;**183**:703–10.
24. Fekadu M, Sumner JW, Shaddock JH, Sanderlin DW, Baer GM. Sickness and recovery of dogs challenged with a street rabies virus after vaccination with a vaccinia virus recombinant expressing rabies virus N protein. *J Virol* 1992;**66**(5):2601–4.
25. Fujii H, Takita-Sonoda Y, Mifune K, Hirai K, Nishizono A, Mannen K. Protective efficacy in mice of post-exposure vaccination with vaccinia virus recombinant expressing either rabies virus glycoprotein or nucleoprotein. *J Gen Virol* 1994;**75**:1339–44.
26. Weyer J, Kuzmin IV, Rupprecht CE, Nel LH. Cross-protective and cross-reactive immune responses to recombinant vaccinia viruses expressing full-length lyssavirus glycoprotein genes. *Epid Infect* 2008;**136**(5):670–8.
27. Moss B. Vaccinia virus: a tool for research and vaccine development. *Science* 1991;**252**(5013):1662–7.
28. Robertson JS. Regulation of poxvirus vectored vaccines. *Vaccine* 2013;**31**:4259–60.
29. Plotkin SA, Cadoz M, Meignier B, Méric C, Leroy O, Excler JL, et al. The safety and use of canarypox vectored vaccines. *Dev Biol Stand (Basel)* 1995;**84**:165–70.
30. Hanlon CA, Niezgoda M, Shankar V, Niu HS, Koprowski H, Rupprecht CE. A recombinant vaccinia-rabies virus in the immunocompromised host: oral, innocuity, progressive parenteral infection and therapeutics. *Vaccine* 1997;**15**(2):140–8.
31. Rupprecht CE, Blass L, Smith K, Orciari LA, Niezgoda M, Whitfield SG, et al. Human infection due to recombinant vaccinia-rabies glycoprotein. *N Engl J Med* 2001;**345**(8):582–6.
32. Berkelman RT. Human illness associated with use of veterinary vaccines. *CID* 2003;**37**:407–14.

33. Dato V, Moose C, Rea N, Fraser G, Seiders J, Rittle C, et al. Human vaccinia infection after contact with a raccoon rabies vaccine bait – Pennsylvania, 2009. *MMWR* 2009;**58**(43):1204–7.
34. Byrd CM, Hruby DE. Construction of recombinant vaccinia virus: using the thymidine kinase locus. In: Isaacs S, editor. *Vaccinia virus and poxvirology: methods and protocols* 1st ed. Totowa New Jersey: Humana Press; 2004. p. 31–9.
35. Isaacs S. Working Safely with Vaccinia Virus: Laboratory Technique and Review of Published Cases of Accidental Laboratory Infections. In: Isaacs S, editor. *Vaccinia virus and poxvirology: methods and protocols. Methods in Molecular Biology Series* 2nd ed. Totowa New Jersey: Humana Press; 2012. p. 1–22.
36. Esposito J, Condit R, Obijeski J. The preparation of orthopoxvirus DNA. *J of Clin Virol* 1981;**2**:175–9.
37. ATCC®: <http://www.atcc.org> [accessed 18.02.14].
38. Green MR, Sambrook J. Polymerase Chain Reaction. In: Green MR, Sambrook J, editors. *Molecular Cloning: A Laboratory Manual* 4th ed. Cold Spring Harbour Laboratory Press; 2012. p. 456–540.
39. Tuchiya K, Matsuura Y, Kawai A, Ishihama A, Ueda S. Characterization of rabies virus glycoprotein expressed by recombinant baculovirus. *Virus Research* 1992;**25**:1–13.
40. Green MR, Sambrook J. Isolation and quantification of DNA. In: Green MR, Sambrook J, editors. *Molecular Cloning: A Laboratory Manual* 4th ed. Cold Spring Harbour Laboratory Press; 2012. p. 1–79.
41. Green MR, Sambrook J. Analysis of DNA. In: Green MR, Sambrook J, editors. *Molecular Cloning: A Laboratory Manual* 4th ed. Cold Spring Harbour Laboratory Press; 2012. p. 81–156.
42. Green MR, Sambrook J. Cloning and transformation of plasmid vectors. In: Green MR, Sambrook J, editors. *Molecular Cloning: A Laboratory Manual* 4th ed. Cold Spring Harbour Laboratory Press; 2012. p. 157–212.
43. Lorenzo MM, Galindo I, Blasco R. Construction and isolation of recombinant vaccinia viruses using genetic markers. In: Isaacs S, editor. *Vaccinia virus and poxvirology: methods and protocols. Methods in Molecular Biology Series* 1st ed. Totowa New Jersey: Humana Press; 2004. p. 15–30.
44. Kotwal GJ, Abrahams M-R. Growing of poxviruses and determining virus titer. In: Isaacs S, editor. *Vaccinia virus and poxvirology: methods and protocols. Methods in Molecular Biology Series* 1st ed. Totowa New Jersey: Humana Press; 2004. p. 101–12.

CHAPTER TWENTY FOUR

Recombinant Rabies Virus Vaccines

Milosz Faber
Department of Microbiology and Immunology, Thomas Jefferson University, Philadelphia, Pennsylvania, USA

Chapter Contents

24.1 Introduction 255
24.2 Materials, Methods, and Results 256
24.2.1 Replication-Competent Recombinant Rabies Virus Vaccines 256
24.2.1.1 Generation of Highly Attenuated RABV Recombinant Viruses using Modified G Genes 256
24.2.1.2 Enhancement of the Immunogenicity of Recombinant Rabies Viruses through Over-Expression of the Rabies Virus Glycoprotein 257
24.2.1.3 Enhancement of the Immunogenicity of Recombinant Rabies Viruses through Expression of Foreign Proteins 259
24.2.2 Replication-Deficient and Replication-Impaired Recombinant Rabies Virus Vaccines 260
24.3 Discussion 260
References 261

24.1 INTRODUCTION

Rabies remains an important public health problem worldwide and causes more than 70,000 human deaths each year. There is no doubt that access to vaccination is a major problem in developing countries. Therefore, more effective and less expensive safe rabies vaccines must be developed. In this respect, reverse genetics technology, which allows the modification of viral elements that account for pathogenicity and immunogenicity, has made the systematic development of safer and more potent rabies vaccines possible. There are several features of rabies virus (RABV) that make the construction of recombinant rabies vaccines feasible: (i) the modular genome of RABV is organized with short transcription stop/start sequences flanking the genes, making it readily amenable to manipulation;[1] (ii) the RABV genome is RNA and the life cycle of RABV is

C. Rupprecht & T. Nagarajan (Eds): Current Laboratory Techniques in Rabies Diagnosis, Research, and Prevention, Volume 1.
Doi: http://dx.doi.org/10.1016/B978-0-12-800014-4.00024-X

exclusively cytoplasmic so that no recombination, reversion, or integration is observed;[2,3] (iii) stable incorporation of large and multiple foreign genes of up to 6.5 kb offers advantages over plus-stranded RNA virus vectors;[2] (iv) RABV is non-cytopathic in infected cells and expresses high levels of foreign proteins over extended periods of time;[2,4] (v) RABV can induce a protective immune response in a variety of animals (e.g. dog and mongoose) following immunization by the oral route,[5,6] and attenuated RABV can target cells in the tonsils and buccal mucosa;[7] and (vi) multiple mutations introduced into the RABV genome that completely abolish the pathogenicity of RABV and render the RABV vector extremely safe,[2] and replication-defective RABVs can be produced that are even safe for completely immunocompromised individuals.[4,8] Using reverse genetics technology, recombinant rabies viruses have been generated that combine the antigenic determinants that render the RABV non-pathogenic with the determinants that are responsible for the elicitation of an effective anti-rabies immune response.

24.2 MATERIALS, METHODS, AND RESULTS

24.2.1 Replication-Competent Recombinant Rabies Virus Vaccines

Most of the recombinant RABVs described here are based on the recombinant RABV SPBN,[9] which was generated from a Street alabama Dufferin (SAD) B19 complementary DNA (cDNA) clone in which the sequence encoding the pseudogene was deleted.[3]

24.2.1.1 Generation of Highly Attenuated RABV Recombinant Viruses using Modified G Genes

The primary requirement for any live vaccine is that the vaccine must be safe for the target species, as well as any other animal species that comes in contact with the vaccine.

The pathogenicity of certain RABV strains (e.g. Evelyn Rokitniki Abelseth, ERA; High Egg Passage, HEP) correlates with the presence of arginine at position 333 of the G protein.[10] An exchange of arginine 333 for glutamine or glycine rendered these viruses nonpathogenic for immunocompetent mammals regardless of the route of infection.[10] Therefore, to reduce the pathogenicity of SPBN, the G gene of SPBN was replaced with a modified G gene encoding a single amino acid exchange, arginine$_{333}$→glutamine$_{333}$ (AGA→GAG).[2] For this approach, the coding sequence of a modified G gene termed GA was amplified by the polymerase chain reaction (PCR) and used to replace the G gene of pSPBN by utilizing the BsiWI and NheI sites flanking the G sequence.[11] The resulting plasmid designated as pSPBNGA was used to rescue the recombinant virus SPBNGA by transfecting BSR-T7 cells[12] with a calcium phosphate transfection kit (Stratagene, La Jolla, CA) containing pSPBNGA and four plasmids encoding the nucleoprotein, phosphoprotein, polymerase, and glycoprotein (N, P, L, and G, respectively) of RABV.[11] Three days after transfection, supernatants were transferred to BSR cells and three days later, the cells were examined for presence of virus by immunostaining with fluorescein isothiocyanate (FITC)-labeled RABV N antibodies (Centocor, Malvern, Pa.). Although SPBNGA was not pathogenic for immunocompetent mice, it became pathogenic after several serial virus passages in newborn mice. The reversion to the pathogenic phenotype was associated with a thymidine to adenosine mutation (T→A) at position 639 of the G gene, which results in an asparagine to lysine exchange at position 194 of G.[13,14] To minimize the possibility for an exchange at amino acid position 194 of G and preventing the reversion to the pathogenic phenotype, the codon at position 637–639 of GA was changed by site-directed mutagenesis to replace asparagine at position 194 by serine, resulting in GAS. After replacement of the GA gene of pSPBNGA with the GAS gene, SPBNGAS was rescued as described above for SPBNGA. SPBNGAS remained non-pathogenic after five suckling mouse brain passages, indicating that the nonpathogenic phenotype of this virus strain was stable.[14]

24.2.1.2 Enhancement of the Immunogenicity of Recombinant Rabies Viruses through Over-Expression of the Rabies Virus Glycoprotein

The RABV G is the major viral antigen responsible for the induction of protective immunity. Because infection with attenuated RABVs results in higher G expression and causes significantly more apoptotic cell death than infection with street RABVs,[15] it was concluded that the G expression level determines the magnitude of the immune response by controlling the speed and extent of apoptosis.[11] Consequently, reverse genetics was used to construct recombinant RABVs that contain two or three identical G genes.[11,16,17] To facilitate insertion of one or two additional GAS genes into pSPBNGAS, a fragment between PacI and BsiWI of pSPBNGAS, containing regulatory and intergenic sequences, was amplified using Deep Vent polymerase (New England Biolabs) and primers

InterG BA(+) (5′-CGA TGT ATA CGT ACG TTT TTG CGA TCG CCG TCC TTT CAA CGA TCC AAG TC-3′) and InterG AN(-) (5′-CTT AGC GCT AGC AAA AAG GCG CGC CGG AGG GGT GTT AGT TTT TTT CAT G-3′). The PCR product was digested with BsiWI and NheI (New England Biolabs) and ligated into RABV vaccine vector pSPBNGAS, previously digested with BsiWI and NheI, resulting in pSPBAANGAS. To insert a second copy of the GAS gene, GAS cDNA was amplified with primers that contain the AscI and NheI sites: SADB19 AscI(+) (5′-CGA ATT TAT TGG CGC GCC AAGATG GTT CCT CAG GCT CTC CTG-3′ and SADB19 NheI(−) (5′-CTT ATC AGC TAG CTA GCT AGT TAC AGT CTG GTC TCA CCC CCA-3′, digested with AscI and NheI, and ligated into pSPBAANGAS, previously digested with AscI and NheI resulting in pSPBAANGAS-GAS. A third copy of GAS gene was introduced in a similar manner into pSPBAAN-GAS-GAS using primers: SADB19 BsiWI(+) (5′-CGA TGT ATA CGT ACG AAG ATG GTT CCT CAG GCT CTC CTG-3′) and SADB19 AsiSI(−) (5′-GAA TCT AGA GCG ATC GCC GTT TAC AGT CTG GTC TCA CCC CCA-3′) resulting in pSPBAANGAS-GAS-GAS.[17] All GAS variants that were rescued from the corresponding full length cDNA clones, as described above for the rescue of SPBNGA, grow in BSR cells to high titers (>10^8 focus-forming units (FFU)/mL). With respect to their immunogenicity, the double G RABV variants SPBNGA-GA and SPBNGAS-GAS are superior to the single G RABV variants SPBNGA or SPBNGAS in inducing protective immunity in Swiss Webster mice, dogs, foxes, raccoons, and mongoose.[11,18–20] No clinical signs or discomfort were observed in these animals after oral, intramuscular (i.m.), and intracranial (i.c.) administration of the double RABV G variants. A single inoculation with this variant resulted in a strong, long-lasting virus neutralizing antibody (VNA) response and protected the animals against a lethal RABV infection. Furthermore, the presence of two nonpathogenic Gs (GA or GAS) within a single RABV genome considerably decreases the possibility of reemergence of a pathogenic phenotype variant because a non-pathogenic RABV G is dominant over a pathogenic RABV G in determining the pathogenic phenotype of the virus.[16]

The triple RABV variant SPBAANGAS-GAS-GAS (termed TriGAS) has been shown to be highly immunogenic, non-neuroinvasive, and nonpathogenic for mice that are either developmentally immunocompromised or have inherited defective immune function, even when administered intracranially.[17,21] After intramuscular administration, TriGAS infects transiently dendritic cells (DCs) and thereby induces a strong and sustained immune response.[22] Most notably, i.c. or i.m. treatment of Swiss Webster mice with TriGAS two and four days after challenge infection with a lethal dose of a street RABV resulted in 100% and 50% protection, respectively.[17]

24.2.1.3 Enhancement of the Immunogenicity of Recombinant Rabies Viruses through Expression of Foreign Proteins

To increase the capacity to induce apoptosis, thereby enhancing the protective immunity against rabies, a recombinant RABV that expresses the pro-apoptotic protein cytochrome c (Cyto c) was constructed.[23] Human Cyto c cDNA was amplified by PCR, and after introduction of BsiWI and Nhe I sites ligated into pSPBN, which resulted in pSPBN-Cyto c(+) [pSPBN-Cyto c(+)]. Following virus recovery from cDNA clones, virus stocks were prepared in BSR cells and used for the immunization of mice. The geometric mean VNA titers induced by SPBN-Cyto c(+) were, on an average, 3.5 times higher as compared to the titers induced by a recombinant RABV containing an inactive cytochrome gene (SPBN-Cyto c(−)). Vaccination of mice with serial dilutions of SPBN-Cyto c(+) and SPBN-Cyto c(−) followed by an intracranial challenge with 10 50% lethal dose (LD_{50}) of Challenge Virus Standard (CVS-24) virus showed that the survivorship of mice immunized with SPBN-Cyto c(+) is significantly higher (on average, 1.7 times, $p = 0.009$) than the survivorship of SPBN-Cyto c(−)-immunized mice. The 50% effective dose (ED_{50}) calculated from the mortality rates in the different vaccine dilution groups revealed that the efficacy of SPBN-Cyto c(+) is 20 times higher than the efficacy of SPBN-Cyto c(−), clearly demonstrating that expression of Cyto c by a recombinant RABV strongly enhances the protective immunity against rabies.[23]

Based on the observation that overexpression of macrophage inflammatory protein-1α (MIP-1α) in mouse brain further decreased RABV pathogenicity, a recombinant RABV expressing MIP-1α (rHEP-MIP1α) was constructed.[24] Intramuscular immunization of BALB/c mice with rHEP-MIP1α resulted in a higher level of expression of MIP-1α at the site of inoculation, increased recruitment of DCs and mature B cells into the draining lymph nodes and the peripheral blood, and higher VNA titers than immunization with the parent rHEP. These data demonstrated that expression of MIP-1α not only reduces viral pathogenicity, but also enhances immunogenicity by recruiting DCs and B cells to the site of immunization.

24.2.2 Replication-Deficient and Replication-Impaired Recombinant Rabies Virus Vaccines

The P protein of RABV is a cofactor for the viral polymerase (L), and essential for the transcription of the viral genome.[25] To generate a replication-deficient RABV, the P gene was deleted by reverse genetics from cDNA clones of the HEP-Flury strain or from the SPBN strain and P gene-deficient virus particles were rescued by transfecting BHK or BSR cells that constitutively express the P protein.[8,26] While the P-deleted RABVs HEP (defP) or SPBN (SPBNΔP) are non-pathogenic for suckling mice or immune-deficient Rag2 mice even when inoculated intracranially, they induce VNA, thereby conferring protective immunity against a lethal infection.[8,26] These results demonstrate the potential utility of P-deficient RABVs for pre- and post-exposure prophylaxis against rabies.

The matrix protein (M) of RABV is responsible for the assembly and budding of the characteristic bullet-shaped RABV particles, and in the absence of M, infectious particles are mainly cell-associated and the yield of cell-free infectious virus is strongly reduced, which indicates a severe impairment of the virus formation process.[27] To investigate the potential utility of M-deficient RABVs as vaccines, a plasmid containing the M gene-deficient SPBN cDNA (pSPBN-ΔM) was constructed using reverse genetics technology. SPBN-ΔM was rescued by transfecting BSR cells stably expressing bacteriophage T7RNApolymerase with pSPBN-ΔM and four different plasmids encoding the RABV proteins N, P, matrix (M), and L.[28] The rescued SPBN-ΔM, which was grown in BSR cells that constitutively express the M protein, was shown to be non-pathogenic for immune-deficient Rag2 mice. Administration of a single dose of SPBN-ΔM induced a potent rabies VNA response in mice and in nonhuman primates, and vaccination with only 10^3 FFUs conferred 100% protection in mice against a lethal RABV challenge.[28] These data support the conclusion that M gene-deleted RABVs are safe and highly immunogenic, and are therefore promising vaccine candidates for the pre- and post-exposure prophylaxis of rabies.

24.3 DISCUSSION

The advantage of a live-attenuated recombinant RABV vaccine over classical modified-live rabies vaccines, or currently used inactivated RABV vaccines for pre- and post-exposure prophylaxis of rabies may not only rest in higher efficacy, but also in significantly lower costs, because only a single dose of live-attenuated vaccine may be necessary, as compared to multiple doses of inactivated RABV vaccine. Furthermore, the use of a live-attenuated vaccine would likely make the need for expensive rabies immune globulin (IgG) dispensable. Although, with respect to safety, replication-deficient recombinant vaccines such as HEP (defP) and SPBNΔP would be preferable to replication-impaired (e.g. SPBN-ΔM) or replication-competent RABV vaccines (e.g. TriGAS), they are much less immunogenic than the replication-impaired[28] or replication-competent RABV vaccines,[11,17] therefore booster immunizations would likely be necessary to induce a long-lasting protective immune response. On the other hand, some of the replication-competent RABV vaccines such as TriGAS are very safe even for immunocompromised mice, regardless of the route of infection.[17] An advantage of replication-competent over replication-deficient or replication-impaired RABVs is that they will grow efficiently in cell substrates such as Vero cells, which have already been approved for the production of RABV vaccines.

A major concern for the use of any live recombinant RABV vaccine is stability. While genetic stability has been demonstrated for several replication-competent RABV vaccines,[13,14,16] these vaccines are stable at 4°C, but may rapidly degrade at higher temperatures (e.g. >25°C).[13] Because it is difficult and probably also not practical to maintain the cold chain during storage and transport, major efforts must be put into the development of vaccine stabilizers that can prevent the inactivation of live recombinant vaccines.

REFERENCES

1. Foley HD, McGettigan JP, Siler CA, Dietzschold B, Schnell MJ. A recombinant rabies virus expressing vesicular stomatitis virus glycoprotein fails to protect against rabies virus infection. *Proc Natl Acad Sci U S A* 2000;**97**(26):14680–5.
2. McGettigan JP, Pomerantz RJ, Siler CA, McKenna PM, Foley HD, Dietzschold B, et al. Second-generation rabies virus-based vaccine vectors expressing human immunodeficiency virus type 1 gag have greatly reduced pathogenicity but are highly immunogenic. *J Virol* 2003;**77**(1):237–44.
3. Schnell MJ, Mebatsion T, Conzelmann KK. Infectious rabies viruses from cloned cDNA. *EMBO J* 1994;**13**(18):4195–203.
4. Dietzschold B, Faber M, Schnell MJ. New approaches to the prevention and eradication of rabies. *Expert Rev Vaccines* 2003;**2**(3):399–406.
5. Meslin FX, Fishbein DB, Matter HC. Rationale and prospects for rabies elimination in developing countries. *Curr Top Microbiol Immunol* 1994;**187**:1–26.
6. Rupprecht CE, Blass L, Smith K, Orciari LA, Niezgoda M, Whitfield SG, et al. Human infection due to recombinant vaccinia-rabies glycoprotein virus. *N Engl J Med* 2001;**345**(8):582–6.

7. Orciari LA, Niezgoda M, Hanlon CA, Shaddock JH, Sanderlin DW, Yager PA, et al. Rapid clearance of SAG-2 rabies virus from dogs after oral vaccination. *Vaccine* 2001;**19**(31):4511–8.
8. Shoji Y, Inoue S, Nakamichi K, Kurane I, Sakai T, Morimoto K. Generation and characterization of P gene-deficient rabies virus. *Virology* 2004;**318**(1):295–305.
9. McGettigan JP, Foley HD, Belyakov IM, Berzofsky JA, Pomerantz RJ, Schnell MJ. Rabies virus-based vectors expressing human immunodeficiency virus type 1 (HIV-1) envelope protein induce a strong, cross-reactive cytotoxic T-lymphocyte response against envelope proteins from different HIV-1 isolates. *J Virol* 2001;**75**(9):4430–4.
10. Dietzschold B, Wunner WH, Wiktor TJ, Lopes AD, Lafon M, Smith CL, et al. Characterization of an antigenic determinant of the glycoprotein that correlates with pathogenicity of rabies virus. *Proc Natl Acad Sci U S A* 1983;**80**(1):70–4.
11. Faber M, Pulmanausahakul R, Hodawadekar SS, Spitsin S, McGettigan JP, Schnell MJ, et al. Overexpression of the rabies virus glycoprotein results in enhancement of apoptosis and antiviral immune response. *J Virol* 2002;**76**(7):3374–81.
12. Schlender J, Bossert B, Buchholz U, Conzelmann KK. Bovine respiratory syncytial virus nonstructural proteins NS1 and NS2 cooperatively antagonize alpha/beta interferon-induced antiviral response. *J Virol* 2000;**74**(18):8234–42.
13. Dietzschold ML, Faber M, Mattis JA, Pak KY, Schnell MJ, Dietzschold B. In vitro growth and stability of recombinant rabies viruses designed for vaccination of wildlife. *Vaccine* 2004;**23**(4):518–24.
14. Faber M, Faber ML, Papaneri A, Bette M, Weihe E, Dietzschold B, et al. A single amino acid change in rabies virus glycoprotein increases virus spread and enhances virus pathogenicity. *J Virol* 2005;**79**(22):14141–8.
15. Morimoto K, Hooper DC, Spitsin S, Koprowski H, Dietzschold B. Pathogenicity of different rabies virus variants inversely correlates with apoptosis and rabies virus glycoprotein expression in infected primary neuron cultures. *J Virol* 1999;**73**(1):510–8.
16. Faber M, Faber ML, Li J, Preuss MA, Schnell MJ, Dietzschold B. Dominance of a nonpathogenic glycoprotein gene over a pathogenic glycoprotein gene in rabies virus. *J Virol* 2007;**81**(13):7041–7.
17. Faber M, Li J, Kean RB, Hooper DC, Alugupalli KR, Dietzschold B. Effective preexposure and postexposure prophylaxis of rabies with a highly attenuated recombinant rabies virus. *Proc Natl Acad Sci U S A* 2009;**106**(27):11300–5.
18. Rupprecht CE, Hanlon CA, Blanton J, Manangan J, Morrill P, Murphy S, et al. Oral vaccination of dogs with recombinant rabies virus vaccines. *Virus Res* 2005;**111**(1):101–5.
19. Blanton JD, Meadows A, Murphy SM, Manangan J, Hanlon CA, Faber ML, et al. Vaccination of small Asian mongoose (Herpestes javanicus) against rabies. *J Wildl Dis* 2006;**42**(3):663–6.
20. Blanton JD, Self J, Niezgoda M, Faber ML, Dietzschold B, Rupprecht C. Oral vaccination of raccoons (Procyon lotor) with genetically modified rabies virus vaccines. *Vaccine* 2007;**25**(42):7296–300.
21. Li J, Ertel A, Portocarrero C, Barkhouse DA, Dietzschold B, Hooper DC, et al. Postexposure treatment with the live-attenuated rabies virus (RV) vaccine TriGAS triggers the clearance of wild-type RV from the Central Nervous System (CNS) through the rapid induction of genes relevant to adaptive immunity in CNS tissues. *J Virol* 2012;**86**(6):3200–10.
22. Schutsky K, Curtis D, Bongiorno EK, Barkhouse DA, Kean RB, Dietzschold B, et al. Intramuscular inoculation of mice with the live-attenuated recombinant rabies virus TriGAS results in a transient infection of the draining lymph nodes and a robust, long-lasting protective immune response against rabies. *J Virol* 2013;**87**(3):1834–41.

23. Pulmanausahakul R, Faber M, Morimoto K, Spitsin S, Weihe E, Hooper DC, et al. Overexpression of cytochrome C by a recombinant rabies virus attenuates pathogenicity and enhances antiviral immunity. *J Virol* 2001;**75**(22):10800–7.
24. Zhao L, Toriumi H, Wang H, Kuang Y, Guo X, Morimoto K, et al. Expression of MIP-1alpha (CCL3) by a recombinant rabies virus enhances its immunogenicity by inducing innate immunity and recruiting dendritic cells and B cells. *J Virol* 2010;**84**(18):9642–8.
25. Schnell MJ, Conzelmann KK. Polymerase activity of in vitro mutated rabies virus L protein. *Virology* 1995;**214**(2):522–30.
26. Cenna J, Tan GS, Papaneri AB, Dietzschold B, Schnell MJ, McGettigan JP. Immune modulating effect by a phosphoprotein-deleted rabies virus vaccine vector expressing two copies of the rabies virus glycoprotein gene. *Vaccine* 2008;**26**(50):6405–14.
27. Mebatsion T, Weiland F, Conzelmann KK. Matrix protein of rabies virus is responsible for the assembly and budding of bullet-shaped particles and interacts with the transmembrane spike glycoprotein G. *J Virol* 1999;**73**(1):242–50.
28. Cenna J, Hunter M, Tan GS, Papaneri AB, Ribka EP, Schnell MJ, et al. Replication-deficient rabies virus-based vaccines are safe and immunogenic in mice and nonhuman primates. *J Infect Dis* 2009;**200**(8):1251–60.

CHAPTER TWENTY FIVE

Adenoviral Vectors as Vaccine Carriers for Prevention of Rabies

Hildegund C.J. Ertl
The Wistar Institute, Philadelphia, PA, USA

Chapter Contents

25.1 Introduction 265
25.2 Basic Characteristics of Ad Viruses 266
25.3 Types of Ad Vectors 266
25.4 Transgene Product for Rabies Vaccines 267
25.5 Construction of Ad Vectors 267
25.6 Ad Virus Rescue and Expansion 268
25.7 Quality Control of Ad Vectors 270
25.8 Discussion 272
References 273

25.1 INTRODUCTION

Symptomatic rabies is nearly always fatal in humans, and only a handful of survivors have been recorded.[1] Although efficacious rabies vaccines are available, more than 50,000 humans residing mainly in less developed countries die each year from rabies.[2] Vaccines are poorly immunogenic and have to be administered 4 to 5 times[3] to achieve adequate titers of virus neutralizing antibodies (VNAs), the primary immune correlate of protection against disease.[4] In cases of severe exposure, active immunization has to be combined with passive immunization with a rabies immune globulin (IgG) preparation of human or equine origin, which is costly and habitually in short supply. Prophylactic rabies vaccination, although mandatory in many countries for dogs and cats, is reserved for humans at high risk of rabies exposure. Prophylactic vaccination of children living in highly endemic areas has been tested and was found to result in sustained B cell responses[5] that in case of exposure rapidly become reactivated by a single booster immunization.[6] Nevertheless,

C. Rupprecht & T. Nagarajan (Eds): Current Laboratory Techniques in Rabies Diagnosis, Research, and Prevention, Volume 1.
Doi: http://dx.doi.org/10.1016/B978-0-12-800014-4.00025-1

prophylactic rabies vaccination as part of childhood vaccination programs is not cost-effective using licensed vaccines.[7]

Some novel rabies vaccines, such as those based on adenovirus (Ad) vectors, are more immunogenic and induce sustained rabies VNA responses after a single immunization.[8,9] Although the actual cost of such vaccines produced under commercial scale-up conditions is hard to estimate, it is expected that a single-dose regimen would be less expensive and in turn make prophylactic vaccination of infants in highly endemic areas economically feasible.

Here we describe some of the basic characteristics of Ad viruses and detail the procedures used to develop replication-defective Ad vectors.

25.2 BASIC CHARACTERISTICS OF Ad VIRUSES

Ad viruses are ubiquitous viruses, and to date 57 different serotypes, defined by numbers as HAdV1-57, have been isolated from humans. These serotypes are divided into seven species (A–G). Additional serotypes have been isolated from non-human primates, dogs, bovines, ovines, and others. Such Ad viruses are largely species-specific. In humans, Ad viruses cause a wide variety of diseases, most commonly upper respiratory infections.

Most humans have neutralizing antibodies to different serotypes of Ad viruses. Seroprevalence rates are universally high to some Ad viruses, such as HAdV-5[10,11] while VNAs to other serotypes, such as HAdV-26, are mainly found in selected geographic regions.[12] A third group of human Ad viruses, such as HAdV-35, can be classified as rare serotypes.[10] Humans also carry T cells to Ad virus, which are mainly directed to conserved regions of Ad proteins, and thereby cross-react between different serotypes.[13]

25.3 TYPES OF Ad VECTORS

Vectors based on Ad viruses are highly suited for gene transfer as they have a broad tropism, high transduction efficiency, and are in general well tolerated. In addition, Ad vectors can be easily produced on appropriate cell lines, and procedures for their purification and quality control are available, thus facilitating clinical development. Vectors have been constructed based on a number of human and non-human Ad viruses.

Initial vectors were based on HAdV-5 virus, the most common and best-characterized serotype of the Adenoviridae.[9] Subsequently, vectors were generated based on HAdV-4,[14] 26,[15] 28,[16] 48, and 35,[17] chimpanzee-derived simian Ad (SAdV) serotypes 3, 23, 24, 25, 63, and others.[18–22]

The Ad genome is packed tightly into the capsid, which limits the size of the genome. Most Ad genomes are 33–35 kb in length, and insertions of an additional 1.8 kb are tolerated while longer insertions prevent viral assembly. In most vectors the expression cassette containing the antigen-encoding gene under the control of a potent promoter, such as the human or mouse cytomegalovirus promoter, are placed into the deleted E1 domain. The E1 deletion allows for insertion of sequences that are up to 5 kb in length. Additional deletion of E3 allows for insertion of up to 8 kb. Such E1-deleted Ad vectors are replication-defective.

25.4 TRANSGENE PRODUCT FOR RABIES VACCINES

Rabies virus (RABV), a member of the lyssavirus genus, is a single negative-stranded RNA virus that encodes five structural antigens. The glycoprotein (G) has receptor binding activity, and is the only antigen that elicits VNAs. Protection against infection is mediated by VNA,[4] and the G-encoding sequence of RABV is therefore the transgene of choice for a recombinant vaccine. Vaccines based on different strains of RABV, such as Pitman Moore (PM) or Pasteur Virus (PV), are available. Other closely related strains are Challenge Virus Standard (CVS) or Evelyn Rokitniki Abelseth (ERA). The G of laboratory strains (such as ERA, PM, PV, etc.) are suitable for the development of recombinant vaccines.

25.5 CONSTRUCTION OF Ad VECTORS

Initially, Ad vectors were constructed by homologous recombination by co-transfecting cells that expressed E1 for transcomplementation with the genome of Ad virus that had been deleted of E1 and a pShuttle vector that contained the expression cassette with the gene of interest, together with E1 flanking sequences.[23] This rather cumbersome method has been replaced by the use of molecular clones. Construction of viral molecular clones requires that the viral genome has been sequenced. Molecular clones are constructed by piece-by-piece cloning of the Ad virus genome into a plasmid vector[24] or a bacterial artificial chromosome.[25] We have used the former method, which is described in more detail below. Viral molecular clones are constructed by isolating DNA fragments upon digestion of the purified viral genome with unique restriction enzymes. Short regions for which such sites are not available can be amplified by polymerase chain reaction (PCR). The fragments are then reassembled by ligation. Deletions are incorporated during this process. In addition, sites for rare restriction enzymes such as I-CeuI and PI-SceI are incorporated flanking potential insertion sites of foreign sequences.

The viral molecular clone is used to develop recombinant virus by inserting the expression cassette containing the gene of interest and suited regulatory elements. Insertion is achieved by cutting the pShuttle vector containing the expression cassette and the viral molecular clone with the rare cutters followed by ligation. Although these procedures are technically straightforward, care has to be taken with long pieces of DNA to avoid DNA damage as can happen upon vigorous pipetting or lengthy exposure to UV light.

25.6 Ad VIRUS RESCUE AND EXPANSION

The recombinant Ad viral molecular clone is used to generate infectious virus. For E1-deleted Ad vectors this requires a cell line, which expresses E1 in trans. Cell lines, which carry the E1 of HAdV-5 virus are available. HEK 293 cells carry a large portion of the 5′ genome of HAdV-5 virus, which can cause homologous recombination of E1 with recombinant viral molecular clones, resulting in outgrowth of replication-competent Ad virus (RCA). This common problem is not encountered in PerC6 cells, which only contain E1.[26] The following descriptions are based on the use of HEK 293 cells. The E1 of HAdV-5 virus can transcomplement the deleted E1 domain of other serotypes, including HAdV-26, SAdV-23, -24, or -25. For other serotypes such as HAdV-35, a packaging cell line that carries the homologous E1 has to be constructed by stable transfection of cells with the E1 gene under the control of a suitable promoter such as the EF-1a promoter.[27] Alternatively, some of the early genes can be replaced with those from a serotype that grows in HEK 293 cells.[28]

For viral rescue, the viral molecular clone is expanded in a competent strain of *Escherichia coli* such as MAX efficiency Stbl2 competent cells (Invitrogen, USA) that tolerates large plasmids. Plasmids are purified, and in our experience plasmid DNA midi or maxi kits from Qiagen, Germany are suitable to provide DNA of sufficient purity. Care has to be taken that purification methods remove endotoxins from the DNA. The purified plasmid is quantified. In addition, we digest a small aliquot with at least two sets of restriction enzymes that cut the Ad genome at 2–6 sites. The digested DNA is separated by agarose gel electrophoresis in the presence of ethidium bromide. Bands are visualized under UV light, and only plasmids that yield the expected banding pattern are used. The plasmid vectors are linearized by digestion with a restriction enzyme such as Pac I that cuts once outside the Ad genome. The linearized DNA is transfected into HEK 293 cells. Transfection can be achieved with calcium phosphate or a number of commercially available transfection kits. Procedures should be optimized, for example, by using molecular clones expressing green fluorescent protein, to ensure high transduction efficiency. HEK 293 cells should be used at a low passage number (below 35), as extensive cell proliferation causes loss of the partial Ad genome. They should be plated into small flasks or 6-well plates, and should have ~80% confluency at the time of transfection. After transfection, medium is replaced and cells are incubated. In general, medium needs to be replaced after 5 days, and then in 2–3 day intervals. Viral plaques tend to become visible after 6–7 days. This depends on the Ad vector and the type of transgene. With some constructs, plaque formation may not occur until days 12–14 after transfection. Once plaques have destroyed at least a quarter of the cell monolayer, cells are harvested and virus is released by freeze thawing (3×). An aliquot of clarified culture supernatant is then used to infect fresh HEK 293 cells. At this stage viral plaques should become visible after 2–3 days, although delays of up to 7–10 days are possible. Ad vectors used for preclinical studies are then gradually expanded on HEK 293 cells until enough is available to prepare a large batch on 10^9 cells (~40 T175 flasks). For clinical lots, the Ad vector is plaque-purified three times on plates with an agarose overlay or by limiting dilution, and several clones are expanded. Virus is again harvested by freeze–thawing. Cell debris is removed by low-speed centrifugation (3–5,000 rpm) at 4°C. The clarified supernatant is further purified by two rounds of ultracentrifugation over cesium chloride (CsCl) gradients at 82,000*g* for 4 hrs, or 65,000*g* overnight. The first round should yield two bands: the upper band contains defective virus particles, and the lower band contains complete Ad vectors. The Ad vectors are then passed over Bio-Gel P-6DG columns (Bio-Rad, Hercules, CA) to remove the CsCl. Purified vector is suspended in 1–2 mL of saline with 10%

glycerol, aliquoted, and stored at −80°C. CsCl banding has the advantage of removing defective Ad virus particles. It is suitable for purification of small- and medium-sized batches, including clinical lots for small-scale trials. For large lots, such as those needed for phase III trials, alternative methods based on gel filtration or chromatography have been developed.[29]

25.7 QUALITY CONTROL OF Ad VECTORS

We conduct a number of *in vitro* studies before Ad vectors are released for animal studies. Content of virus particles (VPs) is determined by spectrophotometry at 260 nm and 280 nm, with the latter determining purity of the preparation. Viral titer in particles is calculated using the formula: OD260 $\times$ dilution $\times$ 1.1×10^{12}. Viral batches expanded on 10^9 HEK 293 cells should yield at least 10^{13} VPs of Ad vectors. Some regulatory agencies, such as the US Food and Drug Administration (USFDA), require dosing of Ad vectors according to VP, as they determine the toxicity of the vectors. Immunogenicity of the vectors on the other hand depends on infectious units (MOI), which is the number of VP per mL that is able to infect cells and transcribe the transgene product. MOI can be measured by plaque assays using an agarose overlay or by an endpoint dilution assay, which determines cytopathic effects. Both assays can be performed on a permissive cell line. Plaque formation is dependent on a number of viral factors, and may not reliably reflect the interactions between the cell line and the E1-deleted Ad vectors. Several alternative assays are available to determine the infectivity of Ad vector batches. One assay is based on staining for hexons using a commercially available antibody to a conserved domain of Ad hexon. Another assay is based on reverse transcription (RT)-PCR using hexon or transgene-specific primers. For both methods, virus is serially diluted and incubated for 7 days on HEK 293 cells using at least six samples for each virus dilution. Cells are then either stained directly with the hexon-specific antibody, or their RNA is isolated for RT-PCR. In either method, untransfected HEK 293 cells are used as controls. VP to MOI ratios of HAdV-5 vectors should not exceed 40:1. Ad vectors based on alternative serotypes, such as some of the SAdV strains that we developed as rabies vaccines, typically have higher MOI to VP ratios, which should not exceed 400:1.

The genetic integrity and identity of large-scale vector batches is assessed by isolation of viral DNA. The recombinant DNA is digested with a set of restriction enzymes and analyzed by agarose gel electrophoresis. The original molecular clones and shuttle plasmids used for generating molecular clones are used in side-by-side restriction digestions to compare signature banding patterns by ethidium bromide-stained agarose gel electrophoresis. In addition, molecular clones of vector backbones without transgene cassettes should be included in the analysis. At least two sets of restriction enzymes are selected: one set focuses on detecting the presence and integrity of the transgene cassettes; the other set cuts the vector backbone. For clinical lots, the entire genome of the Ad vector has to be sequenced.

New vectors that show the appropriate banding pattern are tested for genetic stability of vectors upon 12 serial passages on HEK 293 cells followed by expansion and purification of vectors and restriction enzyme digestion of their DNA. For some Ad vectors, especially those that give poor yields, growth parameters may have to be optimized to achieve appropriate yields and VP to MOI ratios. To this end, different amounts of input virus and different times of incubation in the final expansion of vectors have to be explored.

Vector batches should be checked for RCA on A549 cells. RCA can emerge during the creation and propagation of E1-deleted Ad vectors in HEK 293 cells as a result of recombination between overlapping viral sequences in HEK 293 cells and vectors. Depending on the level of RCA, it could have significant impact on vector performance *in vivo*. Therefore, identification of vector preparations with a high level of RCA contamination is critical for gene transfer and virus-based vaccine applications of Ad vectors. Briefly, cells (in 6-well plates) are treated with 2×10^9, 2×10^{10}, and 2×10^{11} VPs of vector. Control wells are infected with 10 or 50 plaque-forming units (pfu) of the corresponding wild-type Ad virus. In a third set of wells the replication-defective vectors (2×10^{11} VPs) are spiked with infectious virus (1, 10, and 100 MOI) to ensure that at the selected dose of vector, plaque formation is not inhibited. Cells are overlaid with agarose. Plates are read 4 and 8 days later. RCAs commonly contaminate batches of E1-deleted AdHu5 vectors grown on HEK 293 cells, but they have not yet been detected with E1-deleted SAdV vectors transcomplemented with E1 of AdHu5 due to sequence differences in the E1 flanking regions.

Batches to be used in animals should be tested for sterility and endotoxin. Sterility can be tested by placing a small amount of vector in a suitable growth medium followed by an overnight incubation at 37°C.

We measure endotoxin with the Limulus Amebocyte Lysate (LAL) gel-clot method and a commercial kit. Release criteria for vector lots to be used in large animal studies are <5 endotoxin units (EU)/kg of animal weight, which is the parameter for humans, for example, as set by the USFDA.

Transgene product expression can be tested by Western blot or immunoprecipitation upon infection of HEK 293 or CHO cells stably transfected to express the Coxsackie Ad receptor (CAR) with 1,000 and 10,000 VPs/cell of new vectors. Control cells are infected with the same dose of an Ad vector expressing an unrelated transgene product.

For clinical lots, a multitude of additional assays are needed to ensure compliance with regulatory agencies. A detailed description of these assays is beyond the scope of this review, but can be obtained from regulatory agencies such as the FDA in the USA.[30]

25.8 DISCUSSION

Novel vaccines are more immunogenic than currently licensed vaccines and should be explored for preventative rabies vaccination. One class of experimental rabies vaccines is based on recombinant Ad vectors. Ad vectors are genetic vaccines, which transcribe the vaccine antigen once a cell has been transduced and the viral genome has reached the nucleus. This takes time, and as rapid onset of immune responses is crucial for post-exposure prophylaxis, Ad vectors should only be used for prophylactic immunizations.

Ad vectors are very attractive as vaccine carriers, and are easy to construct. Conditions for large-scale production and purification are available. Assays for release testing of clinical lots have been established. Ad vectors are well tolerated by humans and other species at doses that elicit potent immune responses. They trigger strong inflammatory responses[31] and are thus immunogenic without addition of adjuvants. They persist at the injection site for 10–14 days, producing large amounts of antigens. T cells directed to antigens of the Ad vector as well as the transgene product eventually eliminate most of the transduced cells.[32] Nevertheless, Ad vectors similar to Ad viruses persist at low levels mainly in activated T cells.[28] They remain transcriptionally active and thereby maintain transgene product-specific immune responses. The durability of Ad vector-induced immune responses is one of the most attractive features of this class of vaccines.

Developing Ad vectors from new serotypes requires extensive cloning, but is otherwise technically straightforward.[24] Generating new recombinants with available pShuttle vectors and molecular viral clones entails only minimal cloning. However, viral rescue can be challenging. Lack of virus expansion upon transfection of a helper cell line may be caused by overly high passage numbers of the cells or by lack of vector fitness. The latter may be caused by inadvertent modifications of the vector backbone during cloning or more commonly by toxicity of the transgene product, which can be addressed by changing the regulatory elements of the expression cassette to attenuate expression levels. Lack of genetic integrity or stability again most commonly reflects modifications of the vector backbone or transgene toxicity. Poor yields and unduly high VP to MOI ratios can be addressed by optimizing conditions for virus propagation.

REFERENCES

1. Hemachudha T, Laothamatas J, Rupprecht CE. Human rabies: a disease of complex neuropathogenetic mechanisms and diagnostic challenges. *Lancet Neurol* 2002;**1**:101–9.
2. WHO Rabies Fact Sheet: <http://www.who.int/mediacentre/factsheets/fs099/en/> [accessed 20.02.14].
3. Rabies Vaccine Cost: health.costhelper.com/rabies-vaccine.html [accessed 20.02.14].
4. Plotkin SA. Correlates of protection induced by vaccination. *Clin Vaccine Immunol* 2010;**17**:1055–65.
5. Vien NC, Feroldi E, Lang J. Long-term anti-rabies antibody persistence following intramuscular or low-dose intradermal vaccination of young Vietnamese children. *Trans R Soc Trop Med Hyg* 2008;**102**:294–6.
6. Sabchareon A, Chantavanich P, Pasuralertsakul S, et al. Persistence of antibodies in children after intradermal or intramuscular administration of preexposure primary and booster immunizations with purified Vero cell rabies vaccine. *Pediatr Infect Dis J* 1998;**17**:1001–7.
7. Chulasugandha P, Khawplod P, Havanond P, Wilde H. Cost comparison of rabies pre-exposure vaccination with post-exposure treatment in Thai children. *Vaccine* 2006;**24**:1478–82.
8. Chen Z, Zhou M, Gao X, et al. A novel rabies vaccine based on a recombinant parainfluenza virus 5 expressing rabies virus glycoprotein. *J Virol* 2013;**87**:2986–93.
9. Xiang ZQ, Yang Y, Wilson JM, Ertl HC. A replication-defective human adenovirus recombinant serves as a highly efficacious vaccine carrier. *Virology* 1996;**219**(1):220–7.
10. Nwanegbo E, Vardas E, Gao W, et al. Prevalence of neutralizing antibodies to adenoviral serotypes 5 and 35 in the adult populations of The Gambia, South Africa, and the United States. *Clin Vaccine Immunol* 2004;**11**:351–7.
11. Xiang Z, Li Y, Cun A, et al. Chimpanzee adenovirus antibodies in humans, sub-Saharan Africa. *Emerg Infect Dis* 2006;**12**:1596–9.
12. Chen H, Xiang ZQ, Li Y, et al. Adenovirus-based vaccines: comparison of vectors from three species of adenoviridae. *J Virol* 2010;**84**:10522–32.
13. Hutnick NA, Carnathan D, Demers K, Makedonas G, Ertl HCJ, Betts MR. Adenovirus-specific human T cells are pervasive, polyfunctional, and cross-reactive. *Vaccine* 2010;**28**:1932–41.

14. Gurwith M, Lock M, Taylor EM, et al. Safety and immunogenicity of an oral, replicating adenovirus serotype 4 vector vaccine for H5N1 influenza: a randomised, double-blind, placebo-controlled, phase 1 study. *Lancet Infect Dis.* 2013;**13**:238–50.
15. Baden LR, Walsh SR, Seaman MS, et al. First-in-human evaluation of the safety and immunogenicity of a recombinant adenovirus serotype 26 HIV-1 Env Vaccine (IPCAVD 001). *J Infect Dis* 2012;**207**:240–7.
16. Kahl CA, Bonnell J, Hiriyanna S, et al. Potent immune responses and in vitro proinflammatory cytokine suppression by a novel adenovirus vaccine vector based on rare human serotype 28. *Vaccine* 2010;**28**(35):5691–702.
17. Teigler JE, Iampietro MJ, Barouch DH. Vaccination with adenovirus serotypes 35, 26, and 48 elicits higher levels of innate cytokine responses than adenovirus serotype 5 in rhesus monkeys. *J Virol* 2012;**86**:9590–8.
18. Colloca S, Barnes E, Folgori A, et al. Vaccine vectors derived from a large collection of simian adenoviruses induce potent cellular immunity across multiple species. *Sci Trans Med* 2012;**4**:115ra2.
19. Tatsis N, Tesema L, Robinson ER, et al. Chimpanzee-origin adenovirus vectors as vaccine carriers. *Gene Ther* 2005;**13**:421–9.
20. Patel A, Tikoo S, Kobinger GA. Porcine adenovirus with low human seroprevalence is a promising alternative vaccine vector to human adenovirus 5 in an H5N1 virus disease model Randall T.D., ed.. *PLoS ONE* 2010;**5**:e15301.
21. Bru T, Salinas S, Kremer EJ. An update on canine adenovirus type 2 and its vectors. *Viruses* 2010;**2**:2134–53.
22. Seregin SS, Amalfitano A. Overcoming pre-existing adenovirus immunity by genetic engineering of adenovirus-based vectors. *Expert Opin Biol Ther* 2009,**9**(12):1521–31.
23. Mizuguchi H, Kay MA, Hayakawa T. Approaches for generating recombinant adenovirus vectors. *Adv Drug Deliv Rev* 2001;**52**:165–76.
24. Zhou D, Zhou X, Bian A, et al. An efficient method of directly cloning chimpanzee adenovirus as a vaccine vector. *Nat Protoc* 2010;**5**:1775–85.
25. Arii J, Hushur O, Kato K, Kawaguchi Y, Tohya Y, Akashi H. Construction of an infectious clone of canine herpesvirus genome as a bacterial artificial chromosome. *Microbes Infect* 2006;**8**:1054–63.
26. Kovesdi I, Hedley SJ. Adenoviral producer cells. *Viruses* 2010;**2**:1681–703.
27. Gao W, Robbins PD, Gambotto A. Human adenovirus type 35: nucleotide sequence and vector development. *Gene Ther* 2003;**10**:1941–9.
28. Tatsis N, Fitzgerald JC, Reyes-Sandoval A, et al. Adenoviral vectors persist in vivo and maintain activated CD8 + T cells: implications for their use as vaccines. *Blood* 2007;**110**:1916–23.
29. Eglon MN, Duffy AM, O'Brien T, Strappe PM. Purification of adenoviral vectors by combined anion exchange and gel filtration chromatography. *J Gene Med* 2009;**11**:978–89.
30. USFDA, Guidance for Industry: Gene Therapy Clinical Trials. <http://www.fda.gov/BiologicsBloodVaccines/GuidanceComplianceRegulatoryInformation/Guidances/CellularandGeneTherapy/ucm072957.htm > [accessed 20.02.14].
31. Hensley SE, Giles-Davis W, McCoy KC, Weninger W, Ertl HC. Dendritic cell maturation, but not CD8 + T cell induction, is dependent on type I IFN signaling during vaccination with adenovirus vectors. *J Immunol* 2005;**175**:6032–41.
32. Yang Y, Ertl HC, Wilson JM. MHC class I-cestricted cytotoxic T lymphocytes to viral antigens destroy hepatocytes in mice infected with E1-deleted recombinant adenoviruses. *Immunity* 1994;**1**:433–42.

SECTION G

Rabies Antibodies/ Fragments

CHAPTER TWENTY SIX

Production of Human Monoclonal Antibodies using Mammalian Expression Systems

Wilfred E. Marissen
Crucell Holland B.V., Leiden, The Netherlands

Chapter Contents

26.1 Introduction 278
26.2 Materials 278
26.2.1 Reagents 278
26.2.2 Equipment 279
26.2.3 Biological Materials 279
26.3 Methods 279
26.3.1 Cells and Cell Culture 279
26.3.2 Preparation of Expression Vector 279
26.3.3 Preparation of Mammalian Cell Stocks 281
26.3.4 Transfection and Clone Selection 281
26.3.5 Monoclonal Antibody Expression 282
26.3.6 Monoclonal Antibody Purification 282
26.3.6.1 Protein A Chromatography 282
26.3.6.2 Performing Protein A Separation on a Gravity Column 282
26.3.6.3 Performing a Protein A Separation using Liquid Chromatography Systems 283
26.3.6.4 Desalting Columns 284
26.3.6.5 Gravity Desalting Columns 284
26.3.6.6 Desalting using Liquid Chromatography Systems 284
26.3.7 Monoclonal Antibody Product Testing 284
26.3.7.1 Interpretation of Results 288
Acknowledgments 288
References 289

C. Rupprecht & T. Nagarajan (Eds): Current Laboratory Techniques in Rabies Diagnosis, Research, and Prevention, Volume 1.
Doi: http://dx.doi.org/10.1016/B978-0-12-800014-4.00026-3

26.1 INTRODUCTION

The World Health Organization (WHO) recommends that post-exposure prophylaxis (PEP) should be administered to all patients with category III exposure to rabies or to immunodeficient patients with category II exposures.[1] A major component of PEP is passive immunization with rabies immune globulins (RIGs). Due to a lack of supply, many patients die because no PEP/RIG is administered.[2,3] Indeed, access to RIGs in the developing world is particularly poor – a poll in India revealed that only 2.1% of patients bitten by a rabid animal received RIGs.[4] Currently, the antibodies used in PEP are obtained from rabies-immunized donors of human (HRIG) or equine (ERIG) origin.[5,6] As this method means supply is extremely limited, there is a strong need for alternative systems for antibody production.[1]

Producing monoclonal antibodies (MAbs) in large-scale expression systems that can neutralize lyssaviruses such as the rabies virus (RABV) are potentially an ideal alternative to RIGs. Mouse and human MAbs have been shown to protect rodents from lethal RABV challenge.[7–10] The MAbs are also used for postmortem diagnostic testing, such as the direct fluorescent antibody (dFA) test[11] or the direct rapid immunohistochemistry test (dRIT),[12] which is vital to determine whether an animal that has bitten a person has rabies, and as a result, whether the patient will require PEP.

This chapter describes how to express and purify MAbs in mammalian expression systems suitable for use in PEP and diagnosis of rabies. The three major steps in MAb production, such as expression, purification, and product testing, are described. Expression is achieved by transfecting a plasmid coding for light- and heavy-chain immune globulins into mammalian cells and growing them on a large scale, ranging from less than a liter in shaker flasks to much larger volumes in roller bottles and bioreactors. The product can then be purified with Protein A chromatography, which exploits the binding properties of MAbs. Finally, the quality of the end product can be tested with various methods to confirm its molecular weight and isoelectric point (pI), and to determine its ability to bind to and neutralize virus.

26.2 MATERIALS

26.2.1 Reagents

- RABV glycoprotein
- Fluorescein isothio-cyanate (FITC)-labeled anti-rabies and control reagents

- Cell culture medium
- Expression vectors
- Fetal bovine serum (FBS)
- Phosphate-buffered saline (PBS)
- Transfection reagent (e.g. Lipofectamine® or FuGene®)

26.2.2 Equipment

- CO_2 incubator
- Enzyme-Linked Immunosorbent Assay (ELISA) reader
- Fast Protein Liquid Chromatography (FPLC)
- High-Performance Liquid Chromatography (HPLC)
- Fluorescent microscope
- Gel electrophoresis equipment

26.2.3 Biological Materials

- PER.C6® cells
- HEK 293T cells
- Mouse myeloma cells (NS0, SP2.0)
- Chinese Hamster Ovary (CHO) cells
- Challenge Virus Standard (CVS)-11 (or other fixed RABV strain)
- Non-expressing bacteria cells (e.g. DH1, DH5α™, XL1–Blue)

26.3 METHODS

26.3.1 Cells and Cell Culture

Most successful MAb production has been in either CHO or mouse myeloma cells (NS0 or SP2.0). More recently, however, human cell lines such as HEK 293 T and PER.C6 have been explored for MAb production, with the latter having successfully produced rabies MAbs.[13,14]

26.3.2 Preparation of Expression Vector

Most generic expression vectors are suitable for the production of MAbs as the methods of purification do not require any tags, and the purification methods rely on the natural binding properties of the antibodies. Figure 26.1 shows examples of plasmids that may be used. The heavy and light chains are usually cloned into the multiple cloning sites (mcs) of two different expression vectors, but can also be cloned into a dual-expression vector with two mcs, so both the heavy and light chains are expressed by the same construct. For the two-vector method, select two

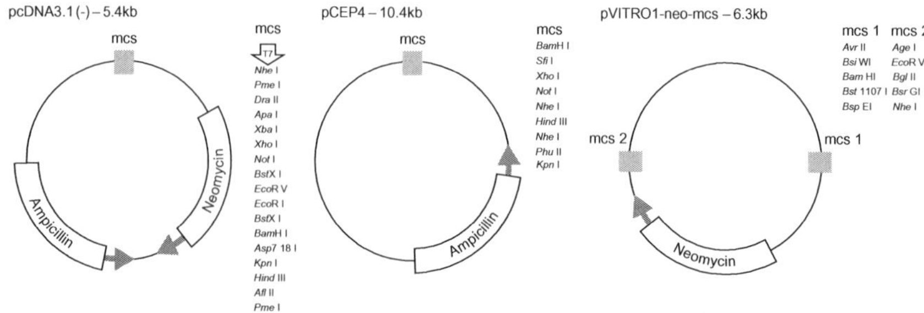

Figure 26.1 Example vectors for MAb expression.

vectors with the same antibiotic resistance before choosing two restriction sites in the mcs for each plasmid. With restriction enzymes chosen, amplify light- and heavy-chain cDNA fragments via polymerase chain reaction (PCR) using primers designed with flanking restriction enzyme sites to produce a product with appropriate sticky ends for ligation into plasmid. Finally, treat plasmids with the same restriction enzymes before adding the PCR product and DNA ligase to form the expression construct. After ligation, transform non-expressing bacteria with 1–2 μL of ligation mix using antibiotic-selective media plates specific to the plasmid, and grow overnight. Single colonies can then be picked to check for appropriate insertion into the parent plasmid, upon which one colony can be amplified by growing in appropriate culture broth before using a plasmid preparation kit to extract plasmid DNA before storing stocks at −20°C. For the dual-expression vector method, apply the same principles; DNA ligation occurs in two steps. First clone the light chain into one mcs with one suitable pair of restriction enzymes before purifying via miniprep, then clone the heavy chain into the other mcs using a different set of appropriate restriction enzymes.

Experimental Tip

Always use high-quality plasmid preparation kits, as low-quality plasmid can affect cell viability and decrease transfection efficiency. Verify the construct with a restriction digest. To select appropriate restriction enzymes that would cut your plasmid into several fragments of known size, use an appropriate tool (e.g. restriction mapper[15]). At least one restriction site in the codon for the MAb itself should be selected for greater accuracy. The sample should then be run on agarose gel electrophoresis to confirm correct-sized bands. Following a successful digest, constructs should be sequenced either in-house or via an external sequencing company to ensure the accuracy of the final product.

26.3.3 Preparation of Mammalian Cell Stocks

Quickly thaw a vial of cells and transfer into pre-warmed media: For PER.C6, HEK 293 T, CHO, grow cells in 10 mL DMEM (with 10% FBS) overnight in a 25 mL flask. For NS0, SP2.0, grow cells in 10 mL MEM at 37°C with 0.5% CO_2 at 37°C with 10% CO_2 overnight in a 25 mL flask. In all cases, after growth, re-suspend cells in PBS and harvest.

Experimental Tip

Use of serum-free media (SFM) during MAb production will decrease background binding in ELISA, and will decrease the chance of background neutralization or interference in rapid fluorescent focus inhibition test (RFFIT) due to co-purification of other antibodies present in serum. It will also lead to a cleaner end product and reduce strain on purification columns.

26.3.4 Transfection and Clone Selection

Plate out adherent cells at 3.5 $\times 10^6$ cells into a 96-mm dish. Before transfection, incubate the plasmid (2 μg of each) with 10 μL transfection reagent in SFM for 30 minutes. Add plasmid to cells for 5 hours before adding DMEM with 10% FBS and incubating overnight. Add growth medium containing appropriate antibiotic. Wash cells twice a week for three weeks to remove dead cells. Pick and transfer clones to a 96-well plate. After several days of further growth under selective pressure, test for MAb presence in cell culture supernatant via ELISA specifically detecting mouse or human MAbs.[16] Alternatively, antibody concentration can be quantified by Protein A HPLC.[17] The results of ELISA can be used to select the highest expressing clones to be expanded further using 24- and 6-well plates before transferring to T25 or T80 culture flasks and growing for several days before flash freezing cells (−80°C). Post-freezing, antibiotic selection pressure is usually no longer used. The process is repeated to select the highest expressing clones and to ensure the construct is stably expressing MAbs.

Experimental Tip

Transfection can be repeated each time you want to produce a MAb batch. However, it is more efficient to lay down a small stock of cells stably expressing the antibody. Be aware that cells have a limit for how long they can be expressed from frozen stocks (usually around 3 months).

26.3.5 Monoclonal Antibody Expression

Quickly thaw a vial of frozen cell clone stocks. Inoculate clones to a target concentration of 1×10^6 viable cells/mL into DMEM containing 10% FBS at 37°C in 10% CO_2 at 125 rpm. This should be performed in shaker flasks, selecting a volume based on required yield (For small-to-medium scale needs, shaker flasks can be used from 20 mL of media in a T125 shaker flask to 1 L of media in a 3 L shaker flask). Cultures should be sampled daily to determine cell concentration and viability (using a cell culture analyzer and ELISA). When cell viability drops below 50–25% (usually after 6–7 days), the culture should be transferred to centrifuge bottles and centrifuged at 5,000 g for 20 min. Supernatant is ready for purification. For larger scale expression, roller bottles or bioreactors can be used instead of shaker flasks. Ensure optimal inoculation cell densities are used according to the cell type.

Experimental Tip

Low levels of expression are usually caused by low cell viability, which is affected by the quality of cell stock and culture medium, incubation temperature, and shaker speed. These should be optimized when low-level expression is experienced. Feed-mixtures containing essential nutrients can be added to increase yield.

26.3.6 Monoclonal Antibody Purification

26.3.6.1 Protein A Chromatography

Protein A has a high affinity for human IgG1/IgG2 and mouse IgG2a/IgG2b, moderate affinity for human IgM/IgA/IgE and mouse IgG3/IgG1, but no affinity for other human/mouse immune globulins (Ig). Protein A chromatography can be performed via gravity columns or liquid chromatography systems. As this is a purification method, the purest water available should be used in all buffers (e.g. HPLC grade).

26.3.6.2 Performing Protein A Separation on a Gravity Column

Each Protein A column will have manufacturer-specific protocols to follow. In short, columns must be equilibrated with binding buffer before applying the sample with a syringe or peristaltic pump. After several wash steps to remove all unbound sample and impurities (protein presence should be determined by UV absorbance at 280 nm), MAb is eluted with elution buffer. As elution conditions are relatively harsh, it helps to collect the fractions into neutralization buffer, which will give a final pH of approximately

7.0. Re-equilibrate the column immediately with 5–10 volumes of binding buffer, wash thoroughly with 20% ethanol, and store at 4–8°C.

Experimental Tip

Samples should be centrifuged (10,000*g* for 10 min) and filtered before loading onto the column with a 0.22 µm filter. NaCl (up to 3 M) may be required in the binding buffer for sufficient binding in some MAbs. In certain cases, an additional wash step after the MAb is bound to the Protein A column may be required to remove all impurities. Only reuse Protein A column with identical MAb samples and be aware that binding ability will reduce with each subsequent use. Sometimes column reuse is not possible.

Binding capacity of columns can be increased by using several columns in sequence or by using larger columns. It is critical that the pH of the elution buffer is set correctly and desalting is done as quickly as possible to minimize aggregate formation.

26.3.6.3 Performing a Protein A Separation using Liquid Chromatography Systems

By using liquid chromatography systems (e.g. FPLC, HPLC), a very similar, yet much faster process, is applied, but is performed in an automated manner. Figure 26.2 shows a chromatogram of a successful human IgG purification.

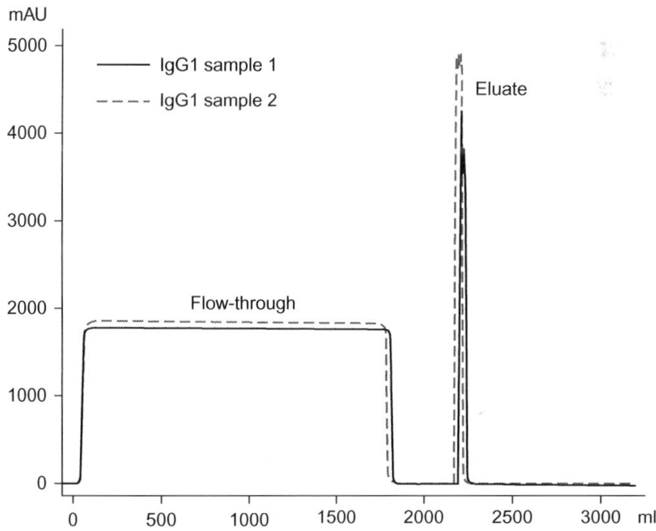

Figure 26.2 Chromatogram from Protein A chromatography purification of IgG. IgG harvest was loaded onto Protein A column and after washing, IgG was eluted. Chromatograms for two MAb samples are shown.

26.3.6.4 Desalting Columns

Purified MAbs inevitably have to undergo buffer exchange to make them suitable for subsequent use. Desalting columns work on similar principles to dialysis, but are much quicker and have much lower risk of sample loss (most manufactures quote a recovery in excess of 95%). Similar to Protein A chromatography, this step can be performed either with bench top gravity columns or using liquid chromatography systems.

26.3.6.5 Gravity Desalting Columns

Each desalting column will have manufacturer-specific protocols to follow. Always note the recommended rates of buffer/sample application. After equilibration with equilibration buffer, remove the stopper from the base of the column and apply the sample via a syringe or peristaltic pump. Make up your sample to the manufacturer's recommended sample volume with buffer if necessary. When the recommended sample volume has been eluted, discard. Elute the protein using the recommended elution buffer volume.

26.3.6.6 Desalting using Liquid Chromatography Systems

Liquid chromatography systems can be used to buffer exchange providing it is performed using a suitable desalting column and 500 mL of required buffer. It is recommended that pre-programmed application templates be applied.

Experimental Tips

Depending on the purpose of the MAb, additional purification steps can be implemented, such as anion- and cation-exchange chromatography. These steps will remove additional impurities such as DNA and protein. All techniques discussed are suitable for research purposes, including the usage of MAbs in animal studies. Sterile filtration through a 0.22 μm filter is highly recommended.

26.3.7 Monoclonal Antibody Product Testing

Several analytical techniques must be used to confirm that the purified protein is the target MAb. Sodium dodecyl sulfate polyacrylamide gel electrophoresis (SDS-PAGE) separates proteins by molecular weight. If the purified MAb has incurred significant insertions or deletions, this should be detected by SDS-PAGE. Figure 26.3 shows an SDS-PAGE of purified

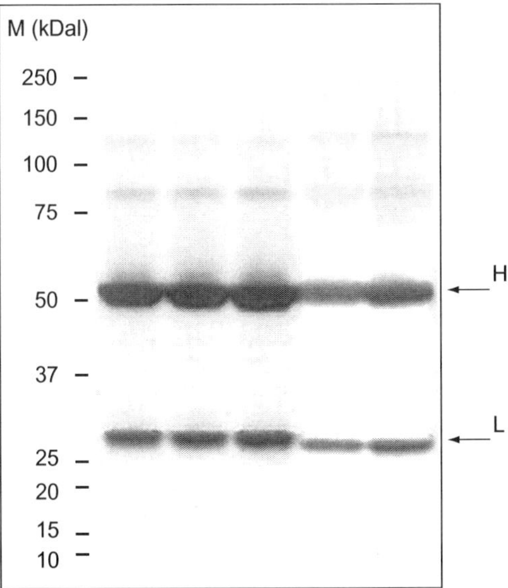

Figure 26.3 SDS-PAGE of purified human IgG MAbs. Purified MAbs were separated by SDS-PAGE (4–12% Bis-Tris Nu-PAGE gel, Invitrogen) under reducing conditions and stained with Sypro Ruby. Light and heavy chains (indicated) are separated by the gel, so this should be taken into account when selecting an appropriate percentage of acrylamide to achieve good separation.

human IgG MAbs. Typically a 4–12% gel yields good results. Isoelectric focusing (IEF) separates proteins by their pI, and will detect if there have been events that altered the overall charge of the purified MAb, including mutations during transfection/expression, protein degradation, or protein deamidation during purification. Figure 26.4 shows an example of an IEF of purified human IgG. The theoretical molecular weight and pI for the MAb you are purifying can be determined using the online tool ExPASy.[18]

Binding ELISA detects the ability of a protein to bind to a ligand. A laboratory strain of inactivated RABV itself (e.g. CVS-11) is a good ligand to use, either prepared in-house or obtained as a rabies vaccine. Alternatively, standard ELISA kit plates are available pre-coated with RABV glycoprotein (e.g. Platelia kit, BioRad). Good binding confirms correct MAb identity and functionality. Poor binding would suggest a loss of tertiary structure, aggregate formation, mutations in the complementarity determining regions, or instability of the MAb. The RFFIT (described elsewhere) tests the potency of an MAb to neutralize RABV,

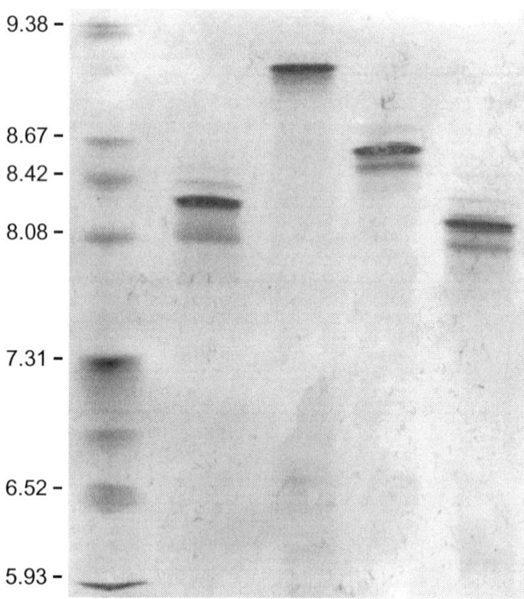

Figure 26.4 Isoelectric focusing in polyacrylamide gel of different purified human IgG MAbs. pI marker is indicated; four different IgG samples were loaded on gel. Note: Several bands are observed for each sample due to multiple protein states with different pIs.

and tests the ultimate functionality of the MAb. However, it is only applicable when producing MAbs both directed against, and known to have a neutralizing capacity against, RABV. For each MAb, comparison of its 50% neutralizing titer to the 50% neutralizing titer (2IU/mL) of a standard reference serum (e.g. standard RIG, lot R3) should be performed. If desired, several purified MAbs can be ranked according to potency. Alternatively, if the appropriate biosafety level laboratory is unavailable, *in vivo* neutralization testing using rabies pseudoviruses can be used as described previously.[19]

High-Pressure Size Exclusion Chromatography (HP-SEC), can be useful in detection of aggregation of the target MAb product. A small sample can be applied to an HP-SEC with an appropriate cut-off. Absorbance at A_{280} can then be used to determine the presence of aggregates. Samples of IgG before and after stress are compared in Figure 26.5. The additional peaks after pH stress at the expense of the major IgG monomer peak are caused by aggregates (i.e. dimers, trimers, tetramers, and oligomers) or MAb fragments. The target monomer proportion is >95%. The latter may also arise over time during storage due to instability of the MAb.

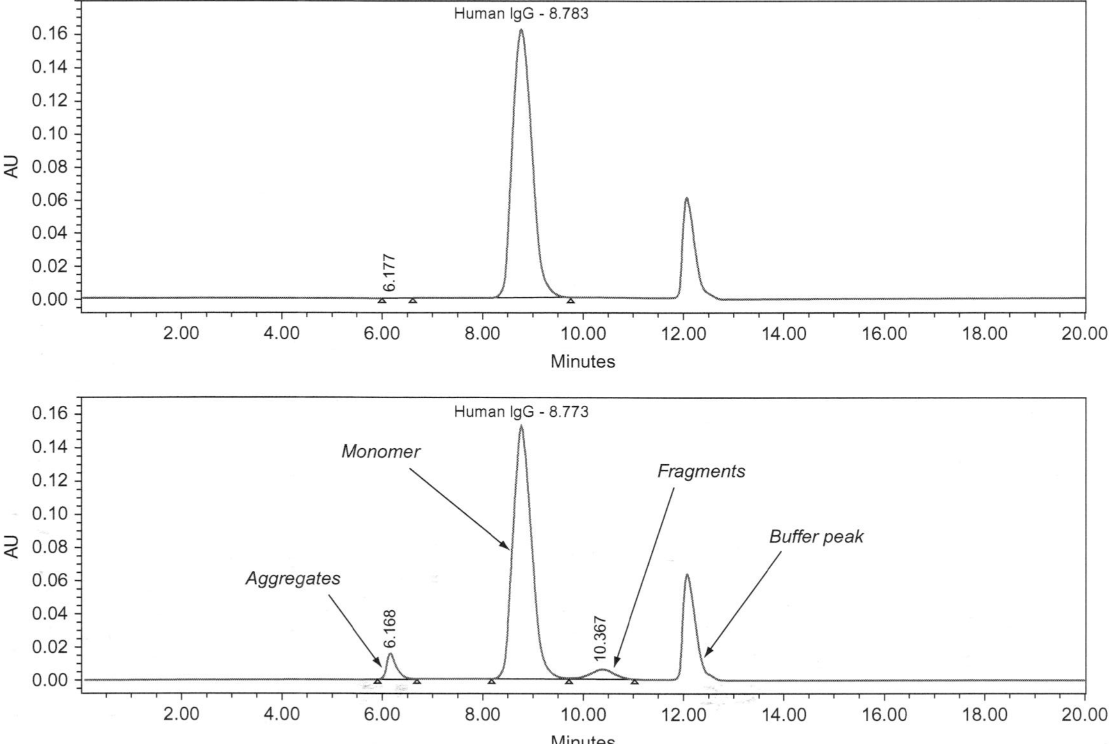

Figure 26.5 Size-exclusion chromatograms of unstressed IgG (A) vs. stressed IgG (B). Aggregates (such as dimers, trimers, oligomers, etc.), IgG monomer, fragment, and buffer peaks are indicated.

Experimental Tips

Many factors can lead to MAb aggregate formation, including extremes of pH, ionic strength, temperature, and other processing conditions.[20] Consequently, aggregation should be monitored throughout the MAb product purification process, including post-expression, where aggregation can be common.

26.3.7.1 Interpretation of Results

The methods presented here allow quick expression and purification of rabies MAbs at a reasonable scale, which can be readily expanded to a much larger scale. These MAbs are suitable for both *in vivo* and *in vitro* research purposes. The quality of the end product is confirmed by SDS-PAGE, IEF, ELISA, RFFIT, and HP-SEC. Because of the methods used, the likelihood of human pathogen contamination is very low. However, in animals, testing of bioburden and endotoxin levels is highly recommended. Although alternative eukaryotic expression systems exist,[21,22] they lack any human-specific post translation modifications that cell lines of human origin (e.g. PER.C6 and HEK 293T) would make. Additionally, both purification methods will be slower and not necessarily available in certain laboratories.

ACKNOWLEDGMENTS

Writing assistance was provided by Chris Whittaker of Gardiner-Caldwell Communications, Macclesfield, UK, and this support was funded by Janssen Pharmaceuticals.

Additional Information

Composition of Media and Buffers

Dulbecco's Modified Eagle's medium (DMEM) is available from many companies, including ATCC, Life Technologies, and Sigma-Aldrich. The composition is very complex and is available on the suppliers' websites. Protein A chromatography: Binding buffer (0.02 M sodium phosphate, pH 7.0), elution buffer (0.1 M citric acid, pH 3–6), neutralization buffer (1 M Tris-HCl, pH 9.0). The composition of buffers from commercially available kits are available from the manufacturer, unless undisclosed due to patenting.

Examples of Major Sources of Supplies/Equipment

PER.C6® cells are only available from Crucell under a specific license agreement. HEK 293T, CHO, NS0, SP2.0 and CHO cells can be obtained from many companies, including ATCC, Aglient, InvivoGen, Promega, and Life Technologies. Recombinant plasmids and restriction enzymes are available from many companies including InvivoGen, Life Technologies, and New England Biolabs. The transfection reagents Lipofectamine and FuGene are only available from Life Technologies and Promega, respectively. Kits for plasmid preparation and gel extraction are available from many companies, including Qiagen, Thermo Scientific, and Sigma-Aldrich. Many companies specialize in protein purification and supply a wide range of Protein A and desalting gravity and HPLC columns, including Thermo Scientific, Sigma-Aldrich, and Life Technologies. For FPLC, GE Healthcare provides the ÄKTA system and appropriate columns. Rabies virus (e.g. CVS-11) can be obtained through ATCC, although availability may be country-specific. A country-specific license may be required to work with lyssavirus, so contact the national regulatory agency for appropriate guidance. ELISA plates coated with RABV glycoprotein or inactivated RABV can be obtained from BioRad or Alpha Diagnostic International. FITC-labeled anti-rabies N antibodies and control reagents are available from Fujirebio or Chemicon.

REFERENCES

1. WHO *Consultation on a Rabies Monoclonal Antibody Cocktail for Rabies Post Exposure Treatment*. Geneva: WHO; 23–24 May 2002. <http://www.who.int/entity/rabies/vaccines/en/mabs_final_report.pdf> [accessed 14.02.14].
2. Dodet B. Asian rabies expert bureau. Preventing the incurable: Asian rabies experts advocate rabies control. *Vaccine* 2006;**24**:3045–9.
3. Wilde H. Failures of post-exposure rabies prophylaxis. *Vaccine* 2007;**25**:7605–9.
4. Sudarshan MK. Assessing burden of rabies in India: WHO sponsored National Multicentric Rabies Survey, 2003. *Indian J Community Med* 2005;**30**:100–1.
5. Rupprecht CE, Hanlon CA, Slate D. Oral vaccination of wildlife against rabies: opportunities and challenges in prevention and control. *Dev Biol* 2004;**119**:173–84.
6. Wilde H, Khawplod P, Hemachuda T. Rabies re-examined. *Lancet Infect Dis* 2002;**2**:327–43.
7. Montano-Hirose JA, Lafage M, Weber P, Badrane H, Tordo N, Lafon M. Protective activity of a murine monoclonal antibody against European bat lyssavirus 1 (EBL1) infection in mice. *Vaccine* 1993;**11**:1259–66.
8. Hanlon CA, Niezgoda M, Morrill P, Rupprecht CE. Oral efficacy of an attenuated rabies virus vaccine in skunks and raccoons. *J Wildl Dis* 2002;**38**:420–7.
9. Hanlon CA, DeMattos CA, DeMattos CC, Niezgoda M, Hooper DC, Koprowski H, et al. Experimental utility of rabies virus-neutralizing human monoclonal antibodies in post-exposure prophylaxis. *Vaccine* 2001;**19**:3834–42.
10. Prosniak M, Faber M, Hanlon CA, Rupprecht CE, Hooper DC, Dietzschold B. Development of a cocktail of recombinant-expressed human rabies virus-neutralizing monoclonal antibodies for postexposure prophylaxis of rabies. *J Infect Dis* 2003;**188**:53–6.

11. CDC Protocol for Postmortem Diagnosis of Rabies in Animals by Direct Fluorescent Antibody Testing. <http://www.cdc.gov/rabies/pdf/rabiesdfaspv2.pdf> [accessed 14.02.14].
12. Lembo T, Niezgoda M, Velasco-Villa A, Cleaveland S, Ernest E, Rupprecht CE. Evaluation of a direct rapid immunohistochemical test for rabies diagnosis. *Emerg Infect Dis.* 2006;**12**(2) <http://dx.doi.org/10.3201/eid1202.050812> [accessed 14.02.14].
13. Jones D, Kroos N, Anema R, van Montfort B, Vooys A, van der Kraats S, et al. High-level expression of recombinant IgG in the Human cell line PER.C6. *Biotechnol Prog* 2003;**19**:163–8.
14. Marissen WE, Kramer RA, Rice A, Weldon WC, Niezgoda M, Faber M, et al. Novel rabies virus–neutralizing epitope recognized by human monoclonal antibody: fine mapping and escape mutant analysis. *J Virol* 2005;**79**:4672–8.
15. RestrictionMapper: <http://www.restrictionmapper.org> [accessed 14.02.14].
16. Boel E, Verlaan S, Poppelier MJ, Westerdaal NA, Van Strijp JA, Logtenberg T. Functional human monoclonal antibodies of all isotypes constructed from phage display library-derived single-chain Fv antibody fragments. *J Immunol Methods* 2000;**239**:153–66.
17. Throsby M, Geuijen C, Goudsmit J, Bakker AQ, Korimbocus J, Kramer RA, et al. Isolation and characterization of human monoclonal antibodies from individuals infected with West Nile Virus. *J Virol* 2006;**80**:6982–92.
18. Compute pI/Mw tool: <http://web.expasy.org/compute_pi/> [accessed 14.02.14].
19. Sloan SE, Hanlon C, Weldon W, Niezgoda M, Blanton J, Self J, et al. Identification and characterization of a human monoclonal antibody that potently neutralizes a broad panel of rabies virus isolates. *Vaccine* 2007;**25**:2800–10.
20. Kükrer B, Filipe V, van Duijn E, Kasper PT, Vreeken RJ, Heck AJ, et al. Mass spectrometric analysis of intact human monoclonal antibody aggregates fractionated by size-exclusion chromatography. *Pharm Res* 2010;**27**:2197–204.
21. Dietzschold B, Gore M, Casali P, Ueki Y, Rupprecht CE, Notkins AL, et al. Biological characterization of human monoclonal antibodies to rabies virus. *J Virol* 1990;**64**:3087–90.
22. Both L, van Dolleweerd C, Wright E, Banyard AC, Bulmer-Thomas B, Selden D, et al. Production, characterization, and antigen specificity of recombinant 62-71-3, a candidate monoclonal antibody for rabies prophylaxis in humans. *FASEB J* 2013;**27**:2055–65.

CHAPTER TWENTY SEVEN

Generation of Immune Globulin Single Variable Domains by Display Technologies

Hans de Haard[1,2], Anna Hultberg[1], Theo Verrips[2], Peter Vanlandschoot[3], Michael Saunders[1], Catelijne Stortelers[3], and Steven Van Gucht[4]

[1]arGEN-X, Zwijnaarde, Belgium
[2]Utrecht University, Utrecht, The Netherlands
[3]Ablynx, Zwijnaarde, Belgium
[4]National Reference Centre of Rabies, Scientific Institute of Public Helath, Brussels, Belgium

Chapter Contents

27.1 Introduction	292
27.2 Materials	293
27.2.1 Reagents	293
27.2.2 Equipment	293
27.2.3 Biological Materials	294
27.2.4 Laboratory Animals	294
27.3 Methods	294
27.3.1 Immunizations	294
27.3.2 Immune Variable Heavy-Chain Fragment Library Construction for Phage Display	295
27.3.3 Phage Display for Selection of Rabies Virus-Specific Llama Variable Heavy-Chain Fragment	296
27.3.4 Periplasmic Expression of Variable Heavy-Chain Fragment	297
27.4 Discussion	298
27.5 Experimental Tips	298
27.5.1 Critical Parameters and Troubleshooting	298
27.5.2 Alternatives	299
27.5.3 Precautions	299
27.5.4 Time Considerations	299
27.5.5 Limitations	299
Acknowledgments	300
References	301

C. Rupprecht & T. Nagarajan (Eds): Current Laboratory Techniques in Rabies Diagnosis, Research, and Prevention, Volume 1.
Doi: http://dx.doi.org/10.1016/B978-0-12-800014-4.00027-5

27.1 INTRODUCTION

Phage display is a well-established technique that uses bacteriophage to connect genotype with phenotype by expression and display of proteins that can be selected from large protein-encoding libraries, and was first described in 1985.[1] Proteins displayed may be peptides, antibody fragments, such as single-chain variable fragments (scFv), antigen-binding fragments (Fab), or as described here, variable heavy-chain fragments (VHH). VHH are the antigen binding, variable part, of "heavy chain only" antibodies present in *Camelidae* family members, such as llamas.[2] The VHHs are small (12–15 kDa), biophysically stable molecules with good solubility characteristics and similar affinities to conventional antibodies.[3] Llama-derived VHHs have proven to be powerful virus neutralizers, including for rabies virus (RABV).[4–6] The single-chain nature of the VHHs allows construction and production of multimeric molecules using the same or different VHH building blocks joined by flexible $Glycine_4Serine$ linkers, thereby targeting either one or two different epitopes on the same molecule.[5,7] These properties make them promising molecules for prophylactic and therapeutic purposes, as well as for diagnostics. Formatted and sequence optimised VHHs are being developed as therapeutic agents by Ablynx, Belgium, under the trade name Nanobodies®.

The principles of phage display and panning are illustrated in Figure 27.1.

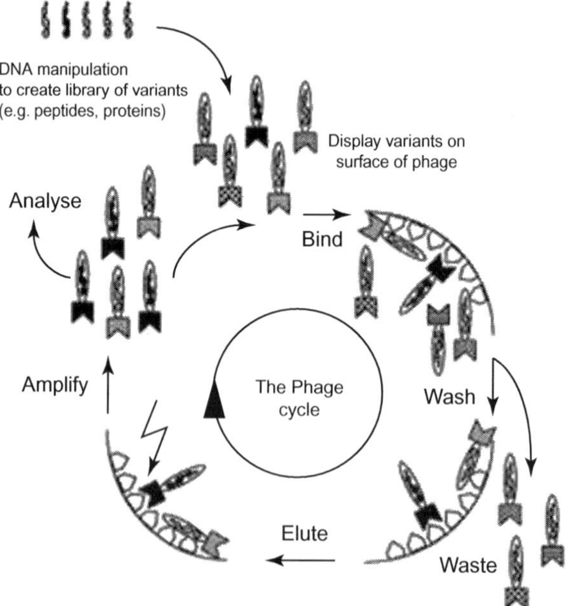

Figure 27.1 Principles of Phage Display and Panning.[8]

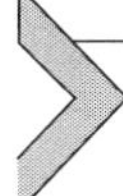

27.2 MATERIALS

27.2.1 Reagents

For total RNA isolation: RNAeasy Midi kit (Qiagen), 96% ethanol (RNase free), 2 M NaAc pH 4 (RNase free), RNase free 1.5 mL microcentrifuge tubes, double autoclaved distilled water, RNase away solution (Molecular BioProducts, Cat. No. 7002).

cDNA synthesis: SuperScript® III First-Strand Synthesis System (Invitrogen).

Primers for amplification of VHH:

- 051 (homology to FR1) 21 bp: GGCTGAGCTGGGTGGTCCTGG
- 052 (homology to FR1) 21 bp: GGCTGAGTTTGGTGGTCCTGG
- 003 (homology to CH2) 23 bp: GGTACGTGCTGTTGAACTGTTCC
- 050 (*Sfi*I containing FR1) 55 bp:
- CATTTGAGTTGGCCTAGCCGGCCATGGCAGAGGTGCAG CTGGTGGAGTCTGGGGG

PCR reagents: 20 mMdNTP, 10 × buffer (+Mg) and Expand High Fidelity enzyme (3.5 U/μL) (Roche Diagnostics GmbH), TAE buffer and agarose.

DNA QIAquick PCR Purification Kit and QIAquick gel extraction kit (QIAGEN).

*Restriction enzymes and ligase: Bst*EII(10 U/μL) and *Sfi*I (20 U/μL, Biolabs), T4 ligase (3 U/μL, Promega).

Antibodies: coupled anti-RABV nucleocapsid IgG – fluorescein isothiocyanate (FITC) conjugate (Bio-Rad Laboratories, France).

Platelia II Rabies kit (Bio-Rad Laboratories, Cat. No. 355-1180), 8-well strips pre-coated with native RABV glycoprotein (G).

Phagemid vector for fusion of VHH with *c-myc*-His_6 and gene3 at sites *Bst*EII 5′ and *Sfi*I 3′,[9] which is identical to pHEN-1,[10] but contains a hexahistidine tail for immobilized metal affinity chromatography (IMAC).

27.2.2 Equipment

Spectrophotometer (UV: preferably 230–310 nm and 600 nm) and cuvettes (if necessary), tabletop centrifuge, thermal cycler, DNA electrophoresis systems cleaned with NaOH and HCl, UV/VIS camera for DNA gels and agar plate imaging, UV camera for slicing out DNA fragments, Incubator at 50°C and 60°C and 16°C (without shaking), electroporator, incubator 37°C with shaker, fluorescent microscope (Olympus IX73).

27.2.3 Biological Materials

Rabies Vaccine, e.g., Mérieux HDCV (Sanofi Pasteur MSD), Baby Hamster Kidney (BHK)-21 cells (ATCC CCL-10), Challenge Virus Standard-11 (CVS-11), (ATCC VR959), *Escherichia coli* (*E. coli*) TG1, Electrocompetent *E. coli* TG1 (Stratagene), M13KO7 Helper phage.

27.2.4 Laboratory Animals

Two llamas (*Lama glama*, purchased from N.V. Neerhofdieren Bocholt) were housed at the animal facilities of the Belgian Scientific Institute of Public Health (WIV-ISP, authorization nr. LA1230177). All experimental procedures were approved by the Ethical Committee of the WIV-ISP and the Veterinary and Agrochemical Research Centre (CODA-CERVA).

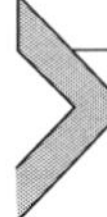

27.3 METHODS

27.3.1 Immunizations

HDCV vaccine was administered through the intramuscular route in the neck region on Days 0, 7, 28, 35, and 57 at 2 sites at 0.5 mL/site, corresponding to 2.5 IU. Collection of blood in EDTA-containing vacutainer tubes was done at different time points as described in Table 27.1.

The neutralizing potency of the immune llama serum can be determined by the Rapid Fluorescent Focus Inhibition Test (RFFIT),[11] using BHK-21 cells and CVS-11 according to the manual of Diagnostic Test and Vaccines for Terrestrial Animals (Office International des Epizooties, 2008). Serial three-fold dilutions of heat-inactivated serum samples were incubated with the CVS-11 in 8-well tissue culture chamber slides for 90 min at 37°C. BHK-21 cells were then added to the serum–virus mixture and incubated for an additional 20–24 h at 37°C, 5% CO_2. The cell sheet was fixed with ice-cold acetone and stained with anti-RABV conjugate and examined under a fluorescence microscope (200 × magnification) to score the number of RABV-infected foci. The number of fields positive out of 20 fields per well was recorded. The neutralization end-point titer was defined as the highest sample dilution at which 50% of the observed microscopic fields contained no infected cells. The *in vitro* neutralizing potency was expressed in International Units (IU)/mL in reference to "The Second International Standard for Anti-Rabies Immunoglobulin" purchased from the NIBSC, UK. A serum titer of 0.5 IU/mL was considered adequate. The titer of neutralizing antibodies

Table 27.1 Immunizations, Blood Collection, and Sero-Conversion (RFFIT titer)

Day	Immunization (vaccine potency IU)	Blood Collection	RFFIT titer Llama 1	RFFIT titer Llama 2
0	2.5	10 mL pre-immune blood	<0.5 IU/mL (1/9)	<0.5 IU/mL (1/9)
7	2.5	–		
27		10 mL immune blood	2 IU/mL (1/66)	6 IU/mL (1/179)
28	2.5	–		
35	2.5	–		
37		10 mL immune blood	22 IU/mL (1/674)	27 IU/mL (1/789)
42		150 mL immune blood (PBL1)	37 IU/mL (1/989)	33 IU/mL (1/896)
49		150 mL immune blood (PBL2)	23 IU/mL (1/674)	15 IU/mL (1/441)
57	2.5	–		
62		150 mL immune blood (PBL3)	22 IU/mL (1/673)	35 IU/mL (1/1071)

in two immune llamas is shown in Table 27.1. They all had a titer in the range of 15–35 IU/mL after repeated immunization with HDCV (Table 27.1).

27.3.2 Immune Variable Heavy-Chain Fragment Library Construction for Phage Display

Immune VHH library construction was divided into five parts: isolation of total RNA, cDNA synthesis, polymerase chain reaction (PCR) amplification of VHH (variable domain fragments, derived from llama heavy-chain immunoglobulins),[2] ligation into phagemid vectors, and expression of VHH containing phage for selections. (Caution: Work on ice as much as possible.)

Isolation of total RNA: Isolate peripheral blood lymphocytes (PBLs) from total blood using routine ficoll gradients (PBL1-3 can be combined or used alone depending on the immune response). Total RNA from PBLs can also be extracted using the RNAeasy Midi kit (Qiagen) following the manufacturer's protocol. Lymph nodes may not be distinguishably enlarged, which makes them difficult to find, and therefore not suitable for use as a source of lymphocytes. Determine the ratio A_{260}/A_{280} as an indication of quality of purified RNA (should be around 2). Estimated total RNA amounts from 150 mL blood are expected to be ≈200–1,000 μg. Determine the integrity of the sample on a 1% agarose gel in a tank (combs and tray) cleaned with 0.1% SDS (at least for 1 h and rinse with double-distilled water, ddH2O) to avoid degradation. You should observe 3 bands of RNA, ~1500 bp, 800 bp, and a band below 200 that correspond to 28S, 18S, and 6S rRNA, respectively. Occasionally a band of 70–100 bp can be visible representing tRNA. Possible DNA contaminations will be above 10,000 bp.

cDNA synthesis: Random primers and the SuperScript III First-Strand Synthesis System for reverse transcription (RT)-PCR are used according to the manufacturer's recommendations to amplify VHH encoding gene fragments. QIAquick PCR Purification Kit was used for clean up, but is an optional step.

PCR amplification: This is divided into two amplifications. The first step amplifies conventional and heavy-chain antibody genes from cDNA, resulting in antibody fragments from FR1 to CH2 of VHH using two framework 1 (FR1)-specific primers 051 and 052 and a CH2-specific primer, 003. DNA fragments, corresponding to 600 bp and separated from the conventional antibody-derived VH of around 900 bp, are extracted from a 1% agarose gel and used as a template in the second nested PCR reaction using primers flanked with *Sfi*I and *Bst*EII restriction sites in the 5′ and 3′ end, respectively (primers 050 and 003). These fragments were cloned into a phagemid vector upstream of a *c-myc* tag and a His6-tag, as well as gene3 for display on filamentous bacteriophage as previously described.[9] Electrocompetent *E. coli* TG1 were transformed, generating library sizes of around 10^8 and phage expressing VHH were prepared as described in the Additional Information section.

27.3.3 Phage Display for Selection of Rabies Virus-Specific Llama Variable Heavy-Chain Fragment

Perform selections for RABV G specific VHH using 8-well strips pre-coated with native RABV G, PV (Pasteur virus) strain (Platelia II Rabies plates). Prepare an overnight culture of *E. coli* TG1 from a single colony taken from a minimal M9 agar plate. Block both the strips and the input phage, prepared using standard protocol (see Additional Information), in superblock (Pierce) or 1% marvel (skimmed milk in PBS) for 1 h at room temperature (RT). Add 10 μL phage to 90 μL superblock per well and incubate with shaking for 2 h at RT. Remove the phage-containing solution from each well with clean filter tips, and wash carefully 20 times with PBS + 0.05% Tween 80, followed 5 times with PBS. Use new tips for every well. Elute with 100 μL/well using trypsin (1 mg/mL) for 15 min at RT with shaking. Stop the trypsin reaction by addition of 5 μL ABSF (4 mg/mL). Infect exponentially growing TG1 for 30 min at 37°C and titrate for determination of output enrichment compared to control as described below. Infect 50 μL of eluted phage into 333 μL TG1 (OD_{600} = 0.5) + 666 μL 2 ×TY. Infect for 30 min at 37°C without shaking. Add 10 mL 2 ×TY amp100 Glu2% and grow overnight at 37°C, 250 rpm. Store grown culture as glycerol stock at −80°C.

Output phage titration: Prepare serial dilutions (10^{-1}–10^{-5}) in a 96-well culture plate (10 μL output phage in 90 μL PBS); add 5 μL of phage dilutions to 95 μL of exponentially growing TG1 (OD_{600} = 0.5). Infect for 30 min at 37°C without shaking. Plate 5 μL drops in duplicate on LB/amp100/gluc2% plates and incubate at 37°C overnight. Calculate the number of input phage when colonies have grown. Store the remaining eluted output phages as glycerol stock at −80°C.

For calculation of input phage: Prepare serial dilutions of input phages in 96-well culture plate (10 μL output phage in 990 μL PBS); typically 1E2, 1E4, 1E6, 1E8, and 1E10. Infect as described for output phage. Repeat the selections if needed to enrich for RABV binding VHH expressed on phage. Use 1 μL of input phage in selection round 2. When enrichment is observed compared to the control (PBS or irrelevant coated protein), dilute the overnight cultures of infected TG1 and plate them on LB/amp100/gluc2% plates. Dilute enough to have individual colonies that are transferred to a 96-well plate (v-shaped) with 85 μL of 2 ×TY amp100/Glu2% and let them grow at 37°C overnight. Store the master plate used for production of VHH, a sequencing at −80°C in 20% glycerol, to be able to trace back the individual clones.

27.3.4 Periplasmic Expression of Variable Heavy-Chain Fragment

From the overnight cultures in the 96-well plate, prepare perisplasmic extracts containing VHH by induction of exponentially growing *E. coli* TG1 with 1 mM IPTG and continue cultivation at 37°C (can be performed in 1–400 mL scale) overnight at 250 rpm for production of VHH. After centrifugation, freeze-thaw the bacterial pellet and re-suspend in PBS. The His_6-tag is used for purifications with Talon Metal affinity resin (Clontech) according to instructions. Determine the concentration of the purified VHH by OD_{280} and the purity (1 μg) by SDS-PAGE under reducing (with DTT) and non-reducing conditions. A band of around 15 kDa corresponds to a VHH.

Purified VHH can be characterized by enzyme-linked immunosorbent assay (ELISA), competition assays, RFFIT, and the VHH encoding genes can be fused into multimeric constructs with flexible G_4S-linkers of different lengths to form bivalent or biparatopic molecules targeting one or more different epitopes.

ELISA: The RABV G precoated on 8-well strips (BioRad) can also be used to identify the G-specific VHH by addition of periplasmic extracts containing VHH, followed by incubation for 2 h at RT. After washing, add rabbit anti-VHH antibody (MCA, The Netherlands), 1/5,000 dilution, and incubate for 1 h at RT. Then add horseradish peroxidase-conjugated goat anti-rabbit IgG (1/10,000, Jackson) and incubate for 1 h at RT. Add TMB substrate and read at 620 nm.

Competition assays can be set up using ELISA to home in on specific epitopes or to evaluate if the selected VHH bind to the same or different epitopes. Label VHH1 with biotin as described by the manufacturer (Thermo Scientific), and determine the concentration at 50% binding. Incubate this fixed amount of VHH1-biotin with a dilution series of VHH2 and evaluate competition by reduction in signal. Use cold, unlabeled VHH1 as a positive control for the competition assay.

27.4 DISCUSSION

VHHs have been shown to be excellent virus neutralizers, including RABV, and the monovalent nature of the molecules allow for multimerization to increase both potency and virus cross-neutralization.[5] The length of the flexible linker can be varied to optimize for intramolecular binding within a trimeric protein like RABV G, or between two different trimeric proteins, potentially preventing viral escape.

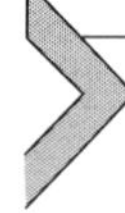

27.5 EXPERIMENTAL TIPS

27.5.1 Critical Parameters and Troubleshooting

All steps should be properly quality controlled before proceeding with the next steps. Good quality RNA is important. Always work with RNase free and on ice. TG1 must be kept on minimal M9 agar plates to maintain the sex pilus on TG1 for infection by the phage when grown to exponential phase. As contamination control, always include TG1 both in culture medium and on plates when spotting.

27.5.2 Alternatives

Elutions can be performed with pH shock using 100 mM Triethylamine (TEA, pH 12) or 0.2 M Glycine-HCl, pH 2.2 for 15 min at RT with shaking. Transfer eluted phage to a new well or tube and neutralize immediately.

Blocking solution can be exchanged with 1% skimmed milk (marvel in PBS), 2% casein, or 2% BSA (all dissolved in PBS). Even if biotinylated antigens are used for selections, milk-derived blocking agents can still be used (but superblock is preferred).

27.5.3 Precautions

- Use gloves when working with RNA.
- Clean bench and pipettes with RNase away (Molecular BioProducts).
- Use a separate set of pipettes for library constructions to avoid cross-contaminations.
- Use QIAquick PCR purification Kit and Quiaquick gel extraction kit dedicated only to library construction.
- Clean gel chambers with 0.2 M NaOH and rinse well with deionized/sterile distilled water before use.
- Work on ice.
- Make sure that there is no excess chromosomal DNA after RNA isolation because it will bind the RNA upon precipitation.
- To avoid contamination of phage, use filter tips for every step and clean the bench carefully with bleach before and after use. Preferably work in dedicated laminar flow hoods.
- Never use pipettes for inoculation of TG1, but sterile serological pipettes to avoid contamination.

27.5.4 Time Considerations

Immunizations take about 6 weeks; RNA extraction and library preparation take 1–2 weeks; Selections take 2–4 weeks depending on complexity and rounds of selection.

27.5.5 Limitations

Displaying VHH by bacteriophage does not allow for affinity selections because more than one copy ofVHH is displayed on each phage, with the exception of monomeric target proteins (should be kept in solution as a biotinylated protein).

ACKNOWLEDGMENTS

This work was funded and executed in conjunction with Ablynx NV, Zwijnaarde, Belgium and supported by an IWT grant (Institute for the Promotion of Innovation by Science and Technology in Flanders, IWT70050) from the Flemish Government. We thank Maria Gonzalez-Pajuelo, previous employee at Ablynx NV, Mireille Koenders, previous employee at Utrecht University, and Valérie Rosseels, previous employee at the National Reference Laboratory of Rabies – Scientific Institute of Public Health, Brussels, Belgium.

Additional Information

Phage Preparation

Precipitate phage with ¼ volume of 20% PEG6000/2.5 M NaCl, for 10–15 min on ice. Centrifuge for 5 min at maximum speed and remove supernatant. The pellet contains the precipitated phage. Repeat the centrifugation, remove remaining supernatant, and leave the tubes head down on tissue paper to remove all PEG. Re-suspend the pellet in ½ volume PBS. Centrifuge for 5 min at maximum speed and take the supernatant into a new eppendorf tube. If there is still a pellet (cellular debris), repeat the centrifugation step and transfer the supernatant to a new eppendorf tube.

Composition of Media, Buffers, Solutions

2 × TY media (1 L): Dissolve 16 g tryptone, 10 g yeast extract, 5 g NaCl, deionized water to 1 L. Autoclave. Add antibiotics (100 µg/mL Ampicillin or 50 µg/mL Kanamycin) when the media has cooled down to 55°C.

LB (Luria broth) media (1 L): Dissolve 10 g tryptone, 5 g yeast extract, 5 g NaCl, deionized water to 1 L. Autoclave. Add antibiotics (100 µg/mL Ampicillin or 50 µg/mL Kanamycin) when the media have cooled down to 55°C.

LB agar (1 L): Dissolve 10 g tryptone, 5 g yeast extract, and 10 g NaCl in 950 mL deionized water. Adjust the pH of the medium to 7.0 using 1N NaOH. Autoclave. Add antibiotics (100 µg/mL Ampicillin or 50 µg/mL Kanamycin) when the media have cooled down to 55°C.

Minimal (M9) agar plates (1 L): Dissolve 15 g Difco Bacto® agar in 888 mL H_2O and autoclave. Prepare 100 mL 10 × M9 salts (60 g Na_2HPO_4, 30 g KH_2PO_4, 10 g NH_4Cl, 5 g NaCl, final volume 1 L autoclaved and stored at 4°C), 1 mL 1 M $MgSO_4$, 0.1 mL 1 M $CaCl_2$, 1 mL 1% thiamine HCL (filter sterilized), 10 mL 20% glucose, and add to the autoclaved agar solution when the temperature reaches 60°C. Fill to final volume of 1 L with deionized water when the temperature reaches 60°C and pour plates.

20% glucose (w/v) (100 mL): 20 g glucose, add to 100 mL with deionized water. Autoclave.

20% glycerol in 2 × TY (90 mL): Add 30 mL 60% glycerol (autoclaved; do not autoclave 100% glycerol) to 60 mL sterile 2 × TY.

1 × TAE (1 L): 20 mL 50 × TAE and fill to 1 L with deionized water.

TE buffer (100 mL): 1 mL 1 M Tris-HCl, 0.2 mL 0.5 M EDTA pH 8.0, 98.8 mL deionized water.

Preparation of M13KO Helper Phage

Grow an overnight culture of TG1 in LB media at 37°C starting from a single colony grown on a minimal M9 plate. Inoculate 1:100 in LB and grow to log phase (OD_{600}: 0.6–0.9).

Make top agar (2 × TY-agar 0.75%) and let it cool to 50°C in a water bath. Streak 1 μL of a helper phage stock (e.g. ~3 × 10^{12} pfu/mL) and 1 μL of a 100 × diluted stock onto the surface of a pre-warmed 2 × TY plate. Add 5 mL of the TG1 culture to 30 mL of top agar (at 50°C). Immediately pour 3 mL of TY top agar containing the TG1 culture across the plate from the end towards the start of the streak. Allow the top agar to solidify for a few minutes and incubate at 37°C overnight. Do not forget to inoculate the new TG1. Pick 4 well-separated single plaques (small ones) with a glass pipette and drop into 4 × 4 mL of 2 × TY with a 1:100 dilution of an overnight TG1 culture. Grow for 2 h at 37°C. Dilute the 4 mL cultures into 100 mL of 2 × TY in a 2-liter flask, preferably baffled for good aeration. Grow for 1 h at 37°C. If baffled flasks are available, add 300 mL of 2 × TY with kanamycin (final conc: 25 μg/mL) and grow overnight. If baffled flasks are not available, just add kanamycin (final conc: 25 μg/mL). Spin down the bacteria in 50 mL tubes for 15 min.

Pour 40 mL of the supernatant in tubes with 10 mL 20% PEG6000/2.5 M NaCl and incubate on ice for at least 30 min. Spin down for 15 min and discard the supernatant, leave the tubes head down on tissue paper to remove all PEG. Re-suspend the pellet (for 4 × 400 mL: 4 times 8 pellets) in 1 mL of filter sterile TE buffer. Centrifuge the bacteria in eppendorf tubes to remove remaining cell debris for 5 min at 4°C. Combine all supernatants (approximately 32 mL) and add an equal volume of 100% sterile glycerol. Determine the titer of the phages by making dilutions of 10^4 to 10^{14} in PBS and plate 100 μL of these dilutions on 2 × TY plates and add top agar with TG1 (see above). Make aliquots and store at −80°C.

REFERENCES

1. Smith GP. Filamentous fusion phage: novel expression vectors that display cloned antigens on the virion surface. *Science* 1985;**228**(4705):1315–7.
2. Hamers-Casterman C, Atarhouch T, Muyldermans S, Robinson G, Hamers C, Songa EB, et al. Naturally occurring antibodies devoid of light chains. *Nature* 1993;**363**(6428):446–8.

3. Harmsen M, De Haard H. Properties, production, and applications of camelid single-domain antibody fragments. *Appl Microbiol Biotechnol* 2007;**77**(1):13–22.
4. Forsman A, Beirnaert E, Aasa-Chapman MM, Hoorelbeke B, Hijazi K, Koh W, et al. Llama antibody fragments with cross-subtype human immunodeficiency virus type 1 (HIV-1)-neutralizing properties and high affinity for HIV-1 gp120. *J Virol* 2008;**82**(24):12069–81.
5. Hultberg A, Temperton NJ, Rosseels V, Koenders M, Gonzalez-Pajuelo M, Schepens B, et al. Llama-derived single domain antibodies to build multivalent, superpotent and broadened neutralizing anti-viral molecules. *PLoS ONE* 2011;**6**(4):e17665.
6. Van der Vaart J, Pant N, Wolvers D, Bezemer S, Hermans PW, Bellamy K, et al. Reduction in morbidity of rotavirus induced diarrhoea in mice by yeast produced monovalent llama-derived antibody fragments. *Vaccine* 2006;**24**(19):4130–7.
7. Roovers RC, Laeremans T, Huang L, De Taeye S, Verkleij AJ, Revets H, et al. Efficient inhibition of EGFR signalling and of tumour growth by antagonistic anti-EGFR Nanobodies. *Cancer Immunol Immunother* 2007;**56**(3):303–17.
8. Hoogenboom HR, de Bruïne AP, Hufton SE, Hoet RM, Arends JW, Roovers RC. Antibody phage display technology and its applications. *Immunotechnology* 1998;**4**(1):1–20.
9. De Haard HJ, Bezemer S, Ledeboer AM, Müller WH, Boender PJ, Moineau S, et al. Llama antibodies against a lactococcal protein located at the tip of the phage tail prevent phage infection. *J Bacteriol* 2005;**187**(13):4531–41.
10. Hoogenboom HR, Griffiths AD, Johnson KS, Chiswell DJ, Hudson P, Winter G. Multi-subunit proteins on the surface of filamentous phage: methodologies for displaying antibody (Fab) heavy and light chains. *Nucleic Acids Res* 1991;**19**(15):4133–7.
11. Smith JS, Yager PA, Baer GM. A rapid reproducible test for determining rabies neutralizing antibody. *Bull World Health Organ* 1973;**48**(5):535.

PART FOUR

Appendix

APPENDIX A

Thirumeni Nagarajan[1] and Charles Rupprecht[2]

[1]Research and Development, Biological E. Limited, Shameerpet, Hyderabad, Andhra Pradesh, India
[2]Ross University School of Veterinary Medicine, Basseterre, St. Kitts, West Indies

Table A.1 List of World Health Organization (WHO) Collaborating Centres for Rabies

S.No	Collaborating Centre	Address
1	The WHO Collaborating Centre for Reference and Research on Rabies	Pasteur Institute of Iran, Tehran, Iran
2	The WHO Collaborating Centre for Rabies Epidemiology	The National Institute of Communicable Diseases [NICD], New Delhi, India
3	The WHO Collaborating Centre for the Characterization of Rabies and Rabies Related Viruses	Veterinary Laboratory Agencies [VLA], Weybridge, UK
4	The WHO Collaboration Centre for Rabies Surveillance and Research	Friedrich – Loeffler Institute [FLI], Germany
5	The WHO Collaboration Centre for Reference and Research on Rabies	Pasteur Institute, France
6	WHO Collaborating Centre for Control, Pathogenesis and Epidemiology of Rabies in Carnivores	Centre of Expertise [COFE] for Rabies, Canadian Food Inspection Agency, Canada
7	WHO Collaborating Centre for Reference and Research in Rabies	National Institute of Mental Health and Neurosciences [NIMHANS], Bangalore, India
8	WHO Collaborating Centre for Research on Rabies Pathogenesis and Prevention	Queen Saovabha Memorial Institute, Thai Red Cross Society, Bangkok, Thailand
9	WHO Collaborating Centre for Reference and Research on Rabies	Centres for Disease Control and Prevention, Atlanta, Georgia, USA
10	WHO Collaborating Centre for Reference and Research on Rabies	The Wistar Institute, USA
11	WHO Collaborating Centre for Neurovirology	Thomas Jefferson University, USA
12	WHO Collaborating Centre for Training and Research in Urban Zoonoses Control	Centro de Controle de Zoonoses, Sao Paulo, Brazil
13	WHO Collaborating Centre on Research and Management on Zoonoses Control	AFSSA-LERPAS, Laboratoire d'étudessur la rage et la pathologie, des animaux sauvages, Malzéville, France
14	WHO Collaborating Centre for Research and Training on Viral Zoonoses	Chulalongkorn University, Bangkok, Thailand

Source: http://www.who.int/rabies/rabies_network/en/ (Accessed March 28, 2014).

APPENDIX B

Thirumeni Nagarajan[1] and Charles Rupprecht[2]
[1]Research and Development, Biological E. Limited, Shameerpet, Hyderabad, Andhra Pradesh, India
[2]Ross University School of Veterinary Medicine, Basseterre, St. Kitts, West Indies

Table B.1 List of World Organization for Animal Health (OIE) Reference Laboratories for Rabies

S.No	Collaborating Centre	Address
1	Centre of Expertise [COFE] for Rabies, Canadian Food Inspection Agency, Canada	Ottawa Laboratory Fallowfield, Animal Diseases Research Institute, 3851 Fallowfield Road, P.O. Box 11300, Station H, Nepean, Ontario K2H 8P9, Canada
2	Diagnostic Laboratory for Rabies and Wildlife Associated Zoonoses	Department of Virology, Changchun Veterinary Research Institute, Chinese Academy of Agricultural Sciences, Liuying Xi Road 666, Jingyue Economic Development Zone, Changchun 130122, People's Republic of China
3	Laboratory for Rabies and Wildlife	French Agency for Food, Environmental and Occupational Health Safety (Anses) Technopôle Agricole et Vétérinaire, CS 40009, 54220 Malzéville, France
4	Institute for Epidemiology, Friedrich-Loeffler Institut	Federal Research Institute for Animal Health, Seestr. 55, D-16868 Wusterhausen/Dosse, Germany
5	Rabies Research Laboratory	Division of Viral Disease, Animal and Plant Quarantine Agency, Ministry of Agriculture, Food and Rural Affairs, 175 Anyang-ro, Manan-gu, Anyang, Gyeongii 430-757, Republic of Korea
6	Onderstepoort Veterinary Institute, Rabies Unit	Private Bag X05, Onderstepoort 0110, South Africa
7	Rabies and Wildlife Zoonoses Group, Virology Department	Animal Health and Veterinary Laboratories Agency, New Haw, Addlestone, Surrey KT15 3NB, Weybridge, United Kingdom
8	Poxvirus and Rabies Branch, Division of High-Consequence Pathogens and Pathology	National Center for Emerging and Zoonotic Infectious Diseases, Centers for Disease Control and Prevention, 1600 Clifton Road, NE, Mail Stop G33, Atlanta, GA 30 333, USA

Source: http://www.oie.int/our-scientific-expertise/reference-laboratories/list-of-laboratories/ (Accessed March 28, 2014).

APPENDIX C: Filter Paper Technology for Sampling, Storage, and Shipment of Rabies Suspect Samples for the Identification of Viral Nucleic Acids

Evelyne Picard-Meyer, Emmanuelle Robardet, and Florence Cliquet
French Agency for Food, Environmental and Occupational Health and Safety (Anses), Nancy Laboratory for Rabies and Wildlife, Community Reference Laboratory for Rabies, Community Reference Laboratory for Rabies Serology, OIE Reference Laboratory for Rabies, Technopôle Agricole et Vétérinaire, WHO Collaborating Centre for Research and Management in Zoonoses Control, Malzéville, France

C.1 INTRODUCTION

The FTA® is a paper-based system designed to fix and store nucleic acids directly on contact from fresh tissues pressed into the pre-treated paper. The FTA paper (e.g. FTA Gene Guard System) is a commercially available product manufactured by the Whatman company.[1] The nucleic acids are entrapped physically, immobilized, and stabilized within the fiber matrix, thus allowing the collection, storage at room temperature, and archiving of the nucleic acids. FTA cards are impregnated with a chaotropic agent under a patented chemical formula to lyse cell membranes and denature proteins on contact. On contact with FTA cards, infectious pathogens such as the rabies virus (RABV) are rendered inactive, allowing shipment of samples via standard postal methods (normal mail routes).[2,3]

Once the FTA cards are prepared and impregnated with the sample to analyze, nucleic acids are recovered by punching the impregnated area. The elution and extraction of RNA are performed with standard reagents and RNA purification kits. The detection of viral RNA is performed by reverse transcription polymerase chain reaction (RT-PCR) with specific pan-lyssavirus primers.[4] We describe the protocol for sampling rabies-suspect tissue and/or body fluid on the FTA card, and the methods for subsequent elution and the detection of viral RNA by conventional RT-PCR.

C.2 MATERIALS

C.2.1 Buffer and Reagents

1. Reagents needed for sampling of rabies virus on FTA cards:
   a. Indicating FTA card (Whatman)
   b. Dessicant pack (Silica gel)

2. Reagents needed for preparing the FTA cards for RNA extraction:
   a. 6.0 mm diameter Harris micro punch
   b. PBS 1 ×
3. Reagents needed for RNA extraction:
   a. RNA purification kits (e.g. Nucleospin RNA virus kit, Macherey Nagel)
   b. 96% ethanol
4. Reagents needed for RT-PCR:
   a. Qiagen One-Step RT-PCR kit (Qiagen)
   b. Platinum Taq polymerase 5 U/μL (Invitrogen)
   c. dNTP 10 mM
   d. RNase free water
   e. RNase inhibitor 40 U/μL
   f. 0.02 μM primer JW12, forward (5′-ATGTAACACCYCTACAATG-3′)
   g. 0.02 μM primer JW6, reverse (5′-CARTTVGCRCACATYTTRTG-3′)
   h. 0.02 μM cocktail of primer JW10, reverse: JW10-1 (5′-GTCATCAAAGTGTGRTGCTC-3′), JW10-2 (5′-GTCATCAATGTGTGRTGTTC-3′), JW10-3 (5′-GTCATTAGAGTATGGTGTTC-3′)
   i. 0.2 mL PCR tubes and sterile tubes (15 mL, 2 mL, 1.5 mL)
   j. Pipettes capable of dispensing volumes 0.5–1,000 μL and sterile tips
5. Reagents needed for the visualization of DNA on agarose gel:
   a. Agarose
   b. TAE (Tris acetate buffer) 10 ×
   c. SYBR Safe DNA Gel Stain 10,000 × (Invitrogen)
   d. Gel-loading buffer
   e. DNA marker

C.2.2 Equipment

1. Equipment needed for the preparation of virus homogenate, sampling of virus on FTA cards, and RNA extraction:
   a. Class II Biological Safety Cabinet
   b. Vortex mixer
   c. Micro-centrifuge
   d. Heating block
2. Equipment needed for RT-PCR and electrophoresis:
   a. PCR thermal cycler
   b. Microwave oven

c. Gel mold
d. 500 mL flask
e. Submarine gel electrophoresis tank
f. Molecular imaging system

C.2.3 Biological Materials

RABV-infected brain tissue.

C.3 METHODS

When working with RABV, precautions need to be taken that follow national and international safety guidelines for working with infectious agents. The three steps described in the first three sections below should be undertaken in a Class II Biological Safety Cabinet to avoid possible exposure. The two steps described in the fourth and fifth sections should be performed in a laboratory equipped to perform PCR by strictly following the typical routine precautions in PCR protocols to avoid cross contamination and false positive results.

To avoid contamination of samples, always wear gloves and proper laboratory clothing when handling biological samples and FTA cards.

C.3.1 Preparation of Brain Tissue Homogenate

1. Homogenize 1 g of brain tissue in 5 mL of PBS 1 × in a 15 mL sterile conical centrifuge tube.
2. Centrifuge for 15 minutes at 2,000 × *g*.
3. Remove 25 μL of supernatant (up to 100 μL) for its application on FTA cards.

C.3.2 Transfer of Rabies Virus to FTA Cards

1. Place the FTA card on a clean, dry, and flat surface. Identify appropriately the FTA card with a unique name and number (Figure C.1).
2. Lift the cover of the FTA card to expose the white sample areas. Drop 25 μL of sample (up to 100 μL) evenly onto the spot, within the sample area, in a concentric circular motion.
3. Let the sample dry at room temperature for 2 hours (the color of the indicating FTA card turns from pink to white once the sample is added).
4. When the FTA card is completely dry, put the impregnated FTA card within a sealable protective pouch with desiccant agents to absorb moisture.

Figure C.1 Example of Whatman FTA card.

5. Store the FTA cards with the desiccant in a cool, dry environment at +4°C until use.
6. As the FTA card inactivates infectious pathogens, the cards can be shipped as regular mail.

C.3.3 Preparation of an FTA Disc for RNA Extraction

1. Use two sample discs (6 mm diameter) from the dried spot with the Punch; cut the disks with a scalpel into little pieces (Figure C.2).
2. Place the little pieces in a 1.5 mL sterile tube.
3. Add 200 μL of PBS 1 ×. Vortex thoroughly and centrifuge 30 sec at 1,000 × *g*.

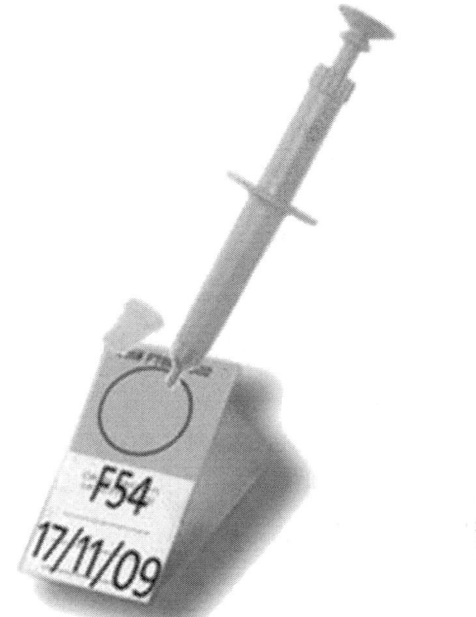

Figure C.2 FTA disc preparation for RNA extraction.

4. Incubate overnight at +4°C.
5. Mix, remove the eluate by pipetting, and transfer to a new sterile tube and store the tube at +4°C until RNA extraction.

C.3.4 RNA Extraction using Nucleospin RNA Virus Kit

The viral RNA extraction is performed as follows, using a NucleoSpin® RNA virus kit from Nacherey Nagel (France).

1. Lysis of viruses: Add 600 μL of Buffer RAV1 containing Carrier RNA to 150 μL of eluate. Pipette the mixture up and down and vortex for 15 sec. Incubate for 5 min at 70°C.
2. Adjust binding conditions: Add 600 μL of 96% ethanol to the clear lysis solution and mix by vortexing for 15 sec.
3. Bind viral RNA:
   a. Place NucleoSpin RNA Virus Columns in Collection Tubes (2 mL) and load 700 μL of lysed sample. Centrifuge for 1 min at 8,000 × *g*.
   b. Use new Collection Tubes (2 mL) for each step. Load the residual lysis solution onto the same NucleoSpin RNA Virus Column.

c. Centrifuge for 1 min at 8,000 × *g*. Discard flow through and place the NucleoSpin RNA Virus Column into a new Collection Tube (2 mL).

4. Wash and dry silica membrane:
   a. Add 500 µL of Buffer RAW to the NucleoSpin RNA Virus Column. Centrifuge for 1 min at 8,000 × *g*. Discard flow through. This washing step removes potential contaminants and PCR inhibitors.
   b. Add 600 µL of Buffer RAV3 to the NucleoSpin RNA Virus Column. Centrifuge for 1 min at 8,000 × *g*. Discard flow through with Collection Tube.
   c. Place the NucleoSpin RNA Virus Column in a new Collection Tube (2 mL) and add 200 µL of Buffer RAV3.
   d. Centrifuge for 5 min at 11,000 × *g* to remove Buffer RAV3 completely.
5. Elute viral RNA:
   a. Place the NucleoSpin RNA Virus Column into a new, sterile 1.5 mL microcentrifuge tube.
   b. Add 50 µL of RNase-free H_2O (preheated to 70°C) and incubate for 1 min.
   c. Centrifuge for 1 min at 11,000 × g.
6. Store the purified RNA at −80°C until use.

C.3.5 Heminested Reverse-Transcriptase Polymerase Chain Reaction

The heminested reverse-transcriptase polymerase chain reaction (hnRT-PCR) method described here allows the specific amplification of all recognized RABV using the validated pan-lyssavirus primers[4], giving a PCR product of 606 bp (first RT-PCR) and of 589 bp when conducting the second round of PCR (Figure C.3). In the first step, the primers JW12 and JW6 are used for the amplification of the nucleoprotein gene (nucleotide positions 55–661 according to the Pasteur virus strain) and primers JW12 and JW10 (nucleotide positions 636–617) in the nested PCR. The RT-PCR is followed by a second PCR to increase the threshold of detection for PCR.

C.3.5.1 Reverse Transcriptase Polymerase Chain Reaction (JW12-JW6 Primers)

1. Thaw template RNA, primer solutions, dNTP Mix, 5 × QIAGEN OneStep RT-PCR Buffer, and RNase-free water, and place them on

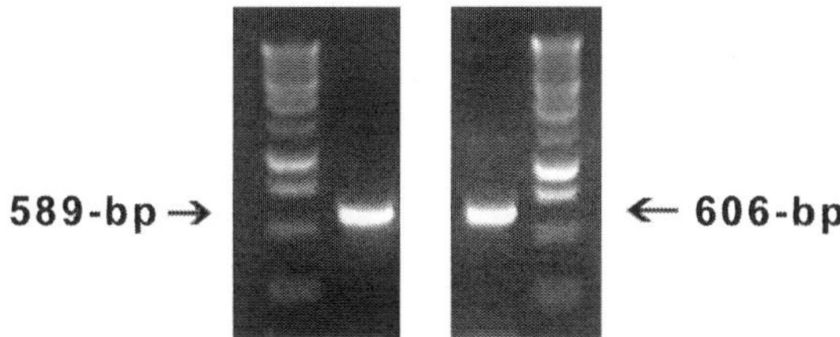

Figure C.3 (A) PCR amplification of rabies virus with primers JW12-JW10 giving an amplicon of 589 bp. (B) PCR amplification of rabies virus with primers JW12-JW6 giving an amplicon of 606 bp.

Table C.1 RT-PCR Mastermix

Reagent	Final Concentration	Volume/ Reaction (in μL)
RNase-free water	–	4.3
5 × Qiagen One step RT-PCR	1×	3
dNTP mix 10 mM	400 μM each dNTP	0.6
Primer JW12 20 μM	0.6 μM	0.5
Primer JW6 20 μM	0.6 μM	0.5
OneStep RT-PCR enzyme mix	–	0.6
RNase inhibitor 40 U/μL	0.4 U	0.5
Total Volume (mix)		10 μL

ice. Mix the solutions completely before use to avoid localized differences in salt concentration.

2. Prepare a mastermix of reagents in RNase-free tubes (0.2 mL) on ice, using the following mix calculations per sample (Table C.1).
3. Mix the mastermix thoroughly and dispense 10 μL into PCR tubes placed on ice.
4. Add 5 μL of template RNA (between 250 and 500 ng/reaction) to the tubes (15 μL of reaction volume).
5. Program the thermal cycler using the protocol in Table C.2.
6. Start the RT-PCR program while PCR tubes are still on ice. Wait until the thermal cycler has reached 50°C, and then place the PCR tubes in the thermal cycler.
7. After amplification, the PCR products are stored at 4°C until required.

C.3.5.2 Nested Polymerase Chain Reaction (JW12-JW10 Primers)

1. Prepare a mastermix of reagents in RNase-free tubes (0.2 mL) on ice, using the composition in Table C.3.

Table C.2 RT-PCR Thermal Cycler Cycle Conditions

Reverse transcription	30 min	50°C
Initial PCR activation step	15 min	95°C
3-step cycling (35 cycles)		
Denaturation	30 sec	94°C
Annealing	30 sec	55°C
Extension	1 min	72°C
Final extension	10 min	72°C

Table C.3 Nested PCR Thermal Cycler Protocol

Reagent	Final Concentration	Volume/ Reaction (in μL)
RNase-free water		13.65
10 × PCR Buffer	1 ×	2
$MgCl_2$ 50 mM	1.5 mM	0.6
dNTP mix 2 mM	0.1 mM	1
Primer JW12 20 μM	0.6 μM	0.25
Primers JW10 20 μM	0.6 μM	0.25
Taq Polymerase 5 U/μL	1.25 U	0.25
Total Volume (mix)		18 μL

Table C.4 Nested PCR Thermal Cycler Cycle Conditions

Initial PCR activation step	5 min	94°C
3-step cycling (30 cycles)		
Denaturation	30 sec	94°C
Annealing	30 sec	55°C
Extension	1 min	72°C
Final extension	10 min	72°C

2. Mix the mastermix and dispense a volume of 18 μL into PCR tubes placed on ice.
3. Add 2 μL of the first PCR product to the tubes (total volume of 20 μL). Run the thermal cycler using the program in Table C.4.
4. Once the PCR is finished, store the PCR product at +4°C before visualization in an agarose gel.

C.3.6 Electrophoresis (Visualization of DNA in Agarose Gel)

Five (5) μL of PCR products are run in a 2% agarose gel along with a DNA marker as described below.

1. Casting of the agarose gel:
   a. Weigh 2.50 g of agarose and put in a 500 mL flask.
   b. Add 125 mL of TAE Buffer to the flask to produce a 2% agarose gel.
   c. Melt the agarose in a microwave until the solution becomes clear.
   d. Let the solution cool to about 50–55°C.
   e. Add 12.5 μL of SYBR Safe DNA Gel stain (10,000 ×).
   f. Seal the ends of the casting tray with two layers of tape and place the combs in the gel-casting tray.
   g. Pour the molten agarose solution into the casting tray and let it solidify.
   h. Place the gel in an electrophoresis chamber.
   i. Add enough TAE Buffer to immerse the gel (over about 2–3 mm of buffer).
2. Loading of the gel:
   a. Mix 1 μL of 6 × sample loading buffer with 5 μL of sample. The DNA ladder must also be mixed with loading buffer to be run alongside the samples.
   b. Load the samples (5 μL) and DNA ladder (5 μL) into the wells.
3. Running of the gel:
   a. Place the lid on the electrophoretic apparatus and connect the electrodes to the power supply.
   b. Run the gel at about 100 volts until the tracking dye reaches the end of the gel.
   c. Remove the lid of the electrophoretic apparatus.
   d. Using gloves, place the gel in the tray and illuminate with the aid of a UV transilluminator.

C.3.7 Interpretation of Results

After electrophoresis, inspection with the aid of a UV transilluminator will allow determination of the following results:

- Whether there is an amplification product(s).
- The quality of the product(s).
- The specific size of the product(s).
- The size of the first-round of PCR should be ≈606 bp, and that of the nested PCR ≈589 bp.

The PCR is considered validated with the following results:

- The size of the PCR product after the first round of PCR is ≈606 bp with primers JW12-JW6, and 589 bp with the nested PCR using primers JW12-JW10.
- The positive and negative controls yielded the expected results (positive controls = positive, and negative controls = negative).

C.4 DISCUSSION

The transport of infectious substances is covered by national and international regulations.[3] Specimens for rabies diagnosis should be shipped within the UN3373 classification (Category B) with triple packaging to avoid any exposure hazards, while for the RABV culture, Category A transport practices should be applied (UN2814 classification). To achieve reliable results, the transit time of specimens for rabies diagnosis should be as short as possible, preferably within 48 hours, and the specimen should be preserved by refrigeration (+4°C) or freezing (with dry ice) during transport to the laboratory. In Europe, the cost of a refrigerated biological transport (UN3373) at +4°C is between 155 € and 365 € (for a distance of 1,200 and 2,400 km, respectively), while the cost of the UN3373 biological transport with dry ice is between 1,245 € and 1,500 € (for a distance of 2,230 km and 1,857 km, respectively).

The chemical reagents impregnated in the FTA paper are designed to inactivate most pathogens and nucleases during drying. The ability of the FTA paper to inactivate the infectivity of RABV when stored on the paper for 2 h at room temperature was demonstrated previously.[2] The result of RABV inactivation is that the sample deposited on the filter paper is no longer infectious, and subsequently does not pose any biohazard. As samples containing DNA or RNA are not covered by the dangerous goods transport regulation,[3] the shipment of these samples can be conducted through normal mail routes. The cost is thus the price of a stamp (0.66 € in France). This method is therefore ideal for shipping specimens for molecular studies and subsequent characterization of viral strains. The genetic characterization of viral strains of positive cases is recommended by international organizations[5–7] to identify the geographical origin of the field strains and to make the distinction between field and vaccine strains in those areas vaccinated orally, thus improving understanding of the epidemiology of rabies.

Advantages of FTA sampling have been demonstrated for RABV-infected samples,[8–10] but also for the detection of various other infectious agents, including avian viruses such as Newcastle disease virus,[11] infectious bronchitis virus,[12] infectious bursal disease virus,[13] or avian influenza virus.[14]

ACKNOWLEDGMENTS

We thank Mélanie Biarnais, Carine Peytavin de Garam, and Jean Luc Schereffer from Anses for their expert technical assistance.

REFERENCES

1. Whatman. FTA™ Nucleic Acid Collection, Storage and Purification. <http://www.whatman.com/FTANucleicAcidCollectionStorageandPurification.aspx>; 2013 [accessed 17.02.14].
2. Picard-Meyer E, Barrat J, Cliquet F. Use of filter paper (FTA) technology for sampling, recovery and molecular characterisation of rabies viruses. *J Virol Methods* 2007;**140**:174–82.
3. OIE Transport of specimens of animal origins. In: OIE *OIE Terrestrial manual*. Paris: OIE; 2013. p. 1–12.
4. Heaton PR, Johnstone P, McElhinney LM, Cowley R, O'Sullivan E, Whitby JE. Heminested PCR assay for detection of six genotypes of rabies and rabies-related viruses. *J Clin Microbiol* 1997;**35**:2762–6.
5. OIE. *OIE Terrestrial Manual*, 22nd ed. Paris: OIE; 2013.
6. WHO. *WHO Expert consultation on rabies*, 2nd report. Geneva: World Health Organization; 2013.
7. European Commission. *The oral vaccination of foxes against rabies*. Brussels: European Commission; 2002.
8. Pant GR, Lavenir R, Wong FY, Certoma A, Larrous F, Bhatta DR, et al. Recent emergence and spread of an arctic-related phylogenetic lineage of rabies virus in Nepal. *PLoS Negl Trop Dis* 2013;**7**:e2560.
9. Nadin-Davis SA, Sheen M, Wandeler AI. Recent emergence of the Arctic rabies virus lineage. *Virus Res* 2012;**163**:352–62.
10. Picard-Meyer E, Robardet E, Moroz D, Trotsenko Z, Drozhzhe Z, Biarnais M, et al. Molecular epidemiology of rabies in Ukraine. *Arch Virol* 2012;**157**:1689–98.
11. Perozo F, Villegas P, Estevez C, Alvarado I, Purvis LB. Use of FTA filter paper for the molecular detection of Newcastle disease virus. *Avian Pathol* 2006;**35**:93–8.
12. Moscoso H, Raybon EO, Thayer SG, Hofacre CL. Molecular detection and serotyping of infectious bronchitis virus from FTA filter paper. *Avian Dis* 2005;**49**:24–9.
13. Moscoso H, Alvarado I, Hofacre CL. Molecular analysis of infectious bursal disease virus from bursal tissues collected on FTA filter paper. *Avian Dis* 2006;**50**:391–6.
14. Abdelwhab EM, Luschow D, Harder TC, Hafez HM. The use of FTA® filter papers for diagnosis of avian influenza virus. *J Virol Methods* 2011;**174**:120–2.

INDEX

Note: Page numbers followed by "*f*" and "*t*" refer to figures and tables, respectively.

A

Adenoviral vectors, as vaccine carriers for rabies prevention, 265
- characteristics of, 266
- construction of, 267–268
- discussion, 272–273
- quality control of, 270–272
- rabies vaccines, transgene product for, 267
- rescue and expansion, 268–270
- types of, 266–267

African lyssavirus RNA demonstration, with RT-PCR, 63, 70*t*–71*t*
- discussion, 68–72
- materials, 64–65
  - biological material, 65
  - equipment, 64–65
  - reagents, 64, 65*t*
- methods, 65–68
  - cycling protocol, 66–68, 66*t*
  - reagent master mix preparation, 65–66

Agarose gel, DNA visualization in, 317

AgPath-ID™ One-Step RT-PCR Kit, 80–81, 81*t*

Amino-ethylcarbazole (AEC), 27–28

Amplification reaction, 91

Antigen distribution, 33

Antigen-binding fragments (Fab), 292

Anti-rabies immune globulin preparation, 39

Anti-rabies serum production, 38–39

Apparent diffusion coefficients (ADC), 225

Aravan virus (ARAV), 7*t*–9*t*, 11

Attenuated vaccines for veterinary use, 233
- discussion, 243–244
- materials, 238–240
  - biological materials, 238–240
  - equipment and facilities, 238, 239*t*
  - reagents, 238, 239*t*
- methods, 240–243
  - downstream process, 242–243
  - master seed lot establishment, 240
  - potency, 243
  - upstream process, 240–242

Attenuation of RABV pathogenicity, 256–257

Australian bat lyssavirus (ABLV), 7*t*–9*t*, 11

B

Baby hamster kidney-21 (BHK-21), 43–50, 135–137, 137*f*, 238–240

Big dye terminator cycle sequencing reactions, 162–163

Biotinylated antibody preparation, 29, 31, 33–36

Blood–brain barrier (BBB), 221–222

Bokeloh bat lyssavirus (BBLV), 7*t*–9*t*, 11

Bovine serum albumin (BSA), 38–41

Brain tissue homogenate reparation, 311

C

C6 cells, 43–50

Canis latrans, 14

Canis lupus, 14

Cell staining, 45–46

Cerebrospinal fluid (CSF), 37–38, 120, 123

CFX96 quantitative PCR system, 76–77

Challenge standard virus (CVS), 68–69, 109, 129, 135, 267

Childhood immunization, 265–266

Clone selection, 281

Colloidal gold conjugation, 55

Control tissues, generation of, 100

Coxsackie Ad receptor (CAR), 272

Cycling protocol, 66–68, 66*t*

Cytomegalovirus, 133–134

D

De Novo project consensus sequence, obtaining, 177–178

Decomposition, of biological materials, 29–30

Deoxynucleoside triphosphates (dNTPs), 160–161

Desalting
columns, 284
gravity, 284
using liquid chromatography systems, 284
DetectiV, 200–201
Deterioration, of biological materials, 29–30
Dideoxynucleoside triphosphates (ddNTPs), 160–163
Diffusion tensor imaging (DTI), 221–222, 225–226
Diffusion-weighted imaging (DWI), 221–222
Digoxigenin (DIG), 97–98
-labeled RNA probe, 98, 186, 190
Direct fluorescent antibody (DFA) method, 49–50
Direct fluorescent antibody test (DFAT), 97–98, 100, 103, 187–189, 278
Direct rapid immunohistochemical test (DRIT), 278
lyssavirus antigens demonstration by
discussion, 33–36
materials, 29–30
deterioration or decomposition, 29–30
equipment, 29
reagents, 29
methods, 30–33
diagnostic assay, performing, 30–32
results, reading and recording, 32–33
test results, reporting, 33
Display technologies, immune globulin single variable domain generation by, 291
discussion, 298
experimental tips, 298–300
alternatives, 299
critical parameters and troubleshooting, 298–299
limitations, 299–300
precautions, 299
time considerations, 299
materials, 293–294
biological materials, 294
equipment, 293
laboratory animals, 294
reagents, 293
methods, 294–298
immunizations, 294–295, 295*t*
phage display, immune variable heavy-chain fragment library construction for, 295–296
rabies virus-specific llama variable heavy-chain fragment selection, phage display for, 296–297
variable heavy-chain fragment, periplasmic expression of, 297–298
DNA intercalating dye, 63–64
DNase digestion, 197–198
Dog saliva, 37–38, 53–54
Dulbecco's Modified Eagle's Medium (DMEM), 148
Duvenhage virus (DUVV), 6, 7*t*–9*t*, 11–12, 63–64
Dye-terminator sequencing, 160–161

E

Electrophoresis, 92–93, 317
Encephalitis, 223–226, 224*f*, 227*f*, 228*f*, 230
Enhancement of RABV immunogenicity
through foreign proteins expression, 259
through glycoprotein over-expression, 257–259
Enzyme-linked immunosorbent assay (ELISA), 16–17, 133–134, 141, 281, 285–286, 288, 298
immune complex demonstration by, 120*f*
discussion, 122–123
materials, 121–122
methods, 122
E-predict, 200–201
Eptesicus fuscus, 13
Equine Rabies Immune Globulin (ERIG), 142
Equine rabies-immunized donors of human (ERIG), 278
European bat lyssavirus, type 1 (EBLV-1), 7*t*–9*t*, 11–12, 63–64
Evelyn Rockitniki Abelseth (ERA), 267
Expression vector preparation, 279–280
EZ1 RNA Universal Tissue Kit, 197–198

F

Flow cytometry (FC), viral antibodies demonstration by, 133
- discussion, 136–139
  - alternative materials and/or methods, 138–139
  - critical parameters and troubleshooting, 138
  - experimental tips, 138
  - future considerations, 139
  - limitations, 139
  - precautions, 138
  - time considerations, 139
- materials, 134–135
  - biological materials, 135
  - equipment, 135
  - reagents, 134–135
- methods, 135–136
  - major step 1, 135
  - major step 2, 135–136
  - results interpretation and rabies virus-neutralizing antibodies levels determination, 136

Flow cytometry, viral antigens demonstration by
- discussion, 47–51
  - alternative materials and methods, 49
  - critical parameters and troubleshooting, 48–49
  - experimental tips, 48
  - future considerations, 50–51
  - limitations, 50
  - precautions, 49
  - time considerations, 49–50
- materials, 44–45
  - biological materials, 45
  - equipment, 45
  - reagents, 44–45
- methods, 45–47
  - cell staining, 45–46
  - infection and incubation, 45
  - infection, flow cytometric analysis of, 46
  - results interpretation, 46–47

Fluorescence antibody test (FAT), 48
Fluorescence detection, 91
Fluorescent antibody virus neutralization (FAVN) test, 127–129, 131
Fluorescent focus inhibition (FFI), 133–134
Formalin-fixed tissue, 97–98, 185–186, 190
454 genome sequences, analyzing and annotating, 177–180
- De Novo project consensus sequence, obtaining, 177–178
- gene and intergenic lengths, checking, 179–180
- genomic termini, obtaining, 180
- Mapper project consensus sequence, obtaining, 178–179

Fractional anisotropy (FA), 225–226, 229*f*
FTA®, 309
- discussion, 318–319
- materials, 309–311
  - biological materials, 311
  - buffer and reagents, 309–310
  - equipment, 310–311
- methods, 311–318
  - brain tissue homogenate reparation, 311
  - electrophoresis, 317
  - heminested RT-PCR, 314–317
  - rabies virus transfer to FTA cards, 311–312
  - results interpretation, 317–318
  - RNA extraction, FTA disc preparation for, 312–313, 313*f*
  - RNA extraction, using Nucleospin RNA Virus Kit, 313–314

Fugene6 (Promega), 149

G

Gel electrophoresis analysis, 109–110, 208–209
GenBank, 109, 206, 208
Gene and intergenic lengths, checking, 179–180
Genetic analyzers, 96-well sample plate(s) preparation for, 166
Genomic termini, obtaining, 180
Glycoprotein (G) gene, 110, 110*f*, 120–121
- attenuation of RABV pathogenicity using, 256–257

Gradient pulse sequence method, 223
Gravity desalting columns, 284

H

HAdV-26, 266, 268
HAdV-35, 266, 268
HAdV-5, 266–268
HAdV1-57, 266
Heminested reverse-transcriptase polymerase chain reaction (hnRT-PCR), 63–64, 314–317
Herpes simplex virus (HSV), 43–44, 133–134
Herpestidae, 15
High amplification efficiency, 85–86, 93–94
High specificity, 93–94
High-pressure size exclusion chromatography (HP-SEC), 286, 287*f*, 288
Homogenous recombinant vaccinia virus, isolation of, 250
Human immunodeficiency virus (HIV), 133–134
Human paralytic rabies, 120–121

I

I-CeuI, 267–268
IdentiFire Library, 109
Ikoma lyssavirus (IKOV), 7*t*–9*t*, 11
Immune complex demonstration, by ELISA, 120*f*
  discussion, 122–123
    alternative materials and/or methods, 123
    experimental tips, 122–123
    limitations and future considerations, 123
    precautions, 123
  materials, 121–122
    biological materials, 121–122
    equipment and other consumables, 121
    reagents, 121
  methods, 122
Immune globulin single variable domain generation by display technologies, 291
  discussion, 298
  experimental tips, 298–300
    alternatives, 299
    critical parameters and troubleshooting, 298–299
    limitations, 299–300
    precautions, 299
    time considerations, 299
  materials, 293–294
    biological materials, 294
    equipment, 293
    laboratory animals, 294
    reagents, 293
  methods, 294–298
    immunizations, 294–295, 295*t*
    phage display, immune variable heavy-chain fragment library construction for, 295–296
    rabies virus-specific llama variable heavy-chain fragment selection, phage display for, 296–297
    variable heavy-chain fragment, periplasmic expression of, 297–298
Immunochromatographic strip test, rabies virus antigen demonstration by, 53
  discussion, 57–59
    experimental tips, 57
    future considerations, 58–59
    technical notes, 57–58
  materials, 54–55
    biological materials, 54–55
    materials and equipment, 55
    reagents, 54
  methods, 55–57
    colloidal gold conjugation, 55
    test strip application, 56–57
    test strip preparation, 55–56
Immunochromatographic strip test, viral antibodies demonstration by, 127
  discussion, 130–131
  materials, 128–129
    equipment, 129
    reagents, 128–129
  methods, 129–130
Immunohistochemical (IHC) diagnostic assay, 27–28
In situ hybridization (ISH)
  rabies viruses genetic characterization by, 185

discussion, 189–191
alternative materials and methods, 189–190
experimental tips, 189
future considerations, 190
materials, 186–187
methods, 187–189
control tissues generation, 187–188
probes, production and evaluation of, 188
results interpretation, 189
section preparation, 188
viral nucleic acids detection by, 97
materials, 98–100
biological materials, 100
equipment, 100
laboratory animals, 100
reagents, 98–100
methods, 100–103
control tissues, generation of, 100
probe production and evaluation, 101–102
results interpretation, 103
section preparation, 100–101
discussion, 103–105
alternative materials and/or methods, 104
critical parameters, troubleshooting, and precautions, 104
experimental tips, 103
future considerations, 105
limitations, 105
time considerations, 104
Incubation, 45
Indirect fluorescent antibody (IFA) test, 16–17
Infection, 45
flow cytometric analysis of, 46
In-house printed microarray, 196
Inkjet printing, 196–197
International Committee on the Taxonomy ofViruses (ICTV), 6
International reference serum (IRS), 135–136
Irkut virus (IRKV), 7*t*–9*t*, 11
Isoelectric focusing (IEF), 284–285, 286*f*, 288

K

Khujand virus (KHUV), 7*t*–9*t*, 11

L

Labeling amplified target, 198–199
Lagos bat virus (LBV), 6–10, 7*t*–9*t*, 12, 63–64
Lateral flow assay, 57
Latex agglutination test (LAT)
rabies antibody demonstration by, 141
discussion, 143–145
experimental tips, 143–144
technical notes, 144–145
materials, 142
biological materials, 142
equipment, 142
reagents, 142
methods, 142–143
Latex coating, 142–143
slide agglutination, 143
rabies virus antigens demonstration by
discussion, 40–42
critical parameters, 41–42
experimental tips, 40
technical notes, 40–41
materials, 38
biological materials, 38
equipment, 38
reagents, 38
methods, 38–40
anti-rabies immune globulin preparation, 39
anti-rabies serum production, 38–39
latex coating, 39
slide agglutination, 39–40
test saliva, 39
Latex coating, 39, 142–143
L-gene, 76, 82
Limulus Amebocyte Lysate (LAL) gel-clot method, 271–272
Liquid chromatography systems
desalting using, 284
protein A separation using, 283
Lleida bat lyssavirus (LLEBV), 7*t*–9*t*, 11

Lyssavirus, 7*t*–9*t*
demarcation criteria for, 6–10
genome organization, 4–5
host range, 12–15
pathobiology of, 15–17
phylogeny of, 6–12
serologic cross-reactivity of, 6–12
speciation, applications for, 193
structure of, 5
taxonomy of, 6
virion, 4–5, 4*f*
Lyssavirus, Next Generation Sequencing of, 171
background, 171–172
methodology, 172–180
discussion, 180–183
coverage analysis, 180
454 genome sequences, analyzing and annotating, 177–180
Roche 454 FLX(+) titanium sequencing, cDNA Preparation for, 174 177
viral RNA extraction, 172–174
Lyssaviruses, Sanger sequencing of, 159
background, 159–161
methodology, 161–169
ABI genetic analyzers, PCR amplicons Preparation for, 161–162
big dye terminator cycle sequencing reactions, 162–163
genetic analyzers, 96-well sample plate(s) preparation for, 166
output data analysis, 168–169
PCR product, purifying, 162
run on 3, starting, 167–168
run on 3130 ×l, starting, 167
run reagents, checking, 166–167
sequencing reactions, ethanol precipitation of, 163–166
96-well sample plate(s), loading, 166
lyssavirus antigens demonstration, by direct rapid immunohistochemical test
discussion, 33–36
materials, 29–30
deterioration or decomposition, 29–30
equipment, 29
reagents, 29
methods, 30–33
diagnostic assay, performing, 30–32
results, reading and recording, 32–33
test results, reporting, 33
lyssavirus nucleic acid demonstration
by pyrosequencing, 205. *See also* Pyrosequencing, genetic characterization via
materials, 206
diagnostic samples, 206
reagents and equipment, 206
methods, 206–210
gel electrophoresis analysis, 208–209
local database, population of, 206–208
One-Step RT-PCR, 208–209
result analysis, 209–210
RNA extraction, 208–209
discussion, 210–214
alternative methods, 213
critical parameters and troubleshooting, 210–213
experimental tips, 210
future considerations, 214
limitations, 213–214
precautions, 213
time consideration, 213
quantitative polymerase chain reaction for, 75
discussion, 82–83
methodology, 76–82
pan lyssavirus RT-PCR, 76–80
viral RNA extraction, 76
using TaqMan probes, 80–82
mastermix preparation, 81
materials/equipment, 80–81
read out, 81
thermal program, 81
real-time polymerase chain reaction for, 76

M

Magnetic resonance imaging (MRI), 221–222
Magnetic resonance perfusion techniques, 225
Mammalian cell stocks preparation, 281

Mammalian expression systems, human monoclonal antibodies production using, 275
materials, 278–279
biological materials, 279
equipment, 279
reagents, 278–279
methods, 279–288
cells and cell culture, 279
expression vector preparation, 279–280
mammalian cell stocks preparation, 281
monoclonal antibody expression, 282
monoclonal antibody product testing, 284–288
results interpretation, 288
monoclonal antibody purification, 282–284
desalting columns, 284
desalting using liquid chromatography systems, 284
gravity desalting columns, 284
protein A chromatography, 282
protein A separation on gravity column, performing, 282–283
protein A separation using liquid chromatography systems, performing, 283
transfection and clone selection, 281
Mapper project consensus sequence, obtaining, 178–179
Mean diffusivity (MD), 225–226, 227*f*, 229*f*
Medulla oblongata, 16
Mega4 software, 109, 208
Meles meles, 15
Melogale moschata, 15
Mephitis mephitis, 14
Metabolite imaging techniques, 223–225
Microarray
hybridization, 199–200
oligonucleotide, 193
MinElute PCR Purification kit, 198–199
Mokola virus (MOKV), 7*t*–9*t*, 11–12, 63–64
Molecular imaging techniques, 225
Monoclonal antibodies (MAbs), 6–10, 27–28, 33–34, 53–54, 120–121, 128, 185–186
expression, 282
product testing, 284–288
purification, 282–284
desalting columns, 284
desalting using liquid chromatography systems, 284
gravity desalting columns, 284
protein A chromatography, 282
protein A separation on gravity column, performing, 282–283
protein A separation using liquid chromatography systems, performing, 283
Mouse inoculation test (MIT), 43–44
Mycophenolic acid resistance screening, 250

N

Nanobodies, 292
Nested polymerase chain reaction, 315–317, 316*t*
Next Generation Sequencing (NGS), 159–160
of lyssavirus, 171
background, 171–172
discussion, 180–183
methodology, 172–180
454 genome sequences, analyzing and annotating, 177–180
coverage analysis, 180
Roche 454 FLX(+) titanium sequencing, cDNA Preparation for, 174–177
viral RNA extraction, 172–174
NMR imaging, 219
discussion, 230
materials, 222–223
human and animal subjects, 222
magnetic resonance imaging examination, 222
methods, 223–230
early and late stages, utility and limitations during, 226–230
magnetic resonance perfusion techniques, 225

NMR imaging (*Continued*)
metabolite imaging techniques, 223–225
molecular imaging techniques, 225
structural data, 223
susceptibility imaging techniques, 223
voxel-based group analysis, 226
Nucleic acid extraction, 197–198
Nucleoprotein, 120–121
Nucleoprotein (N) gene, 76, 108
Nyctereutes procyonoides, 14

O

Oligonucleotide microarray, 193
discussion, 201–202
methodology, 195–201
data analysis, 200–201
in situ synthesis (inkjet printing) technology, 196–197
in-house printed microarray, 196
labeling amplified target, 198–199
microarray hybridization, 199–200
nucleic acid extraction and DNase digestion, 197–198
oligonucleotide design, 195–196
post hybridization washes and image capture, 200
reverse transcription-polymerase chain reaction, 198
process and analysis, 195*f*
Oligonucleotide primers, 87*t*–88*t*
One-Step Reverse Transcription-Polymerase Chain Reaction, 76–77, 80–81, 109–110, 205–206, 208–209
Oral rabies vaccination (ORV), 108

P

Pan lyssavirus RT-PCR, 76–80
internal controls, 77, 78*t*–79*t*
mastermix preparation, 77, 77*t*
materials/equipment, 76–77
read out, 80
thermal program, 80, 80*t*
Pasteur virus (PV), 44–45, 135, 267
Phosphate buffered saline (PBS), 38–39
PI-Scel, 267–268
Pitman Moore (PM) virus, 267
Polyclonal antibodies (PAbs), 27–28, 33–34, 54
Polyethylenimine (PEI), 149
Polymerase chain reaction (PCR), 37–38
product, purifying, 162
real-time. *See* Real-time polymerase chain reaction
Post hybridization washes, 200
Post-exposure prophylaxis (PEP), 278
Postmortem diagnosis, of rabies, 33
Pox viral vectored vaccines, for rabies, 245
discussion, 251–252
materials, 246–248
cells, viruses, and other biological materials, 247
equipment, 247–248
reagents, 246–247
methods, 248–251
animal studies, culturing of recombinant vaccinia virus for, 251
biological safety considerations, 248
expression analysis, 250–251
homogenous recombinant vaccinia virus, isolation of, 250
recombinant vaccinia viruses, preparation of, 249–250
vaccinia virus transfer vector, preparation of, 248–249
Pre-clinical studies, 268–270
Pre-exposure prophylaxis (PrEP), 127, 130–131
PrimerExplorer V4 software, 86–89
Probabilistic tractography, 226, 229*f*
Probe detection, 102–103
Probe hybridization, 102
Procyon lotor, 13–14
Production batch, 241
Protein A
chromatography, 282
separation
on gravity column, performing, 282–283
using liquid chromatography systems, performing, 283
Pseudotype virus neutralization, viral antibodies demonstration by, 147
materials, 148–149

biological materials, 149
equipment, 149
reagents, 148–149
methods, 149–153
PNA, setting up, 151
pseudotyped virus stock generation, 149–151, 150*t*
pseudotyped virus stock titration, 151
reporter gene assay, 151–152
results interpretation, 152–153
discussion, 153–155
experimental tips and critical parameters, 153
future considerations, 154
limitations, 154
Pseudotype virus-based neutralization assays (PNA)
setting up, 151
Pseudotyped virus stock generation, 149–151
Pseudotyped virus stock titration, 151
Pyromark ID, 110–111
Pyrophosphate (PPi), 205–206
Pyrosequencing, genetic characterization via, 107
discussion, 111–115
alternative methods, 113–114
critical parameters and troubleshooting, 111–113, 112*f*
experimental tips, 111
future considerations, 115
limitations, 114–115
precautions, 113
time consideration, 114
materials, 108–109
diagnostic samples, 109
reagents and equipment, 108–109
methods, 109–111
gel electrophoresis analysis, 109–110
local database, population of, 109
One-Step RT-PCR, 109–110
result analysis, 110–111
RNA extraction, 109–110

Q

QIAamp Viral RNA Kit (QIAGEN), 90, 197–198
Quality control, of adenoviral vectors, 270–272
Quantitative polymerase chain reaction (qPCR)
for lyssavirus nucleic acid demonstration, 75
discussion, 82–83
methodology, 76–82
pan Llyssavirus RT-PCR, 76–80
viral RNA extraction, 76
using TaqMan probes, 80–82
mastermix preparation, 81
materials/equipment, 80–81
read out, 81
thermal program, 81

R

Rabies immune globulins (RIGs), 278
Rabies prevention, adenoviral vectors as vaccine carriers for, 265
characteristics of, 266
construction of, 267–268
discussion, 272–273
quality control of, 270–272
rabies vaccines, transgene product for, 267
rescue and expansion, 268–270
types of, 266–267
Rabies serogroup, 6
Rabies virus (RABV), 3, 6, 7*t*–9*t*, 11–15, 17, 27–28, 30, 33, 63–64
antigens demonstration, by latex agglutination test
discussion, 40–42
by immunochromatographic strip test, 53
materials, 38
methods, 38–40
challenge virus standard strain of, 68–69, 109, 129, 135, 267
detection, 49–50
using RT-LAMP system, 85
isolation, 43–44
postmortem diagnosis of, 33
transfer to FTA cards, 311–312
transgene product for, 267
in vitro binding using inactivated, 130*f*

Rabies viruses genetic characterization, by *in situ* hybridization, 185
discussion, 189–191
alternative materials and methods, 189–190
experimental tips, 189
future considerations, 190
materials, 186–187
methods, 187–189
control tissues generation, 187–188
in situ hybridization, 188–189
probes, production and evaluation of, 188
results interpretation, 189
section preparation, 188
Rabies virus-neutralizing antibodies (RVNA), 133–134
Rabies-immunized donors of human (HRIG), 278
Rabies-related viruses (RRVs), 6
Rapid fluorescent focus inhibition test (RFFIT), 39, 127–131, 133–134, 138, 141, 143–144, 281, 285–286, 288, 294–295, 295*t*
Rapid neutralizing Antibody (RAPINA) test, 128–131
Reagent master mix preparation, 65–66
Real-time polymerase chain reaction (RT-PCR), 109–110, 185–186, 198, 314–315
African lyssavirus RNA demonstration with, 63
heminested, 63–64, 314–317
for lyssavirus nucleic acid demonstration, 76
One-Step RT-PCR, 76–77, 80–81, 109–110, 205–206, 208–209
Pan lyssavirus, 76–80
internal controls, 77, 78*t*–79*t*
mastermix preparation, 77, 77*t*
materials/equipment, 76–77
read out, 80
thermal program, 80, 80*t*
Recombinant rabies virus vaccines, 255
discussion, 260–261
enhancement of RABV immunogenicity
through foreign proteins expression, 259
through glycoprotein over-expression, 257–259
highly attenuated virus generation, using modified G genes, 256–257
replication-deficient recombinant rabies virus vaccines, 260
replication-impaired recombinant rabies virus vaccines, 260
Recombinant vaccinia viruses, preparation of, 249–250
Replication-deficient recombinant rabies virus vaccines, 260
Replication-impaired recombinant rabies virus vaccines, 260
Reporter gene assay, 151–152
Restriction mapper, 280
Results interpretation, 152–153
Reverse genetics, 255–258, 260
Reverse transcription-loop-mediated isothermal amplification (RT-LAMP) system
rabies virus detection using, 85
discussion, 93–94
materials, 86–90
biological materials and sample preparation, 90
equipment, 89–90
primer set, 86–89
reagents, 86
methods, 90–93
amplification reaction, 91
detection and results interpretation, 91–93
master mix and sample solution, mixing of (on ice), 91
master mix preparation, 90–91, 90*t*
Reverse transcription-polymerase chain reaction, 97–98, 104–105
Rhabdoviridae, 3
Ribonucleic acid (RNA)
extraction, 109–110, 208–209
FTA disc preparation for, 312–313, 313*f*
using Nucleospin RNA Virus Kit, 313–314
messenger, 15–16
nucleoprotein, 4, 15–16
single-stranded, 3

Ribonucleoprotein (RNP), 4
Roche 454 FLX(+) titanium sequencing, cDNA Preparation for, 174–177
emulsion polymerase chain reaction, 176–177
library preparation, 174–176
Roche 454 FLX + sequencing, 177
Run reagents, checking, 166–167

S

Sanger sequencing, 111–114, 213
of lyssaviruses, 159
background, 159–161
methodology, 161–169
ABI genetic analyzers, PCR amplicons Preparation for, 161–162
big dye terminator cycle sequencing reactions, 162–163
genetic analyzers, 96-well sample plate(s) preparation for, 166
output data analysis, 168–169
PCR product, purifying, 162
run on 3, starting, 167–168
run on 3130 ×l, starting, 167
run reagents, checking, 166–167
sequencing reactions, ethanol precipitation of, 163–166
96-well sample plate(s), loading, 166
Seed train, 240–241
Sequence analysis (SQA), 108, 205–206
Sequencing reactions, ethanol precipitation of, 163–166
Serum-free media (SFM), 281
Shimoni bat virus (SHIBV), 7*t*–9*t*, 11
Simplified focus inhibition (SFIMT), 133–134
Single nucleotide polymorphism (SNP) genotyping, 205–206
Single-chain variable fragments (scFv), 292
Single-stranded RNA (ssRNA), 3
Slide agglutination, 39–40, 143
Slide preparation, 102
Sodium dodecyl sulfate polyacrylamide gel electrophoresis (SDS-PAGE), 284–285, 285*f*, 288
Staining intensity, 32
Structural data, 223
Susceptibility imaging techniques, 223
Susceptibility-weighted imaging (SWI) technique, 223
SYBR® Green, 64, 80*t*

T

Tadarida brasiliensis, 13, 15
TaqMan probes, RT reverse transcription PCR for lyssavirus species using, 80–82
mastermix preparation, 81
materials/equipment, 80–81
read out, 81
thermal program, 81
Test saliva, 39
Thermal cycler protocol, 316*t*
Tragelaphus strepsiceros, 14
Transfection, 281
Transgene product, for rabies vaccines, 267
Tris phosphate buffered solution (TPBS) buffer, 30–31
TRIzol® Reagent, 90, 173–174, 182, 197–198
Turbidity detection, 91

U

U.S. Centers for Disease Control and Prevention (CDC), 27–28, 33–34

V

Vaccines, attenuated
for veterinary use, 233
Vaccinia virus transfer vector, preparation of, 248–249
Variable heavy-chain fragments (VHH), 292, 295–296
library construction, for phage display, 295–296
periplasmic expression of, 297–298
selection, phage display for, 296–297
Vesicular stomatitis virus (VSV), 148, 153
Veterinary use, attenuated vaccines for, 233
discussion, 243–244
materials, 238–240
biological materials, 238–240
equipment and facilities, 238, 239*t*
reagents, 238, 239*t*

Veterinary use, attenuated vaccines for (*Continued*)
methods, 240–243
downstream process, 242–243
master seed lot establishment, 240
potency, 243
upstream process, 240–242
Viral RNA extraction, 76, 172–174
tissue samples, 173
virus culture samples, 173–174
Virus neutralizing antibodies (VNAs), 127–128, 148, 265–266
Voxel-based group analysis, 226
Vulpes lagopus, 14
Vulpes vulpes, 14

W
"Walking the Genome", 160
96-Well sample plate(s)
ABI genetic analyzers, PCR amplicons preparation for, 161–162
loading, 166
preparation, for genetic analyzers, 166
West Caucasian bat virus (WCBV), 7*t*–9*t*, 11–12
World Health Organization (WHO), 128
World Organization for Animal Health (OIE), 128

CPI Antony Rowe
Eastbourne, UK
February 02, 2015